Practical Biotransformations

Postgraduate Chemistry Series

A series designed to provide a broad understanding of selected growth areas of chemistry at postgraduate student and research level. Volumes concentrate on material in advance of a normal undergraduate text, although the relevant background to a subject is included. Key discoveries and trends in current research are highlighted, and volumes are extensively referenced and cross-referenced. Detailed and effective indexes are an important feature of the series. In some universities, the series will also serve as a valuable reference for final year honours students.

Practical Biotransformations

A Beginner's Guide

Gideon Grogan

*York Structural Biology Laboratory, Department of Chemistry,
University of York, UK*

A John Wiley and Sons, Ltd., Publication

This edition first published 2009
© 2009 Gideon Grogan

Registered office
John Wiley & Sons Ltd, The Atrium, Southern Gate, Chichester, West Sussex, PO19 8SQ, United Kingdom

For details of our global editorial offices, for customer services and for information about how to apply for permission to reuse the copyright material in this book please see our website at www.wiley.com.

Library of Congress Cataloging-in-Publication Data

Grogan, Gideon.
  Practical biotransformations : a beginner's guide / Gideon Grogan.
    p. cm. – (Postgraduate chemistry series)
  Includes bibliographical references and index.
  ISBN 978-1-4051-9367-2 (cloth) – ISBN 978-1-4051-7125-0 (pbk. : alk. paper)
  1. Biosynthesis. 2. Enzymes–Biotechnology. 3. Microbial biotechnology.
  4. Organic compounds–Synthesis. I. Title.
  QD415.5.G76 2009
  660.6–dc22

                              2009004187

A catalogue record for this book is available from the British Library.

ISBN Cloth: 9781405193672 (HBK)
ISBN Paper: 9781405171250 (PBK)

Typeset in 10/12 Minon by Laserwords Private Limited, Chennai, India
Printed and bound in Great Britain by TJ International Ltd, Padstow, Cornwall.

Contents

Foreword

'Biotransformations' is a fundamentally practical discipline in which the objective is to use enzymes and whole cells as catalysts for conversions of organic substrates. This subject is growing very rapidly, finding ever increasing applications ranging from pharmaceuticals and fine chemicals to food additives, cosmetic ingredients and more recently biofuels. Against this backdrop of an expanding demand for this technique, it is clearly important that we continue to train the next generation of 'biotransformers', hence the timely nature of this book by Gideon Grogan. Dr. Grogan has set out to provide a step-by-step guide to acquiring the basic techniques that are required to carry out practical biotransformations in the laboratory. He has put himself in the position of a scientist who possesses good basic laboratory skills but needs to be led by the hand in terms of how to set up a laboratory and how to acquire basic techniques (e.g. culturing microorganisms, using isolated enzymes, gene cloning) necessary for practical biotransformations. Crucially, in addition to the excellent text, the book is illustrated with many helpful pictures and diagrams. Books of this nature are not freely available and hence the availability of this manual will serve to provide an unmet need, particularly for organic chemists wishing to avail themselves of this emerging technology.

Professor Nicholas J. Turner,
Director
Centre of Excellence for Biocatalysis,
Biotransformations and Biocatalytic Manufacture
University of Manchester, UK

Preface

The development of new asymmetric catalytic methods is of fundamental importance to industrial synthetic chemistry. In addition to the importance of generating optically pure synthetic intermediates, the recent drive to adopt greener methods of synthesis has stimulated a growing interest in biologically catalysed reactions as a means of selective and environmentally benign synthesis. However, even when a biocatalytic method may be the best solution for a synthetic problem, there are obstacles that exist which bar it from being considered as a real option. Many of these are due to a lack of understanding of the basic techniques associated with biocatalysis, or a want of facilities or trained personnel.

This book is intended as a beginner's guide to microbes and enzymes and how to use them to do synthetic organic reactions in the laboratory. Rather than a list of reactions, or an overview of the literature, it is intended as a laboratory manual that can be readily referred to in an everyday experimental environment. It assumes very little knowledge of biochemical reactions or microbiology, but seeks, with appropriate advice on aspects of microbiological practice and associated safety, to help the uninitiated to begin to understand how biocatalysts work and how they can be used safely and efficiently for the generation of valuable intermediates and metabolites. It would therefore be suitable for undergraduate or postgraduate students of chemistry with little or no experience of biochemistry, microbiology or molecular biology. As a book intended primarily for those with a knowledge of synthetic organic chemistry, knowledge of synthetic reactions and techniques, analytical methods, such as TLC, NMR, UV spectroscopy, mass spectrometry and standard laboratory techniques, such as solvent extraction and column chromatography, has been assumed.

The book is not an exhaustive treatment of biochemistry, molecular biology or their techniques. Only those techniques, which are directly relevant to laboratory-scale biotransformations, or those that will help those unfamiliar with biocatalysis to engage with the relevant literature, which may include subjects such as microbiology or protein purification, are included. Different individuals will already have varying levels of exposure to the methods described herein, so it should be possible to dip in and out of the book depending on the level of previous experience.

Where more specialised techniques are required, e.g. more advanced molecular biology, or process improvement techniques such as immobilisation, the reader will be referred to more specialist texts in these areas. However, it is hoped that this book will serve as a useful primer to new workers in laboratories that already undertake biotransformations but, more importantly, to encourage the uptake of biotransformations in synthetic organic laboratories that have until now not appreciated the benefits or inherent experimental simplicity of biocatalytic solutions.

Acknowledgements

I would like to thank Dr Mark Fogg and Dr David Nelson (University of York), Dr Robert Speight (Ingenza Ltd), Dr Andrew Wells (Astra Zeneca), Dr Richard Lloyd (Dr Reddy's) and Professor Nicholas Turner (Centre of Excellence for Biocatalysis, Biotransformations and Biocatalytic Manufacture (CoEBio3), University of Manchester) for their critical reading of the manuscript and their comments. I would also like to thank Phil Roberts for photography and assistance with graphics and Mark Thompson, Florian Fisch, Claudia Szolkowy, Sam Johnston and Laila Roper for additional photographs and protocols.

Chapter 1
Biotransformations, Microbes and Enzymes

1.1 Introduction

A biotransformation, as understood by the growing community of chemists and bioscientists who practise in the area, is the conversion of one chemical entity to another by the action of a biological system which, in our case will be primarily micro-organisms or enzymes derived from them. Micro-organisms and their extracts have been exploited for thousands of years for chemical reactions, largely for fermentation reactions leading to alcoholic beverages, vinegar or the production of foodstuffs such as yoghurt. The history of biotransformations shares much with the history of the study of the chemistry of the brewing process in the nineteenth century, on which an excellent essay is to be found in 'Introduction to Biocatalysis using Enzymes and Micro-organisms' by Professor Stanley M. Roberts and colleagues [1]. In the twentieth century, early interest in biotechnology was stimulated both by the use of organisms to produce bulk chemicals, such as sugars and amino acids, and also in the exploitation of microbes for pharmaceutical production for the chemical industry, after the discovery of penicillin. In terms of biotransformations science, such developments were crucial, as they encouraged research into the optimisation of large-scale fermentations of micro-organisms and the relevant enzyme biochemistry, and also encouraged the development of microbial culture collections that still serve as a valuable reservoir of biocatalysts today. However, it is in the last twenty or thirty years or so that microbes and enzymes have been recognised as very real solutions to many of the challenges facing synthetic chemistry in terms of their capabilities as catalysts of single-step reactions that are analogous to standard organic reactions. The first wave of applications in organic synthesis was neatly summarised in reviews by Professor J. Bryan Jones, one of the pioneers in preparative biocatalysis, in 1976 and 1986 [2], the latter having been cited over 740 times at the time of writing

Practical Biotransformations: A Beginner's Guide   © 2009   Gideon Grogan

this book. This review summarises over 350 reports of both applied studies of enzymes and related biochemistry that incorporated early developments including microbial hydroxylations, the use of lipase enzymes for kinetic resolutions and alcohol dehydrogenases for the asymmetric reduction of ketones. The early work of Jones, Professor George Whitesides [3] and others inspired a whole new generation of organic chemists to investigate the possibility of applying enzymes in organic synthesis and important developments were made in several laboratories around the world. The involvement and expertise of microbiologists, molecular biologists and enzymologists would prove crucial and the development of biotransformations science in the last two decades represents one of the best examples of synergistic relationships across disciplines in the biological and chemical sciences. Biocatalytic reactions are now reported in the literature and used in industry on a routine basis. Although it is not perhaps quite correct to say that there is an enzymatic equivalent for every type of reaction applied in abiotic chemical synthesis, the breadth of chemical catalytic diversity exhibited by enzymes is astonishing, particularly given that they are all made of the same, fairly small repertoire of twenty amino acids. Enzymes have been described that catalyse carbon and heteroatom oxidations, reductions of carbonyls and nitro groups, carbon-carbon bond formation, group transfer leading to the production of, for example, oligosaccharides and the hydrolysis of a number of functionalities including esters, amides, epoxides and even carbon-carbon bonds. All of these activities and more are now employed routinely in industrial synthetic chemistry by some of the world's largest chemical companies.

As both the level of interest and the amount of research in biotransformations science has grown, a number of textbooks exists that summarise the relevant literature and give examples of the application of enzymes in synthesis. In addition to the early volume by Professors Whitesides and Wong [3], notable amongst these are 'Biotransformations in Organic Chemistry' [4] by Professor Kurt Faber of the University of Graz, which is a convenient single volume that constitutes an excellent summary of the current state of the area, and 'Enzyme Catalysts in Organic Synthesis' [5], a larger volume edited by Professor Karlheinz Drauz of Degussa and Professor Herbert Waldmann of the University of Dortmund, with contributions from various authors on specific aspects of biocatalysis, with detailed sections on different types of enzymatic reaction and their application. In addition there are specific texts on certain enzyme classes, such as 'Hydrolases in Organic Synthesis' by Professors Bornscheuer and Kaslauskas [6], and in techniques such as high-throughput methodology in 'Enzyme Assays. High Throughput Screening, Genetic Selection and Fingerprinting', edited by Professor Jean-Louis Reymond of the University of Berne [7]. The interest in biotransformations has also fuelled the formation of several research Centres of Excellence throughout the world, notably in Iowa, USA, at the University of Graz in Austria, the University of Delft in the Netherlands and recently at the University of Manchester, UK. Many of these run training courses in biocatalysis methodology for those in the synthetic organic chemistry industry. Several companies have also been established that offer biocatalytic solutions for industrial chemistry and these grow in number annually.

This recent and continuing appreciation of biocatalysis has probably arisen from three major considerations. First, the emerging recognition that the synthesis of single enantiomer forms of chiral drugs was going to be increasingly important in industrial chemistry in the late twentieth century and beyond, and that enzymes would be able to accomplish an important role as chiral catalysts. Second, that industry was to come under increasing pressure to develop environmentally benign methods of synthesis – 'green chemistry' criteria – that would be fulfilled by many of the natural characteristics of enzymes and microbes. Third, a revolution has been undergone in gene and protein engineering and techniques of molecular biology, ensuring that the manipulation of enzymes has become a standard technique in laboratories worldwide.

1.1.1 Biocatalysts catalyse selective reactions

The properties of enzymes as chiral catalysts has been appreciated for decades, and it is with the increase in demand for enantiopure drugs that there has been the growth of interest in enzymes in fine chemical synthesis. Enzymes are themselves chiral, being comprised of L-amino acids that go to make up their polypeptide chain. Within the active site of an enzyme, the three-dimensional, chiral environment ensures that the enantiomeric constituents of racemates are discriminated, and enzymes are therefore capable of catalysing kinetic resolutions, and also the desymmetrisation or chiral functionalisation of *meso-* or prochiral compounds. In addition, owing to their exquisite specificity, enzymes are also capable of catalysing *regioselective* reactions, functionalising only one of many chemically equivalent sites in a substance. An excellent example of this is the hydroxylation of progesterone by the fungus *Rhizopus arrhizus*, which occurs only at the 11 position of the steroid nucleus, and only in the α-configuration (Figure 1.1).

Biocatalysts are also capable of catalysing *chemo-* selective reactions, discriminating between chemically similar yet structurally different functions that may be

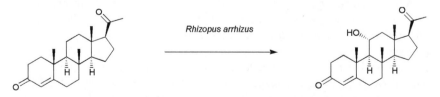

Figure 1.1   Regio- and stereoselective hydroxylation of progesterone by *Rhizopus arrhizus*

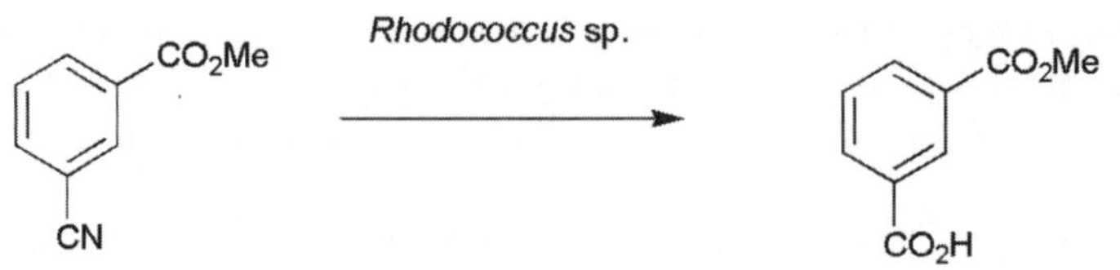

Figure 1.2   Chemoselective hydrolysis of an aromatic nitrile by a *Rhodococcus* species

labile under the same chemical conditions, such as in the preferential hydrolysis of a nitrile over a carboxy-ester, two acid-labile groups, by a strain of bacterium known as *Rhodococcus* (Figure 1.2).

1.1.2 Biocatalysts catalyse Green Chemistry

As both industrial and academic chemical laboratories come under increasing pressure to adopt environmentally benign methods of chemical synthesis, biocatalytic methods of synthesis offer several advantages. The twelve principles of Green Chemistry, proposed by Anastas and Warner in 1998 [8], are listed in Table 1.1.

Table 1.1 also highlights those principles to which enzymes and microbes conform either in part or totally. The most notable of these include the catalytic nature of biological catalysts, their derivation from renewable resources and inherent biodegradability and their ability to operate at neutral pH and ambient temperature and pressure. Enzymes are also often able to work in aqueous environments, thus removing the use of organic solvents. Their selectivity can also permit the direct functionalisation of molecules at one of many chemically identical sites, thus removing the need for protection/deprotection strategies. Whilst not providing a panacea for green chemical solutions, biocatalysts are certainly worthy of consideration where environmentally clean chemistry is an issue, and indeed, the Presidential award for Green Chemistry (2006) was awarded to a biocatalytic process, developed by Professor Roger Sheldon at the University of Delft in collaboration with Codexis Ltd in the USA, for the production of the side-chain of one of the world's most commercially significant pharmaceuticals, atorvastatin (Lipitor^{TM}) [9].

Another advantage of enzymes in the context of Green Chemistry is that they often provide a route to natural-equivalent materials where these may be required in, for example, the food or cosmetics industries. If a natural, i.e. non-genetically modified organism, is used to transform a natural substance, the product of that reaction may be labelled as 'natural', as having exploited both natural source materials and reagents in its production. Hence, despite the wealth of interest in recombinant biocatalysts, enzyme engineering and the associated advantages, there is still a great deal of interest in biotransformations catalysed

Table 1.1 The twelve principles of Green Chemistry, as proposed by Anastas and Warner [8]

	Green Chemistry principle	Notes
1	It is better to prevent waste than to treat or clean up waste after it is formed	
2	Synthetic methods should be designed to maximise the incorporation of all materials used in the process into the final product	
3	Wherever practicable, **synthetic methodologies should be designed to use and generate substances that possess little or no toxicity to human health and the environment**	Biocatalysts are natural products with no inherent toxicity issues
4	Chemical products should be designed to preserve efficacy of function while reducing toxicity	
5	**The use of auxiliary substances (e.g. solvents, separation agents, etc.) should be made unnecessary whenever possible and innocuous when used**	Biocatalysts are often employed in aqueous media.
6	Energy requirements should be recognised for their environmental and economic impacts and should be minimised. **Synthetic methods should be conducted at ambient temperature and pressure**	Most biocatalysts work at both ambient temperature and pressure
7	A **raw material feedstock should be renewable** rather than depleting whenever technically and economically practical	Biocatalysts are being intensively investigated for their application in renewables processing
8	**Unnecessary derivatisation (blocking group, protection/deprotection, temporary modification of physical/chemical processes) should be avoided whenever possible**	Biocatalysts are often both regio- and stereoselective, obviating the need for protection/deprotection steps
9	**Catalytic reagents** (as selective as possible) are superior to stoichiometric reagents	Biocatalysts are, of course, catalytic

(continued overleaf)

Table 1.1 (*continued*)

	Green Chemistry principle	Notes
10	Chemical products should be designed so that at the end of their function they do not persist in the environment and break down into innocuous degradation products	Biocatalysts are, by their nature, biodegradable after use
11	Analytical methodologies need to be further developed to allow for real-time in-process monitoring and control prior to the formation of hazardous substances	
12	Substances and the form of a substance used in a chemical process should chosen so as to **minimise the potential for chemical accidents, including releases, explosions, and fires**	Again, the use of biocatalysts reduces the risk of these hazards

by naturally occurring organisms and exploiting the enzymatic systems within them.

Associated with the context of clean and sustainable chemical technology is the rapidly growing area of biorefining, in which renewable materials from sustainable resources will be transformed using either chemical or biocatalytic methods in biorefineries, to high-value products. The natural abilities of biocatalysts to degrade bulk natural materials, notably lignin and cellulose, will lend themselves well to the production of valuable platform chemicals such as sugars and phenolics that might themselves be transformed into high-value products using further enzymatic elaboration. Such industrial processes are often intensive and are hence dependent on engineered micro-organisms. One excellent example is in the microbial production of the polymer precursor propane-1,3-diol by DuPont in the USA (http://www2.dupont.com/Sorona/en_US/). In this process, corn starch is broken down to glucose, and then converted into the polymer precursor propane-1,3-diol by the action of an engineered biochemical pathway in the bacterium *Escherichia coli* (Figure 1.3). The propane-1,3-diol derived by biotransformation is then used to make the polymer Sorona™. Such biotransformations by engineered biochemical pathways may in future offer routes to many bulk chemicals from feedstocks that offer an alternative to non-renewable petrochemical sources.

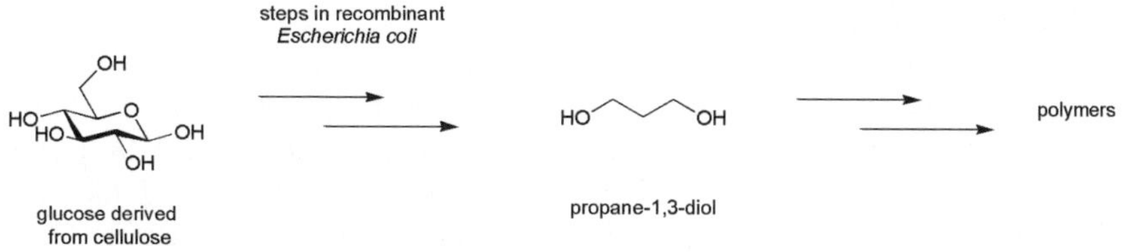

steps in recombinant
Escherichia coli

glucose derived
from cellulose

propane-1,3-diol

polymers

Figure 1.3 Production of propane-1,3-diol from corn-derived cellulose by DuPont

1.1.3 The impact of molecular biology on biotransformations chemistry

In the last twenty years, the ability to acquire, study and exploit the vast wealth of genetic diversity offered by the microbial world has extended to previously unimaginable proportions. The genomes of organisms are sequenced routinely and may be analysed and 'mined' for biocatalysts of interest, as well as the DNA of organisms that have never been isolated using the techniques of 'metagenomics' (see Chapter 2). The speed of genome sequencing is sure to increase over the coming years as the relevant technologies gather pace, offering rapid access to new enzymes and biochemical pathways. Allied to this is the development of bioinformatics, which, with the exponential increase in computer processor power, is offering user-friendly tools for the analysis of genomes and, in some cases, the prediction of the function of enzymes that are encoded within genomes. Laboratory tools for gene cloning and expression are becoming more accessible, affordable and user-friendly and are to be found in an increasing number of laboratories previously working in synthetic chemistry. Last, even the tools for protein engineering, including those that are dependent on structure ('rational engineering'), or random methods of mutagenesis dependent on so-called 'directed evolution' techniques, are becoming increasingly accessible to those not directly associated with the fields of molecular biology and structural biology. Whilst collaboration between groups in different disciplines is now common, there is a much greater appreciation of molecular biology and enzymology by those trained in synthetic chemistry, and the increasing significance of the now-established field of chemical biology should ensure that generations of new chemists are equipped with the knowledge and skills necessary to incorporate biological systems into their work.

Having reviewed the background to the current growth in biotransformations science, we will now examine biotransformations themselves, and the biocatalysts that are responsible for effecting the reactions.

1.2 Biotransformations

A working definition of a biotransformation for the purposes of this book is 'the transformation of a substrate to a recoverable end-product by either microbial or enzymatic means'. The term 'recoverable' makes an important distinction that separates the preparative biotransformations of interest from those 'biotransformations' that are described in the literature that refers largely to drug metabolism studies *in vivo*. The biotransformations that will be considered will be simple analogues of single-step synthetic organic reactions that convert substance A to product B through the action of a single primary catalyst, either enzyme or microbe, and will not therefore, consider multi-step, single-pot enzyme reactions (apart from where cofactor regeneration is an issue) or any aspects of 'pathway engineering'.

Before beginning a more detailed discussion on the nature of biotransformations and the types of reaction that can be catalysed, it would be useful to address briefly the characteristics of biocatalysts themselves, in order to familiarise the worker new to the area with some general considerations of micro-organisms and the enzymes that are derived from them. The range of natural biocatalysts is not restricted to the microbial world of course – there has been much work on catalytic antibodies and ribonucleic acids, for example – but our discussion will focus on micro-organisms and their enzymes, as these will be used most routinely in research and applications.

1.3 Microorganism

Whilst an extended discussion of microbial taxonomy or biochemistry is outside the scope of this book, we have attempted to provide below some information that will be useful in approaching the use of microbes for applications in biotransformations. For more detailed information, useful general introductions to microbiology and microbial biochemistry are provided by both Hans G. Schlegel's book 'General Microbiology' [10] and 'Brock Biology of Microorganisms' [11].

Micro-organisms (eubacteria, archaea, fungi, algae, viruses) are ubiquitous, being able to exploit a huge range of substances for growth in order to survive or thrive in environments that are diverse in terms of growth substrates, temperature, pressure and salinity. It is this naturally evolved diversity that renders their biochemistry so fascinating and amenable to the catalysis of useful reactions and their application in chemistry. The micro-organisms typically used for biotransformations are *eubacteria*, unicellular micro-organisms such as *E. coli, Pseudomonas* or *Lactobacillus* that are encountered in everyday life, *yeasts* such as *Saccharomyces cerivisiae* (common baker's yeast) and microscopic *fungi* such as *Aspergillus* and *Mucor*. There are fundamental differences in the biology and biochemistry of prokaryotic micro-organisms or prokaryotes (the eubacteria and archaea) and eukaryotic micro-organisms or eukaryotes (yeasts and fungi) that have a profound impact, particularly on the genetic manipulation of these organisms or the extraction and application of their genetic material.

1.3.1 Prokaryotes

The prokaryotes are single celled organisms typified by well known species such as *E. coli*. They are distinguished from the other major group, eukaryotes, in that they possess no membrane bound vesicles, known as *organelles*, within the cell but carry out their biochemical functions either at the cell membrane that surrounds the cell or in the soup of proteins, carbohydrates and other biochemicals that constitutes the *cytoplasm*. Their genetic material usually consists of a circular chromosome, sometimes also with a number of smaller circular pieces of DNA known as *plasmids*. For our purposes, prokaryotic organisms can be divided into two groups: the *archaea* and the *eubacteria*.

1.3.1.1 Archaea

The Archaebacteria or Archaea are a kingdom of life in their own right and consist of evolutionarily primitive bacteria that are thought to be the common ancestor of both prokaryotes and eukaryotes. They are distinguished from the other major kingdom of bacteria, the eubacteria, by a number of physiological features including an absence of peptidoglycan in their cell walls and their use of ether rather than ester linkages in their lipid chemistry. Archaea are often found to thrive in extreme environments, such as at high temperatures (thermophiles), cold temperatures (psychrophiles) or pressures (barophiles). Whilst their rather exotic growth requirements dictate that it is unlikely that whole cell preparations of Archaea would be used as biocatalysts, many of the enzymes that one finds in Archaea are, for example, highly thermostable, and are hence extremely interesting from the biotechnology perspective. It is therefore common to find applications of biocatalysts from thermophiles in the biocatalysis literature, from organisms such as *Aquifex, Thermotoga, Thermoanaerobium, Pyrococcus* and *Sulfolobus*, but these enzymes have almost always been prepared using recombinant techniques of the type described in Chapter 7.

1.3.1.2 Eubacteria

Prokaryotic organisms that have been exploited for biotransformations are drawn almost exclusively from this kingdom of life. The eubacteria can be divided into two sub-groups based on their response to a type of staining used for microscopic visualisation called Gram staining. Strains of eubacteria are referred to as either Gram positive or negative, depending on whether the Gram stain can be removed by washing with ethanol. The difference in response is due to a difference in the structures of cell walls in these two groups – Gram-positive bacteria show a purple coloration and Gram-negative bacteria are red. Of the seventeen or so major classes of eubacteria or *phyla*, most biocatalysts have been drawn from only a few of these. Notable amongst these are the Gram-negative Pseudomonad group, which includes organisms such as *Pseudomonas, Burkholderia, Zymomonas* and *Agrobacterium* and the acetic acid bacteria, such as *Acetobacter* and *Gluconobacter*. The Pseudomonads (Figure 1.4) are distinguished by their ability to grow on a large range of carbon sources and possess many enzymes that are capable of processing both natural and xenobiotic compounds of different structural types. Such biodegradative capacity exploiting a wide range of enzymes is of course extremely useful when it comes to preparative biotransformations.

Acetic acid bacteria are capable of oxidising primary alcohols to carboxylic acids and have hence been used for the preparative oxidation reactions in the mode of potassium permanganate or Jones reagent. Gram-positive bacteria are often separated into those which are 'high-GC' or 'low-GC' – a reference to the proportion of the nucleobase cytosine or guanosine that features in their DNA sequence. Of the Gram-positive bacteria, the enteric bacterium *E. coli* is one of the most used bacteria in laboratories worldwide, providing both an early model for bacterial

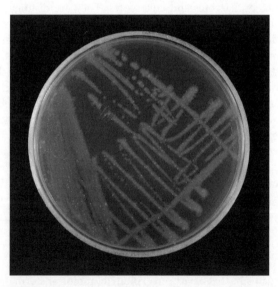

Figure 1.4   Bacteria of the genus *Pseudomonas*, growing on an agar plate

genetics and routinely acting as a host strain for simple cloning experiments. It was also the first to have its entire genome sequenced [12]. *E. coli* is most often used as the host for enzymes from other organisms, expressing their genes in heterologous fashion using techniques described in Chapter 7. The low-GC spore forming Gram-positives such as *Bacillus*, and high-GC organisms such as *Corynebacterium* and *Rhodococcus* have biodegradative abilities that rival the Pseudomonads in enzymatic scope and substrate range. The high-GC Gram-positive bacteria actinomycete group that includes *Streptomyces* and *Rhodococcus* have also provided a host of valuable biocatalysts. The genus *Streptomyces* (Figure 1.5), which is most well-known for its ability to form a large amount of structurally diverse antibiotics, presents an impressive reservoir of enzymes that are used biosynthetically to make these valuable secondary metabolites. Such enzymes are often used to elaborate core structures such as macrocylic polyketides that have been assembled by large, processive enzyme complexes. Some of these interesting enzyme activities have been recruited for single-step biotransformations such as for preparative hydroxylation reactions.

In Chapter 3 we provide more detail about the growth requirements and characteristics of the bacteria used in biotransformations and give examples of preparative biotransformations catalysed by them.

1.3.2 Eukaryotes

The other major kingdom of life, the eukaryotes, provides other major examples of biocatalysts – these predominantly being drawn from the filamentous fungi

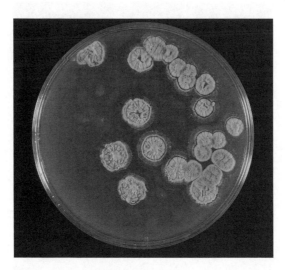

Figure 1.5   Bacteria of the genus *Streptomyces*, growing on an agar plate

and yeasts. The kingdom of eukaryotes covers many organisms from yeast to plants and mammals, but the 'higher' eukaryotes have of course seen a more limited application in biocatalysis. There are examples of the use of algae, plant cell cultures and indeed, cultured mammalian cells, and these have been shown to possess useful enzymatic activities. Our discussion will be mostly limited to the yeasts and fungi.

Eukaryotes, in contrast to prokaryotes, possess subcellular organelles, some of which are responsible for such biochemical processes as respiration (mitochondria) or, where relevant, photosynthesis (chloroplasts) and their genetic material is contained within the nucleus of the cell, bounded by a nuclear membrane. They can be single celled (yeasts) or multicellular (fungi).

1.3.2.1 Yeasts

Yeasts are single-celled eukaryotic organisms and, as one of the simplest forms of eukaryote, have been used as the model for many eukaryotic biochemical processes and genetics, including the first genome sequence of a eukaryotic organism, namely, that of *Saccharomyces cerivisiae* (Figure 1.6) [13].

They have also been hugely significant in biotechnology, initially as the active ingredient in both breadmaking and brewing (*S. cerivisiae; Schizosaccharomyces pombe*). Yeasts are easy to grow in the laboratory and, indeed are available not only from chemical suppliers but are also available in freeze-dried form from high-street

Figure 1.6   Freeze-dried *Saccharomyces cerivisiae*, a yeast often used for biotransformation reactions

supermarkets, so the applications of, for example *S. cerivisiae* or baker's yeast have been widespread, notably for their capacity to catalyse enantioselective reductions of carbonyl groups using alcohol dehydrogenase activities. Other yeasts, notably non-pathogenic strains of *Candida*, have also been used for reduction reactions, and have also provided enzymes such as formate dehydrogenase, one of the most important enzymes in cofactor recycling techniques.

1.3.2.2 Filamentous fungi

Filamentous fungi, sometimes called moulds, are multicellular organisms that grow often on dead or decaying matter, notably wood, which they are well-equipped to degrade using a powerful arsenal of ligninolytic and cellulolytic enzymes. On agar plates (Figure 1.7), the filamentous fungi usually appear as mats of *mycelia*, which are in turn made up of intertwining filaments called *hyphae*, which are extended filaments made up of many cells.

Some hyphae project upwards from the mat, and bear sporulation structures – the spores that are produced by such fungi are often visible as dark spots dusting the top of the fungal mat. Their capacity for biodegradation makes them, like many species of bacteria, extremely useful in the biotransformation of xenobiotic substrates, and they have been particularly highly prized as catalysts of *hydroxylation* reactions. Fungi such as species of *Penicillium* were shown in early studies in the 1950s at Eli Lilly and Upjohn to hydroxylate the steroid nucleus both

Figure 1.7 A filamentous fungus (*Mucor fragilis*) growing on an agar plate, showing characteristic mat made up of the fungal mycelium

regio- and *stereo*-specifically, and a wide range of substrates for hydroxylation by fungi have been described since. Other filamentous fungi commonly encountered in the biotransformations literature are *Aspergilllus, Rhizopus, Fusarium, Mucor* and *Beauveria*. Many of these have an established history in biotechnology as antibiotic producers, foodstuffs and as producers of commodity chemicals such as citric acid. Fungi, as eukaryotes, have more complex genetics than simple prokaryotes such as bacteria, but progress is being made on the genomics of fungi, and the genomes of *Neurospora crassa*, and a number of species of *Aspergillus* have been sequenced [14]. Macroscopic fungi such as ordinary mushrooms can also be a source of enzymes, such as tyrosinase, an oxidase which converts catechol to benzoquinone, but such organisms are of course less convenient to culture using standard laboratory techniques.

1.3.3 Plants and animals as sources of enzymes for biotransformations

There have been a number of reports of the use of plant cell cultures as biocatalysts [15], notably, tobacco and geranium, and also of useful enzymes from plants such as spinach and potato. It is certainly true that plants possess a fantastic biosynthetic capability that can be recruited for applied biocatalytic reactions, and indeed the sequenced genomes of organisms such as *Arabidopsis thaliana* have revealed many useful enzymes, such as glycosyltransferases, that can be exploited using biocatalysis [16]. However, plant cells in culture are comparatively difficult to grow reliably, and the exploitation of plant enzymes such as the glycosyltransferases of *Arabidopsis* have been accessed using recombinant biotechnology in easy-to-handle laboratory hosts such as bacteria and fungi.

Whilst animals were once the source of useful biocatalysts, notably the esterase from porcine liver (PLE) and alcohol dehydrogenase from equine liver (HLADH), the advent of recombinant biotechnology and the emergence of a huge and ever-increasing diversity of microbial biocatalysts of comparable or superior activity has reduced the use of these enzymes; indeed, the use of enzymes from animal sources is often precluded, particularly in the preparation of substances for human consumption and for which the use of animal products is now prohibited.

1.4 Organism Nomenclature

The nomenclature of micro-organisms is important as it enables us to distinguish between species of related bacteria/fungi and strains that may be available from different sources. As molecular biological techniques evolve, so do the rules of taxonomy and nomenclature and it is not unusual to find that an organism has been renamed. The name of a micro-organism is usually constructed from a prefix that defines its *genus* and a suffix that denotes the *species*. This name is italicised. There is also usually a 'strain descriptor' which should be specific to one strain, or one that is genetically and biochemically distinct. This is often in the form of a mixed number-letter code at the end of the organism's name. In the context of biotransformations, it is the strain descriptor that is also vitally important, as it may only be one certain strain of, for example, *Pseudomonas putida* that possesses the activity required.

Hence, using a bacterial example, the genus *Pseudomonas* includes species such as *Pseudomonas putida, Pseudomonas fluorescens* and *Pseudomonas aeruginosa*. The strain of the organism is usually specified with a suffix that often denotes either the accession number of that strain to a culture collection (see below) such as *Pseudomonas putida* ATCC (American Type Culture Collection) 17453 or, the code given to a strain by the researchers who discovered it, such as *P. putida* M10. Microscopic fungi and yeasts bear equivalent names, *Aspergillus niger, Mucor fragilis, Beauveria bassiana, S. cerivisiae* and so on; again qualified in each case by a strain descriptor such as *B. bassiana* ATCC 7159.

In addition, genetically modified organisms, particularly strains of *E. coli* that are routinely used for recombinant gene expression and thence as biocatalysts, become new strains once the foreign gene, located on a circular piece of DNA called a plasmid, has been taken up by the bacterium. Hence, the commonly used and commercially available expression strain of *E. coli* BL21 (DE)3, which has been transformed with (or taken up) a plasmid pXYZ, may be referred to as *E. coli* BL21 (DE3) pXYZ. An introduction to simple recombinant DNA techniques for creating recombinant or 'designer' biocatalysts is provided in Chapter 7.

Ultimately, it is important to stress that the enzyme complements of two different strains of the same species for example, *P. putida*, could be very different. The biocatalytic properties of an organism may well be restricted to the specific *strain* of an organism described by the literature and that strain must be obtained in an effort to repeat published results.

As we shall see, both whole cells of the organisms described above, and the enzymes that exist within those cells, may be used as biocatalysts. We will now consider the nature of enzymes and the possible useful chemical reactions that may be exploited in preparative applications.

1.5 Enzymes

Enzymes are the catalysts of the majority of biochemical reactions in living cells. They are proteins and as such are polymers made up of combinations of twenty amino acids, usually up to hundreds of amino acids in length, and typically have molecular weights in the range of 10 000–100 000 Da, although the polymeric chains can associate to form larger functional complexes.

The twenty amino acids (Appendix 1) each possess a distinct side-chain, conferring chemical and electronic properties which, when combined within the active site of an enzyme, perhaps with a small coenzyme or metal ion can confer the ability to catalyse a wide range of chemical catalytic processes including oxidations, reductions, hydrolyses and carbon-carbon bond formation. It is common to find these amino acids referred to by either a three-letter code (e.g. Tyr for tyrosine) or a single letter code (e.g. Y for tyrosine) when reading about enzymes in biocatalysis publications. Whilst each amino acid is itself relatively simple in structural and chemical terms, their combination to form proteins gives rise to several levels of *hierarchical structure* which are summarised briefly below:

- *Primary structure:* This refers to the sequence of the amino acids in the protein chain, which are connected together by amide (peptide) bonds formed in a condensation reaction on the ribosome – the cellular organelle of protein synthesis. It is common to see the primary structure of enzymes displayed as sequences of their amino acids in a single letter code (Figure 1.8).
- *Secondary structure:* This refers to local folding of the polypeptide chain, into discrete, three-dimensional structures, such as *alpha-helices* and *beta-strands*, the latter of which associate to form *beta-pleated sheets*. These structures are themselves often connected by loops and turns, which may themselves be ordered or disordered.
- *Tertiary structure:* The collection of helices, sheets, and other turns and loops is folded into a globular structure in enzymes that is termed the tertiary structure (Figure 1.9). This structure is stabilised by a number of usually weak binding interactions between amino acids and their side-chains, or the peptide backbone, including van der Waals contacts, extensive hydrogen bonding and electrostatic interactions ('salt bridges') between positively and negatively charged residues (such as lysine and aspartate, respectively). The structure is also stabilised by a number of interactions with water molecules.
- *Quaternary structure:* Should more than one polypeptide chain associate in order for an enzyme to function, this structure of associated proteins is known as the quaternary structure (Figure 1.10). For example, the quaternary structure of the lyase enzyme shown in Figure 1.10 is made up of three identical monomeric subunits.

```
MKQLATPFQEYSQKYENIRLERDGGVLLVTVHTEGKSLVWTSTAHDELAY
CFHDIACDRENKVVILTGTGPSFCNEIDFTSFNLGTPHDWDEIIFEGQRL
LNNLSIEVPVIAAVNGPVTNHPEIPVMSDIVLAAESATFQDGPHFPSGIV
PGDGAHVVWPHVLGSNRGRYFLLTGQELDARTALDYGAVNEVLSEQELLP
RAWELARGIAEKPLLARRYARKVLTRQLRRVMEADLSLGLAHEALAAIDL
GMESEQ
```

Figure 1.8   The primary structure of an enzyme, or its sequence of amino acids

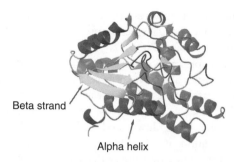

Figure 1.9 The tertiary structure of an enzyme (pdb code 1ein) illustrating both of the secondary elements. The alpha-helices are shown in red and the beta-strands in yellow. The strands can be seen to lie together to form a beta-pleated sheet. Together with a variety of loops and turns, these elements make up the *tertiary* structure of the protein monomer as shown

The weak interactions that operate in enzymes are crucial for structural and functional integrity. The polypeptide will also often interact with non-proteinaceous molecules, cofactors or coenzymes, which may be as simple as metal ions such as zinc, or small, yet complex organic molecules such as vitamins. These cofactors can be crucial for stability and/or activity and considerations of these elements are as fundamental in the study of biochemistry and applications

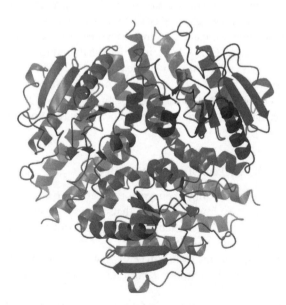

Figure 1.10 The quaternary structure of a protein as represented by the trimeric structure of a lyase enzyme (pdb code 1o8u). The constituent monomers are shown in red, blue and green

of enzymes as the proteins themselves. An excellent introduction to the structure and function of proteins and enzymes is provided by 'Protein Structure and Function' by Petsko and Ringe [17].

Enzymes are remarkable catalysts, in some examples raising the catalytic rate of reaction some 10^{18} times above the non-catalysed reaction and in some cases are certainly 'perfect' catalysts, catalysing reaction rates at limits only imposed by rates of molecular diffusion. They catalyse an extraordinary range of chemical reactions, both catabolic (break-down) and anabolic (synthetic) and as such are responsible for the creation of a bewildering array of molecular architectures that can be observed in the extraordinary diversity of known natural products. Whilst it may not be entirely true to say that there is a biochemical equivalent for every known synthetic chemical reaction, many of the reactions catalysed by Nature have direct equivalents in synthetic chemistry. We offer a brief overview of the range of chemical reactions catalysed by enzymes below – a more detailed introduction is offered by Professor Tim Bugg's book 'An Introduction to Enzyme and Coenzyme Chemistry' [18].

1.6 Types of Enzymatic Reactions

In 1956, the Enzyme Commission, under the auspices of the International Union of Biochemistry, organised the types of enzymatic reaction into six classes, by which enzymes can be described using their so-called 'E.C. numbers' (http://www.chem.qmul.ac.uk/iubmb/enzyme/). Each enzyme is accorded a four-figure designation (E.C. X.X.X.X), in which the first number is the number of the representative class of six into which the enzyme falls. The next two numbers subdivide the enzyme classes further, perhaps on the basis of substrate class, or the type of cofactor employed, and the last is the number of that enzyme in its subclass. The six classes of enzymes are listed below, with some simple examples given in each case.

1.6.1 E.C. 1.X.X.X. Oxidoreductases

Oxidoreductases are a vast family of enzymes that catalyse oxidation and reduction reactions. In this class are found enzymes that catalyse, for example, hydroxylation reactions, ketone-alcohol interconversions, the reduction of carbon-carbon double bonds and the oxidation of amines.

One example is cytochrome P450cam (E.C. 1.14.15.1), an enzyme that hydroxylates the terpene molecule camphor **1** both regio- and stereoselectively to yield 5-*exo* hydroxy camphor (Figure 1.11).

Another example of an oxidoreductase is lactate dehydrogenase (E.C. 1.1.1.27), that catalyses the interconversion of lactic acid and pyruvic acid (Figure 1.12).

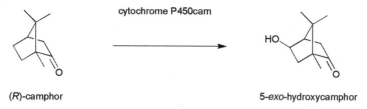

cytochrome P450cam

(R)-camphor   5-exo-hydroxycamphor

Figure 1.11   Hydroxylation of camphor by cytochrome P450cam

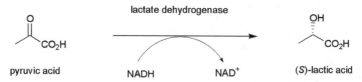

lactate dehydrogenase

pyruvic acid   NADH   NAD$^+$ (S)-lactic acid

Figure 1.12   NADH-dependent reduction of lactate by lactate dehydrogenase

Amine oxidases (E.C. 1.4.3.4) catalyse the oxidation of amines to imines, which then react with water in the aqueous environment of enzymes to form carbonyl groups, with the release of ammonia. Some of these enzymes are dependent on copper, whereas some, like the enzymes from *Aspergillus niger*, use a flavin cofactor (Figure 1.13).

The oxidation of molecules in chemistry is of course extremely important as it offers a route into the functionalisation of non-activated chemicals such as hydrocarbons. Oxidoreductase enzymes have therefore been a subject of great interest for biotransformations research, but they also present somewhat complex problems. Oxidoreductases are very often dependent on non-proteinaceous cofactors such as nicotinamide cofactors (NAD, NADP – see below), flavins (FAD, FMN) or haem, or even perhaps the presence of other proteins acting as part of an electron transport chain with the enzyme of interest (e.g. cytochrome P450cam above). As such, they have been perceived as challenging from the perspective of applied biocatalysis, as the cofactors may require recycling, having been used up in one catalytic cycle, or introduce extra considerations of stability and expense, as they may have to be added as separate reagents to the enzymatic reaction.

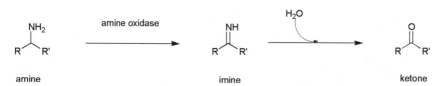

amine oxidase   H$_2$O

amine   imine   ketone

Figure 1.13   Oxidation of amines to imines, and then to ketones, by amine oxidases

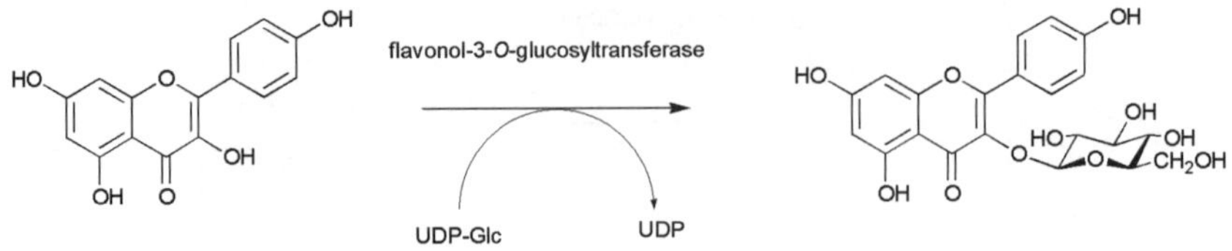

Figure 1.14   Reaction catalysed by flavonol-3-*O*-glucosyltransferase

1.6.2 E.C. 2.X.X.X. Transferases

Transferases are enzymes that catalyse group transfer reactions. In this class are found glycosyl transferases, that transfer sugar units from activated sugar donors to acceptor molecules to form glycosides, and kinases, which catalyse phosphorylation reactions.

One example is flavonol-3-*O*-glucosyltransferase (E.C. 2.4.1.91). This enzyme transfers glucose from the activated sugar donor UDP-glucose to a flavonoid acceptor substrate (as shown in Figure 1.14).

Sugar transferases are also perceived as challenging owing to their dependence on activated sugar donors, themselves being composed of sugar-base-phosphate nucleotides which are expensive to add in stoichiometric amounts. Other examples of transferases include kinases, which transfer phosphate groups and are involved in cell signalling processes, but can also be exploited in biotransformation reactions for the synthesis of phosphate esters.

1.6.3 E.C. 3.X.X.X. Hydrolases

Hydrolases are enzymes that catalyse the cleavage of bonds using water. The most widely used examples are lipases and proteases, that cleave C-X bonds, but C-C bond hydrolases are also known.

A simple example is offered by the protease chymotrypsin (3.4.21.2), which has seen many applications in organic synthesis, including the hydrolysis of peptides, but can also be used for the cleavage of ester bonds in enantioselective hydrolysis reactions to yield optically pure alcohols and carboxylic acids (Figure 1.15). This

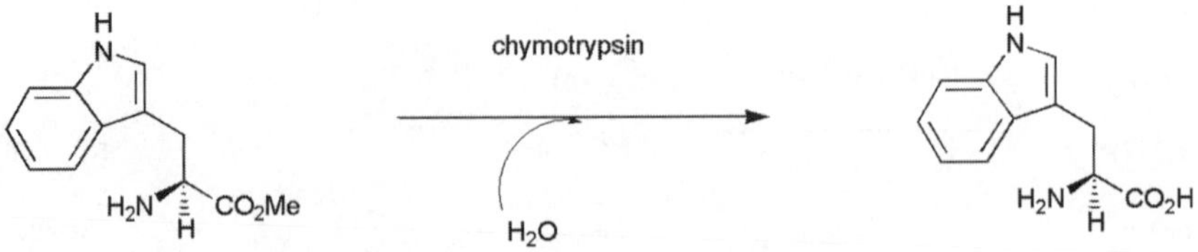

Figure 1.15   Hydrolysis of an ester by chymotrypsin

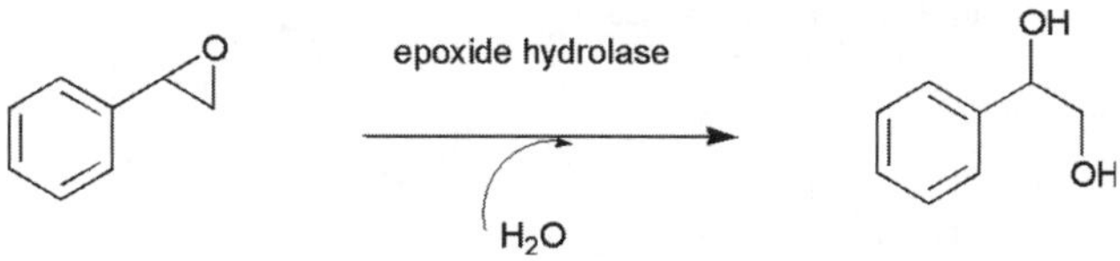

Figure 1.16 Hydrolysis of an epoxide by an epoxide hydrolase

reaction is also routinely performed by esterases (e.g. E.C. 3.1.1.13) and lipases (E.C. 3.1.1.3) which cleave the carboxy ester function in Nature.

Epoxide hydrolase enzymes (E.C. 3.3.2.X) perform the ring-opening of epoxides to form vicinal diols. If the epoxide is chiral, these enzymes are also often capable of performing kinetic resolutions of the substrates, to give residual epoxides and product diols, each with high optical purity (Figure 1.16).

Glycosidases (3.2.1.X) perform the hydrolysis of glycosidic bonds between sugars. The enzyme can be highly regio- and stereoselective and, in some cases, using either medium or protein engineering, can be exploited to catalyse the *synthesis* of glycosidic bonds (Figure 1.17).

The use of hydrolases has become even more widespread once it was realised that in many cases they would catalyse their natural reaction in the *reverse* direction if suitable conditions, notably low-water conditions in organic solvents, were applied. We will briefly examine the mechanistic basis for this reversal of activity in lipase enzymes in Chapter 4. An example of a reverse hydrolysis, an esterification reaction using the lipase B enzyme (E.C. 3.1.1.3) from the yeast *Candida antarctica* is shown in Figure 1.18.

1.6.4 E.C. 4.X.X.X. Lyases

Lyases are enzymes that catalyse the addition of water or ammonia across double bonds. This can often be catalysed with excellent selectivity, giving rise to enantiomerically pure alcohols or amines. A good example is that of the industrial enzyme fumarase (E.C. 4.2.1.2), which catalyses the addition of water across the double bond in fumaric acid to give malic acid (Figure 1.19).

Another example of an enzyme with lyase activity is phenylalanine ammonia lyase (E.C. 4.3.1.5), which catalyses the reversible addition of ammonia to a

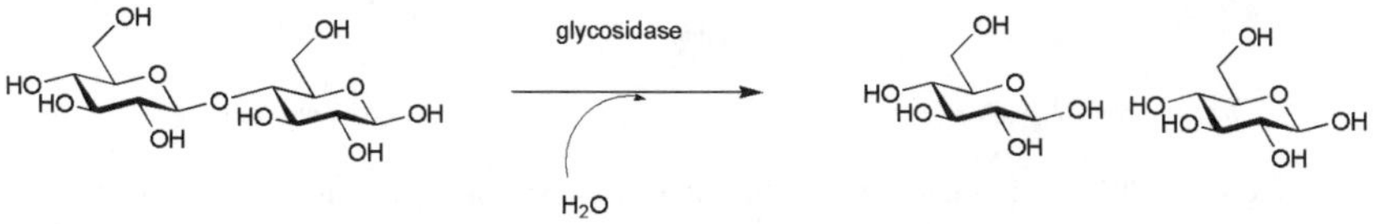

Figure 1.17 Hydrolysis of a glycosidic bond by a glycosidase

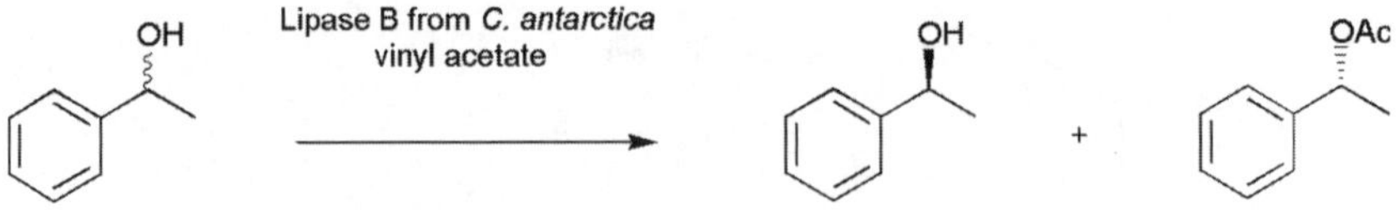

Figure 1.18   An esterification reaction catalysed by *Candida antarctica* lipase B

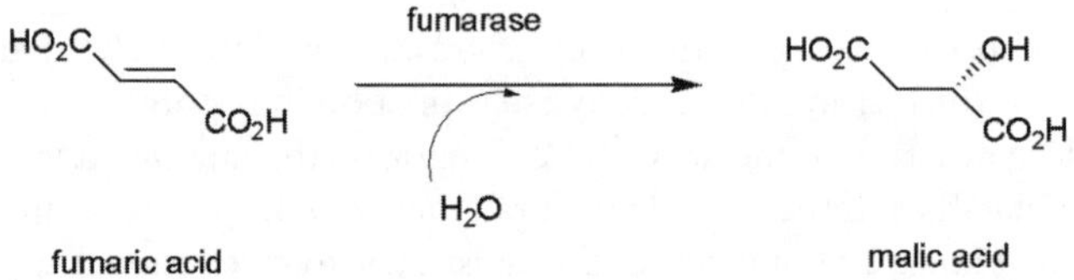

Figure 1.19   Addition of water to fumaric acid catalysed by fumarase

cinnamic acid derived substrate. These lyases can be useful in preparative organic chemistry for the synthesis of chiral amines (Figure 1.20).

1.6.5 E.C. 5.X.X.X. Isomerases

Isomerases are enzymes that catalyse isomerisation reactions, such as hexose-pentose isomerisations. The isomerisation of glucose-6-phosphate to fructose-6-phosphate catalysed by glucose-6-phosphate isomerase (E.C. 5.3.1.9) is one of the most economically significant biotransformations (Figure 1.21).

Inversions of absolute configuration are also possible, using racemase enzymes such as mandelate racemase (E.C. 5.1.2.2) (Figure 1.22). These enzymes have recently attracted interest because of their potential application as catalysts of stereo inversions and for dynamic kinetic resolution experiments.

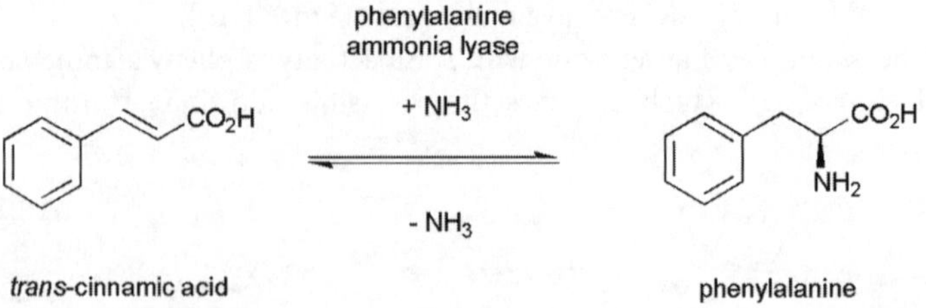

Figure 1.20   Addition of ammonia to a double bond by phenylalanine ammonia lyase

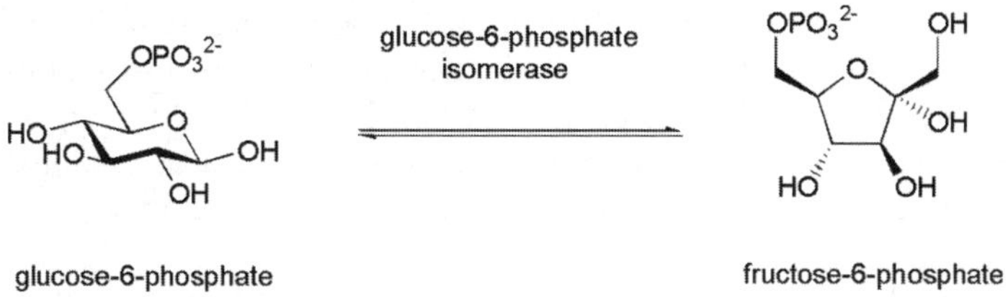

Figure 1.21 Isomerisation of glucose-6-phosphate by glucose-6-phosphate isomerase

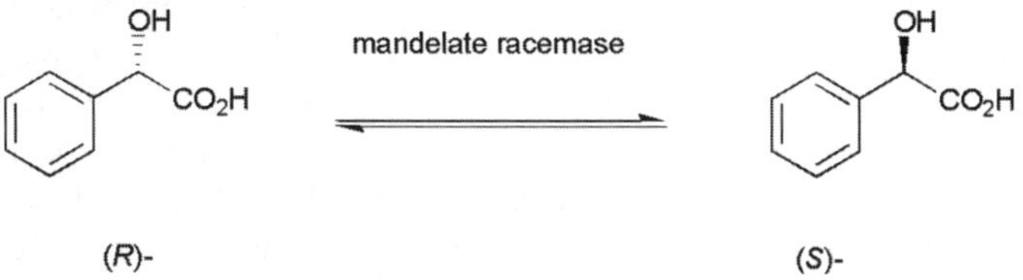

Figure 1.22 Racemisation of mandelic acid catalysed by mandelate racemase

1.6.6 E.C. 6.X.X.X. Ligases

Ligases are enzymes that catalyse the formation of bonds, with the concomitant hydrolysis of the coenzyme adenosine triphosphate (ATP) to adenosine diphosphate (ADP) and phosphate. Whilst potentially very useful, these enzymes have seen limited applications in the biotransformations literature. One example of a ligase is aspartate-ammonia ligase (E.C. 6.3.1.1), which catalyses the ATP-dependent reaction of ammonia with L-aspartate to form L-asparagine (Figure 1.23).

1.6.7 Web-based resources for enzymes – BRENDA

There are a number of useful web-based resources that provide details of enzyme function based on the E.C. number classification. One of these is BRENDA, 'The Comprehensive Enzyme Information System' (http://www.brenda.uni-koeln.de/), which is administrated from the University of Köln in Germany. It provides details

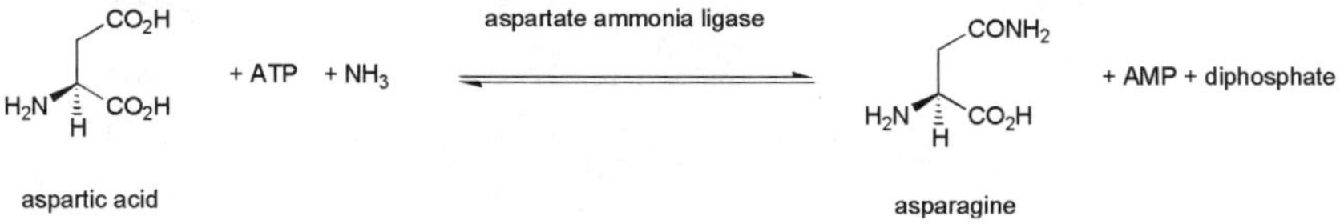

Figure 1.23 Transformation of L-aspartate to L-asparagine catalysed by aspartate-ammonia ligase

of enzyme function, information on substrate specificities, stability and other characteristics and links to literature and other websites such as structure-related databases which can be accessed by text-based searching or using the E.C. number of the enzyme of interest. The database also includes sections on the ability to cross-reference with organism names, and sections on applications in biotechnology.

Given the changing nature of enzymology in the last fifty years, with genome sequencing, bioinformatics and molecular biology giving information on the evolutionary relationship of enzymes as dictated by either tertiary structural fold or amino-acid sequence, it may be that the E.C. system is a somewhat simplistic method of classifying enzymes. However, it is still extremely useful in distinguishing between six broad areas of chemical transformation catalysed by enzymes, and gives researchers a point of entry for the study of a new enzyme system.

It is telling that of all the six classes of enzyme, it is the hydrolases (E.C. 3.X.X.X) that have found the greatest application in organic synthesis. If we review the papers on applied biotransformations that have appeared in the major twenty or so synthetic organic chemistry journals in 2007, we observe that just over half of these papers detail the application of hydrolase enzymes (Figure 1.24). This is because they are easy to use, being often stable, cheap, readily available from a variety of commercial sources, and depend only on water for activity. The proportional use of hydrolases would be much higher if only isolated enzyme examples were represented. The proportion of oxidoreductase reactions is also high, but many

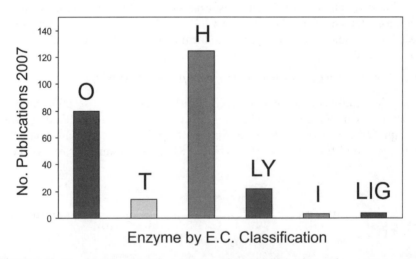

Figure 1.24 Bar chart illustrating the proportion of use of different classes of enzymatic reaction reported in papers in the major synthetic organic chemistry journals in 2007. (O) Oxidoreductase E.C. 1.X.X.X; (T) transferase E.C. 2.X.X.X; (H) Hydrolase E.C. 3.X.X.X; (LY) Lyase E.C. 4.X.X.X; (I) Isomerase E.C. 5.X.X.X; (LIG) Ligase E.C. 6.X.X.X

of these will describe the use of oxidoreductase enzymes within whole cells. Many other transformations throughout the six classes would be highly prized chemical transformations, yet are often complicated by the requirement for enzymatic cofactors or auxiliary electron transport proteins, non-proteinaceous molecules such as small organic molecules or metal ions, that are essential for catalysis.

1.7 Enzymatic Cofactors

We have already alluded to the fact that whilst some enzymes are structurally and chemically simple, merely requiring the folded protein to effect catalysis in the presence of perhaps water, many other chemical transformations catalysed by enzymes require small molecule cofactors that may be as simple as a metal ion, or be more complicated such as one or other kind of small organic vitamin molecule. In whole cell biotransformation reactions, these cofactors are most usually produced and recycled in the organism. However, in isolated enzyme biotransformations, they may pose an additional consideration. They may be covalently or non-covalently bound to the protein molecule, and if the interaction is weak or transient, it may be necessary to add this cofactor, with the additional considerations of expense and complexity that this would imply.

This brief discussion will be restricted to a few of the common coenzymes and cofactors found in synthetically useful enzymes that will feature in this book. For a useful introduction to many of the specifics for enzyme–cofactor interactions and their mechanisms, the reader is referred to reference [18].

1.7.1 Nicotinamide cofactors [NAD(P)/H]

The nicotinamide cofactors such as (nicotinamide adenine dinucleotide phosphate) will be routinely encountered when using oxidoreductase-catalysed reactions, and must be added to the reaction in either stoichiometric amounts, which is prohibitively expensive at scale, or more usually, recycled by the action of an auxiliary enzyme, acting on an alternative auxiliary substrate as detailed in Chapter 5. Nicotinamide cofactors will most commonly act as either the donor or acceptor of chemical reducing equivalents, for example, by adding hydride to the substrate itself in much the same mode as sodium borohydride, or, donating/accepting reducing equivalents from/to another cofactor (perhaps a flavin) or a different protein cofactor as part of an electron transport chain.

The structure of NADPH is shown in Figure 1.25. In the case of a simple reduction of a carbonyl group by an alcohol dehydrogenase (Figure 1.26) (E.C. 1.1.1.1), hydride is delivered stereospecifically to one enantiotopic face of the prochiral ketone from the dihydronicotinamide ring, which is consequently aromatised. In the example reaction, the reduced cofactor must be regenerated (see Chapter 6) for use in another catalytic cycle.

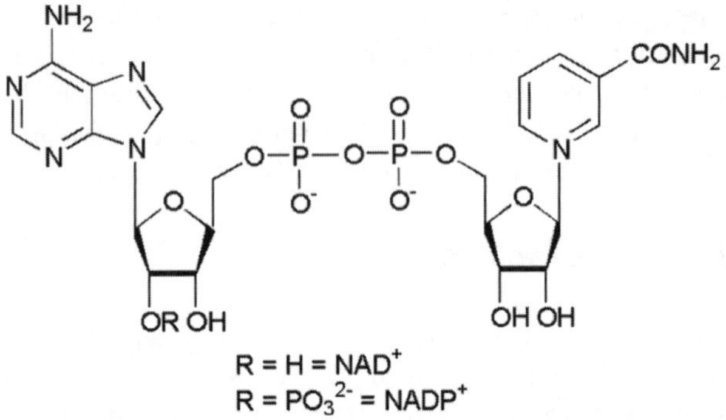

Figure 1.25   Structure of NAD(P)H

NADPH retails for approximately £ 600 g^{-1}, but its analogue, NADH, which differs only in the removal of the 2' hydroxyl on the ribose, retails at about one-tenth of this price, offering a more economic solution to coenzyme-dependent biocatalysis in some cases. However, unfortunately, it is rarely the case that enzymes dependent on NADPH will also use NADH as efficiently, so unless an enzyme that catalyses a reaction equivalent to the one required can be found that employs NADH, the researcher may either have to optimise conditions for the use and recycling of NADPH, or resort to protein engineering in an effort to change the coenzyme specificity.

1.7.2 Flavins [flavin mononucleotide (FMN) and flavin adenine dinucleotide (FAD)]

Flavins are also often found in oxidoreductase enzymes, and also mediate electron transport, but are able to participate in single electron transfer steps through the

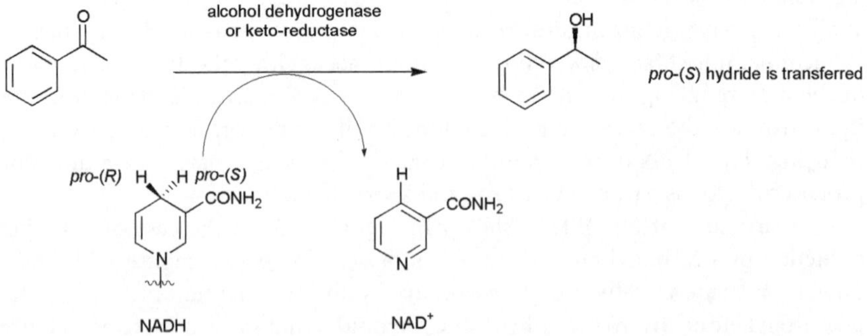

Figure 1.26   Reduction of a prochiral ketone by an alcohol dehydrogenase showing stereospecific transfer of hydride from the nictonamide ring of NADH.

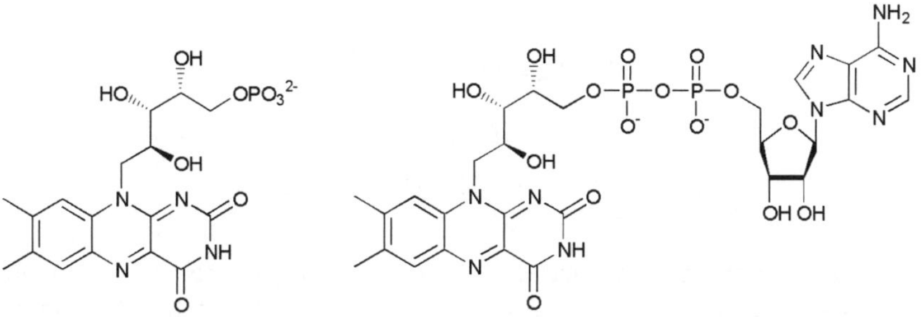

flavin mononucleotide (FMN)

flavin adenine dinucleotide (FAD)

Figure 1.27 Structures of flavin mononucleotide (FMN) and flavin adenine dinucleotide (FAD)

generation of radical species. In some cases, the flavin, FAD, must be reduced to $FADH_2$, which then reacts with molecular oxygen to generate an oxidising species such as a hydroperoxy flavin that acts in a similar fashion to peracid reagents such as *m*CPBA on alkene substrates, to give epoxides, or ketone substrates to give esters or lactones in an equivalent of the Baeyer–Villiger reaction.

The structure of FAD is shown in Figure 1.27. In many cases, FAD is covalently bound to the enzyme of interest and thus does not need to be added to isolated enzyme reactions. However, this is not always the case, and FAD and FMN must sometimes be added to either ensure stability during a purification protocol or to reconstitute full enzyme activity in assays or biotransformation reactions.

An example of a flavin dependent reaction is the transformation of the aromatic substrate vanillyl alcohol by vanillyl alcohol oxidase (VAO) (Figure 1.28).

1.7.3 Pyridoxal phosphate (PLP)

PLP is most often associated with enzymes that catalyse the transformation of amino acid substrates, including decarboxylations, racemisations and transamination reactions. The structure of PLP is shown in Figure 1.29. The aldehyde

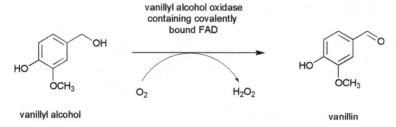

vanillyl alcohol oxidase
containing covalently
bound FAD

O_2 H_2O_2

vanillyl alcohol

vanillin

Figure 1.28 The oxidation of vanillyl alcohol by an oxidase that uses FAD as a coenzyme

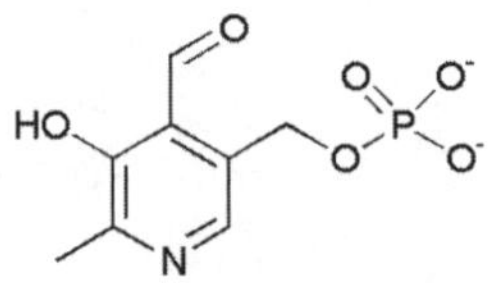

Figure 1.29 Structure of pyridoxal phosphate (PLP)

moiety in the pyridine ring reacts with the amino group of amino acid substrates to yield a covalent adduct, which, through a variety of chemical transformations, gives rise to the diverse chemical reactions mentioned. PLP is not routinely added to isolated enzyme biotransformations that use PLP dependent enzymes as the coenzyme is covalently bound to an active site lysine residue that is displaced by the amino acid group on substrate binding, and then reconstituted after the catalytic cycle is complete.

An example of a PLP dependent reaction is the isomerisation of the amino acid glutamate by glutamate racemase (Figure 1.30). In this reaction, the amino acid forms a complex with PLP in the active site of the enzyme, and a proton is abstracted from the C-alpha of the amino acid from one face of the resulting covalent intermediate. This intermediate is then reprotonated from the other face to give an inversion of stereochemistry.

1.7.4 Thiamine diphosphate (ThDP)

ThDP is a vitamin cofactor that is used by a number of enzymes, often in lyases that catalyse carbon-carbon bond formation. The structure of ThDP is shown in Figure 1.31.

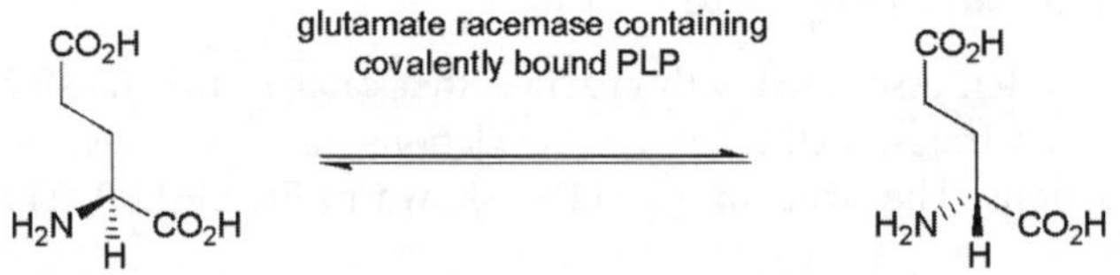

Figure 1.30 Isomerisation of glutamic acid by the PLP-dependent enzyme glutamate racemase

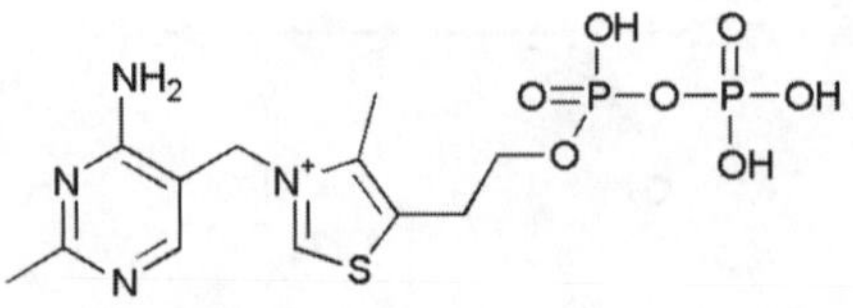

Figure 1.31 The structure of thiamine diphosphate (ThDP)

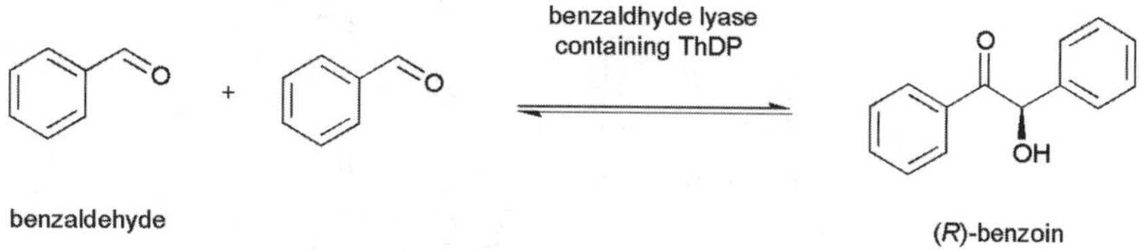

Figure 1.32   The ligation of two benzaldehyde molecules by ThDP-dependent benzaldehyde lyase

Carbon-carbon bond forming reactions by this enzyme are usually dependent on the generation of a carbanion at the C-2 position of the thiazolium ring – this acts as a nucleophile for attack at the carbonyl group of, say, an aldehyde to give a covalent enzyme-bound intermediate that is attacked by another molecule to form a C-C- bond. One simple example is illustrated by benzaldehyde lyase (Figure 1.32) (4.1.2.38), an enzyme that catalyses the ligation of two benzaldehyde molecules to form the optically active benzoin. The binding of ThDP is usually associated with the binding of magnesium ions, and these and the coenzyme itself, which is relatively inexpensive, are often added to the relevant isolated enzyme reactions.

1.7.5 Haem

Haem is a coenzyme known best as being the active moiety in haemoglobin and other oxygen-binding or oxygen transport proteins. It is also the basis for the central reactive species in a number of oxidative biotransformations such as hydroxylation, epoxidation and O- or N-demethylation reactions. Haem, as shown in Figure 1.33, consists of a tetracyclic protoporphyrin ring containing four pyrrole rings that are coordinated to a central iron atom through ring nitrogens. The haem is usually covalently bound to the enzyme via an iron-sulfur

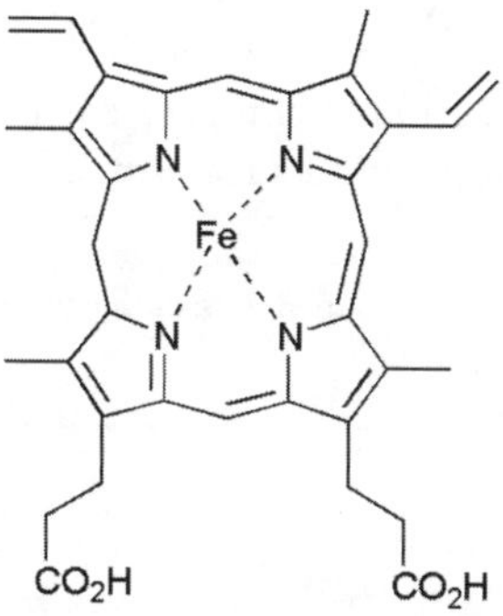

Figure 1.33   Structure of haem

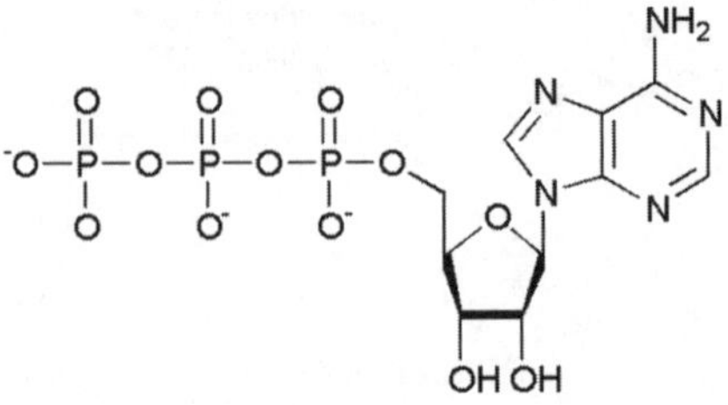

Figure 1.34   Structure of adenosine triphosphate (ATP)

bond to a cysteine residue in one of the other coordination positions. The last coordination position is usually taken either by a water molecule or a molecule or atom of oxygen. It is the iron-bound oxygen that is the oxidising species in haem-dependent enzyme catalysed biotransformations. As cytochromes P450 are most often exploited as catalysts in whole cells, the addition of haem is not necessary and stability considerations do not arise.

1.7.6 Adenosine triphosphate (ATP)

ATP has been described as the energy currency of biochemical systems. Many biochemical reactions and processes are dependent on the energy that is released by the hydrolysis of ATP to ADP. The structure of ATP is shown in Figure 1.34.

Some enzymes used in biotransformations require the addition of ATP, particularly ligases, for instance in DNA manipulations but also in preparative synthetic reactions such as pyruvate carboxylase (E.C. 6.4.1.1), which catalyses the ligation of pyruvate and carbonic acid to give oxaloacetate (Figure 1.35). Phosphorylation reactions, which are becoming more widely used in preparative biocatalysis, also employ ATP as cofactor. ATP may also be recycled using auxiliary enzyme–substrate systems.

1.7.7 Coenzyme A (CoA)

CoA (Figure 1.36) is a structurally complex coenzyme that features an ADP residue connected to a pantotheinic acid substituent, and then a cysteamine that

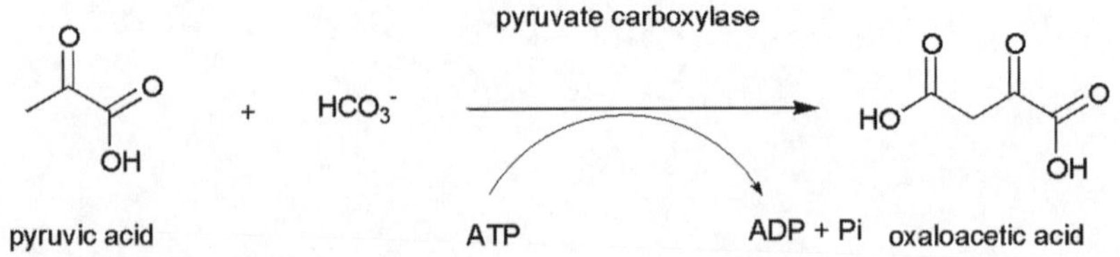

Figure 1.35   Ligation of pyruvic acid and carbonic acid catalysed by pyruvate carboxylase

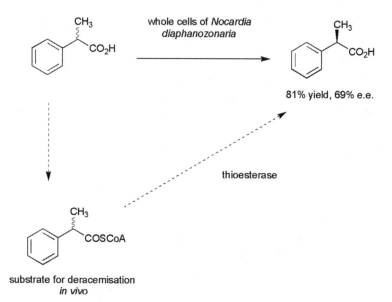

Figure 1.36 Structure of Coenzyme A

terminates in a thiol function. CoA ligases catalyse the formation of CoA thioesters which are then used to transfer acyl groups to other chemical entities through the action of acyl CoA transferases.

CoA is also often involved in the transformation of carboxylic acid substrates – it is sometimes necessary that these are activated for catalysis by the formation of the CoA thioesters. The requirement for CoA in a biocatalytic process may be obscured, perhaps by the use of whole cells, and will therefore constitute an issue should the reaction need to be moved outside the cell. An example is given by α-methyl carboxylic acid racemisation reactions. In the strain of *Nocardia* shown in Figure 1.37, the acid is first transformed to the CoA ester, which is the substrate for deracemisation; the CoA is then removed by a thioesterase [19].

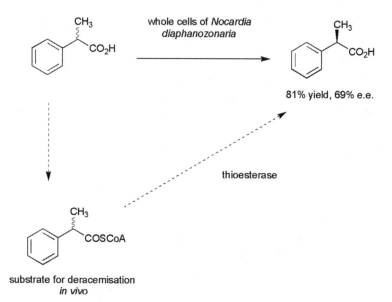

Figure 1.37 Deracemisation of α-methyl carboxylic acids catalysed by a Coenzyme-A-dependent isomerase from *Nocardia* sp.

1.7.8 Sugar nucleotides

The formation of glycosidic bonds in Nature is catalysed by glycosyltransferase enzymes. These enzymes transfer a sugar residue from a sugar donor to a sugar (or other, such as amino acid) acceptor, and form a glycosidic bond between the two. The donor sugar must be activated, which in the majority of cases means that it is ligated to a nucleotide, for example UDP-glucose (Figure 1.38) or GDP-mannose.

Such activated sugar donors are available commercially, but again, they are expensive and not amenable to preparative glycosylation reactions. In the reaction which couples the sugar to the aglycone (curcumin in Figure 1.39), the nucleotide is lost as a leaving group. In order to participate in another catalytic cycle, this must be joined to another sugar by the action of a sugar nucleotidyl transferase. In many cases, this recycling of the nucleotide sugar may occur within a whole-cell biocatalyst, but *in vitro* cofactor regeneration systems have also been adopted.

The chemical, steric and electronic diversity afforded by a combination of just twenty amino acids and a handful of coenzymes and cofactors is immense and gives rise to a wealth of organic chemical reactions that can be applied in synthesis. Many further examples of these reactions will be given in later chapters, where the use of various types of biocatalyst including whole cells, commercially available and 'home-grown' isolated enzymes, and engineered enzymes is described.

1.8 Some Basic Characteristics of Enzyme Catalysis

1.8.1 Enzyme kinetics

Enzyme-catalysed reactions conform to many of the chemical and physical principles that govern 'conventional' chemical catalysis, and indeed many of the terms applied to those techniques are found in biocatalysis nomenclature. However, a brief perusal of biocatalysis papers will reveal some terms that may be unfamiliar to those inexperienced with enzymes and there are some definitions and descriptions of these that would be valuable at this stage.

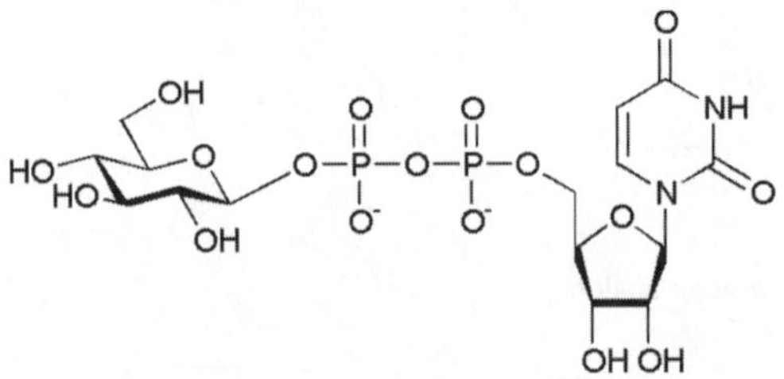

Figure 1.38   Structure of an 'activated' sugar substrate for a glycosyltransferase – UDP-glucose

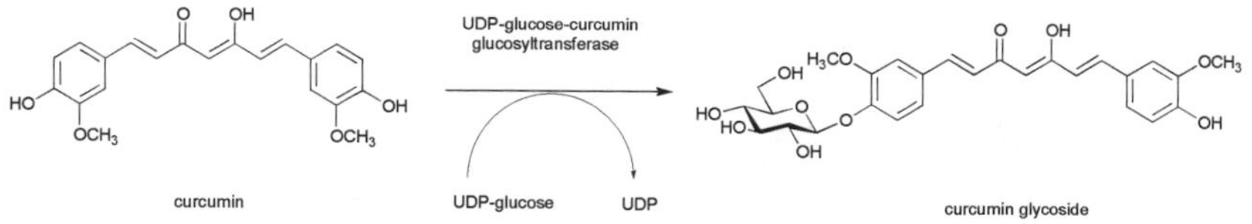

Figure 1.39 Coupling of curcumin to glucose catalysed by a UDP-Glc-dependent glycosyltransferase (from reference [20])

Let us consider the kinetics of a simple non-cofactor dependent enzyme such as a protease or esterase. When the enzyme E and substrate S come into contact, an enzyme-substrate complex [ES] is formed as the substrate binds within the active site (Scheme 1.1).

The chemical transformation catalysed by that enzyme then occurs, via the relevant transition state [ES]*, resulting in an enzyme-product complex [EP], which dissociates to give free enzyme and product in solution, and allowing another substrate molecule to bind. A simple energy level diagram of the reaction coordinate (Figure 1.40) reveals that, as with other forms of catalysis, the enzyme is able to lower the energy of the transition state for the reaction and the reaction proceeds at a faster rate than the uncatalysed process. The molecular determinants of transition state stabilisation are discussed at length in both references [18] and [21].

The kinetics that describes enzyme activity is characteristic and is often, but not always, observed to conform to so-called 'Michaelis–Menten' behaviour (Figure 1.41).

Figure 1.41 illustrates the behaviour of an enzyme when challenged with increasing concentrations of its substrate and plots substrate concentration on the x-axis against the rate of reaction on the y-axis. At low concentrations, it is observed that there is an increase in rate that is proportional to the concentration of the substrate. This linear portion of the graph corresponds to the fact that there are enzyme sites available for the transformation of molecules of substrate up to a point at which no more are available and the enzyme active sites are said to be saturated – that is, all of them within the sample are occupied and are working at their maximum rate. Hence, on reaching a certain concentration the rate of

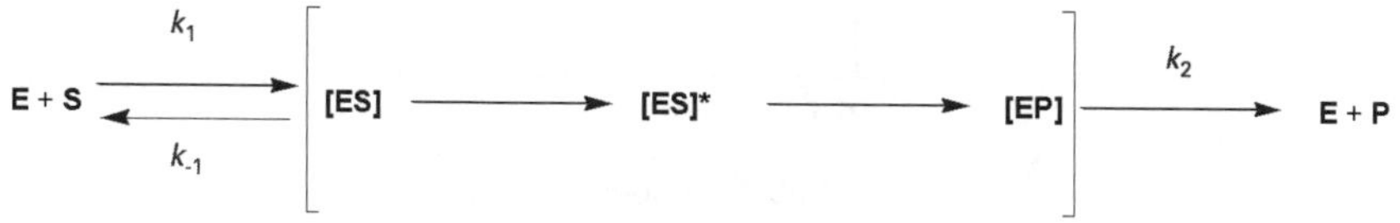

Scheme 1.1 Reaction coordinate for a simple enzyme catalysed reaction. Processes in brackets occur within the enzyme active site. k_1, k_{-1} and k_2 refer to the rate constants for the respective steps.

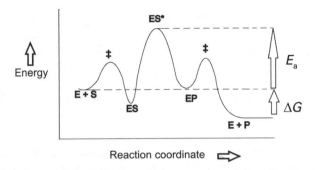

Figure 1.40   An energy level diagram depicting the reaction coordinate for a simple enzyme-catalysed process. ΔG, the energy change for the reaction (which has a negative value); E_a, the activation energy for the reaction

enzymatic reaction reaches a plateau beyond which it cannot increase – this is known as the maximum velocity of the enzyme or V_{max}, measured in units of moles (micromoles) or substrate converted (or product evolved) per unit time, and can be easily determined using any number of the enzyme assays described in Chapter 6 conducted using increasing concentrations of substrate. The Michaelis–Menten graph also allows the determination of another important kinetic parameter – the K_M or Michaelis–Menten constant – for an enzyme with a particular substrate. The K_M is the concentration of substrate at which the enzyme catalyses its reaction at half the maximal rate and hence has units of mol dm^{-3}. In terms of the kinetic

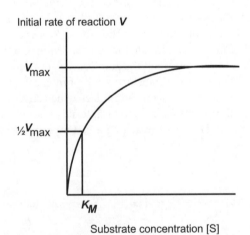

Figure 1.41   Graph depicting Michaelis–Menten kinetics exhibited by a simple enzyme illustrating the kinetic quantities V_{max} and K_m

$$K_m = \frac{k_{-1} + k_2}{k_1}$$

Scheme 1.2   Equation for the Michaelis constant K_m

constants for the enzyme-catalysed reaction introduced in Scheme 1.1, the value of K_m is described by the equation in Scheme 1.2.

This value of K_m is specific for any enzyme working with a particular substrate and may be thought of as an association constant for the formation of the enzyme-substrate complex as shown in Scheme 1.1. The K_m provides a value for the affinity of an enzyme for its substrate. A lower number indicates that the enzyme has a greater affinity; a higher number indicates that the affinity is lower. The kinetic constants V_{max} and K_m are related by the Michaelis–Menten equation, the derivation of which can be found in all good biochemistry textbooks (Scheme 1.3, where v is the initial rate of reaction and [S] the concentration of the substrate). The equation is often represented in reciprocal form as this allows the relevant data to be plotted in a straight line (for example Lineweaver–Burk and Eadie–Hofstee plots – see reference [21]) and allows the more simple manual calculation of the kinetic constants. However, in practise, these days the constants can be calculated from the raw data using efficient chemical graphics programs such as Grafit (http://www.erithacus.com/grafit) or Enzyme Kinetics in SigmaPlot (http://www.systat.com/products/enzyme_kinetics/).

Two other kinetic terms are frequently encountered in the literature reports of the characterisation of enzymes or their mutants. k_{cat} is the 'turnover number' of an enzyme, and corresponds to the number of molecules of substrate converted to product per number of molecules of enzyme, and hence has units of s^{-1}. This is equivalent to the kinetic constant for the breakdown of the enzyme-product complex [EP] given as k_2 in Scheme 1.1. Tables of kinetic data found in reports of enzyme characterisations often report both K_M and k_{cat} values for an enzyme and its mutants, which may be prepared in order to investigate the role of a single amino acid residue within an active site, for example (see Chapter 8) and also another value, that of k_{cat}/K_M, with units of $s^{-1}\ mol^{-1}\ dm^3$ often called the *specificity constant* for an enzyme. The specificity constant for an enzyme is thought to be the best measure of the catalytic efficiency of an enzyme and would be used to compare the catalytic performance of two forms of the same enzyme with different mutations perhaps, or to compare the efficiency of an enzyme working on two different substrates.

$$v = \frac{V_{max}\,[S]}{K_m + [S]}$$

Scheme 1.3   The Michaelis-Menten equation

It is important to note that, whilst the concepts described above are most easily understood for an enzyme that operates on a single substrate, many enzymes use more than one substrate, and in some cases, the binding and release of substrates and products is consequently much more complex in many cases, sometimes occurring sequentially as in 'ping-pong' mechanisms giving rise to very complex kinetic descriptions of their activity indeed. A flavin-dependent monooxygenase, for example (see Chapter 5), will bind a nicotinamide coenzyme NADPH, a ketone substrate and oxygen and the reaction will also involve a flavin coenzyme FAD which is bound to the monooxygenase. The detailed study of such mechanisms can be a significant challenge therefore.

The calculation of simple kinetic parameters such as the V_{max} also allows the determination of the amount of enzymatic activity present in an extract. This is usually quantified by the use of units of activity (U), wherein a unit of activity denotes the transformation of 1 μmol of substrate (or evolution of 1 μmol of product) per minute. This is sometimes seen as units mL^{-1} of enzyme preparation. The specific activity of an enzyme preparation refers to units of activity mg^{-1per} of enzyme and will increase as an enzyme increases in purity during, say, a purification procedure such as that described in Chapter 6. In recent years, the international community has agreed on the implementation of the katal (kat) or nanokatal (nkat) as a means of quantifying catalytic activity, wherein 1 katal corresponds to the amount of catalyst required to transfer 1 mol of substrate per second, although use of this terminology in the biocatalysis literature is still limited.

1.8.2 Physical parameters affecting enzyme activity

As organic molecules, enzymes are subject to the same physical and chemical stresses that result from different reaction conditions and a thorough investigation of these is part and parcel of the process of optimising a preparative enzyme-catalysed reaction as seen in Chapters 5 and 6. These include simple parameters such as pH and temperature.

1.8.2.1 pH

As many of the molecular interactions that govern both stability of enzymes and the chemical mechanisms of their action are governed by electrostatic interactions and hydrogen bonding, in the latter case often involving proton transfer steps that accompany general acid and base catalysis, it is no surprise that the optimum performance of enzymes varies greatly with pH. Whilst some enzymes are known to have pH optima at acidic or alkaline pH, it is most common to find enzymes with optimum activity at around physiological pH 5–9, with activity tailing decreasing around that optimum, resulting in a characteristic bell-shaped curve

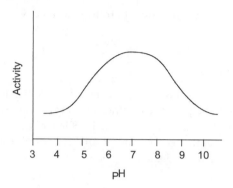

Figure 1.42 Graph depicting enzyme activity versus pH for a typical enzyme-catalysed reaction and showing the pH optimum

(Figure 1.42). The pH optimum of the enzyme may well dictate consideration of the reaction medium when either using the enzyme in a preparative reaction (Chapter 5) or when attempting a purification (Chapter 6).

1.8.2.2 Temperature

The bonding interactions that are responsible for protein stability will also be greatly affected by temperature. Enzymes work best at physiological temperatures largely, but there are of course examples of enzymes that function best at very high temperatures, notably those derived from thermophiles (see above). The rate of an enzymatic reaction is usually observed to increase in line with principles that affect increases of catalytic rates in conventional chemical reactions but this is only true up to the point at which the protein structure itself begins to denature, i.e. the bonding interactions that hold the structure together begin to fall apart (Figure 1.43). Again, an investigation of the optimum temperature for

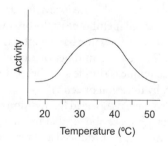

Figure 1.43 Graph depicting enzyme activity versus temperature for a typical enzyme-catalysed reaction

an enzyme-catalysed reaction is usually routine when developing a biocatalytic process.

1.8.2.3　Other factors

Other physical factors affecting enzyme activity include the amount of water in the reaction mixture, the presence of other solvents or solutes such as salts, and even, in some cases, the pressure in the system. Some of these considerations are addressed specifically in the sections on the applications of isolated enzymes and their purification in Chapters 5 and 6.

1.9　Types of Biocatalyst - Biotransformations by 'Whole Cells' or 'Isolated Enzymes'

In the single-step biotransformations mentioned above, a clear distinction can be made between two separate techniques: those of *whole-cell* biotransformations and *isolated-enzyme* biotransformations. There will be many implications, in both technological and practical terms, from the choice of type of biocatalyst, which in addition to facilities and cost, will include safety and containment requirements.

1.9.1　Whole cells

In whole-cell biocatalytic reactions, organisms (eubacteria, yeast or filamentous fungi) are cultivated in one or other type of growth medium and the substrate, or starting material is added after the biomass has reached an appropriately substantial level (a *growing-cell biotransformation*). Alternatively, the biomass may be harvested by centrifugation or filtration after a suitable period of growth, washed with water or a buffered solution, and resuspended in water or a buffer of choice and substrate added to the resultant cell suspension (a *resting-cell biotransformation*). In each case, the technique is exploiting the necessary enzyme for transformation in something approaching its 'natural' state *in vivo* inside the cells. Substrate must therefore cross the cell wall or cell membrane of the organism in order to be transformed, and products must cross the same barrier after transformation, back into the surrounding medium. There are a number of advantages to using whole-cell biotransformations: the technique is simple and cheap, the removal of the catalyst is straightforward (usually by filtration or centrifugation) and reactions take place in an aqueous medium. Also, there is no need for addition of the expensive and possibly fragile enzymatic cofactors (see above) that may be necessary for enzymatic activity, as these will be in most cases produced and recycled in the cell. The major disadvantages are usually due to competing reactions – either over-metabolism of the required product, or side reactions that transform the substrate into chemical entities other than that required or expected. In addition, the use of whole-cell preparations of micro-organisms implicates training and facilities requirements which are dealt with in detail later in this book. Such organisms will be 'viable' – that is to say living, and must be handled using the appropriate techniques and containment procedures. They must also, unlike enzymes, be deactivated prior to disposal, by either treatment with a chemical disinfectant or by sterilisation using an autoclave.

1.9.1.1 Recombinant or 'designer' biocatalysts

When considering whole-cell biotransformations, it is worth mentioning a relatively new development that may be commonly encountered in modern-day biotransformations. It is commonplace that the genes encoding enzyme of interest from one organism may be cloned and expressed in a different, more user friendly organism such as the common laboratory bacterium *E. coli* or the yeast *S. cerivisiae* (baker's yeast). This new 'recombinant' strain will then produce the enzyme of interest, often in much higher amounts than the organism from which the original gene was derived. The result is a 'designer' biocatalyst, a whole cell biocatalyst that, although possessing many of the same characteristics as wild-type whole cell biocatalysts, will have other problematic aspects removed, such as over-metabolism or side reactions, as the enzyme of interest has been removed from its natural context. Recombinant biocatalysts, whilst introducing a genetically modified element into the research, also enable workers to exploit enzymes from an organism that would be otherwise difficult because the source organism may be pathogenic or have a requirement to be grown on an expensive or toxic carbon source in order to elicit the required enzyme activity. The genetics, biochemistry and behaviour under industrial fermentation conditions of generic host systems such as *E. coli, Pseudomonas fluorescens, S. cerivisiae, Pichia pastoris* and species of *Aspergillus* are well understood, facilitating the optimisation of biotransformations at scale. Such designer biocatalysts are hence becoming the standard for many biotransformation processes both in academic and industrial settings.

1.9.1.2 Growing cells or resting cells

Biotransformations catalysed by whole cells may be further subdivided into two sections: *growing cells* or *resting cells*. In the first, cells are grown on a suitable liquid medium in a shake-flask, and after a period of growth, the substrate for transformation is added directly to the cell culture. The advantages of this method are that it is not necessary to harvest the cells by centrifugation or filtration and so the whole growth and biotransformation process can be conducted in 'one pot'. It is also true to say that some biocatalytic reactions, notably hydroxylations, work much better using growing cell protocols rather than resting cell ones. In a resting-cell biotransformation, the cells are grown in an equivalent medium, and then removed from it once a suitable amount of biomass has been accumulated. They are then washed with water or a buffered solution and resusupended in another amount of buffer for purposes of reaction. This allows the pH of the reaction to be controlled to some extent, but also results in cleaner transformations, the organism having been removed from the complex mixture of chemicals that accumulates during growth, and thus can simplify downstream processing of reaction products.

1.9.2 Isolated enzymes

In this case, the enzyme required for biotransformation is liberated from the cells by cell disruption and then purified to the required degree using a range of chromatographic techniques. Of course, in many instances, the isolation and purification has been performed by an enzyme supplier and the isolated enzyme is provided to the researcher in a usable powdered form that may be employed in much the same way as a conventional chemical reagent from the shelf. The advantages inherent in isolated enzyme reactions are of course this ready usability and clean reactions, free from either over-metabolism or unwanted side reactions. The disadvantages are usually associated with enzyme stability and the possible need for non-enzymatic cofactors that are too expensive to add in stoichiometric amounts. In addition, if the researcher wishes to explore the possibility of using an isolated enzyme on the back of a successful whole-cell investigation, there will need to be an investment in equipment and expertise in order to equip the laboratory with the necessary facilities for cell disruption, protein chromatography and protein analysis. An appreciation of the potential utility of biotransformations in the synthetic organic community has led to more companies offering a wider range of 'off- the-shelf' isolated enzymes than was previously available.

It is certain that both of these general techniques have their uses and are widely applied in both laboratory research and in industry. The choice of which biocatalyst to use in a certain situation will be affected by a great many factors dependent on the nature of the desired chemical transformation, but also in many cases, on the economics of the process. These considerations will become apparent in Chapters 4 and 5, when we examine the protocols associated with implementing different types of biocatalyst.

1.10 Conclusion

Biotransformations are becoming the method of choice for an increasing number of processes in the chemical industry, from the production of bulk chemicals to pharmaceuticals, flavours and materials. This is primarily because they offer advantages of reaction selectivity and Green Chemistry credentials that are often not supplied by conventional abiotic chemistry. In the following chapters, we will explore the sources of biocatalyst available to the researcher, and then examine how facilities might be set up and used for preparative biotransformations in the laboratory.

References

1. S. M. Roberts, N. J. Turner, A. J. Willetts and M. K. Turner (1995) Introduction to Biocatalysis Using Enzymes and Micro-organisms, Cambridge University Press, Cambridge.
2. J. B. Jones, C. J. Sih and D. Perlman (Eds) (1976) Applications of Biochemical Systems in Organic Chemistry; Parts 1 and 2, John Wiley & Sons, Ltd, New York; J. B. Jones (1986) Enzymes in organic synthesis. *Tetrahedron*, **42**, 3351–3403.
3. C.-H. Wong and G. M. Whitesides (1994) Enzymes in Organic Synthesis, Tetrahedron Organic Chemistry Series Volume 12, Academic Press, Oxford.
4. K. Faber (2004) Biotransformations in Organic Chemistry, 5th Edn, Springer-Verlag, Berlin.
5. K. Drauz and H. Waldmann (Eds) (2002) Enzyme Catalysts in Organic Synthesis, Wiley-VCH, New York.
6. U. T. Bornscheuer. and R. J. Kalauskas (2005) Hydrolases in Organic Synthesis. Regio and Stereoselective Biotransformations, 2nd Edn, Wiley-VCH, New York.
7. J.-L. Reymond (Ed.) (2006) Enzyme Assays. High Throughput Screening, Genetic Selection and Fingerprinting, Wiley-VCH, Weinheim.
8. P. T. Anastas and J. C. Warner (1998) Green Chemistry Theory and Practice, Oxford University Press, New York.
9. R. J. Fox, S. C. Davis, E. C. Mundorff, L. M. Newman, V. Gavrilovic, S. K. Ma, L. M. Chung, C. Ching, S. Tam, S. Muley, J. Grate, J. Gruber, J. C. Whitman, R. A. Sheldon and G. W. Huisman (2007) Improving catalytic function by ProSAR-driven enzyme evolution. *Nature Biotechnol.*, **25**, 338–344.
10. H. G. Schlegel (1993) General Microbiology, 7th Edn, Cambridge University Press, Cambridge.
11. M. T. Madigan and J. M. Martinko (2006) Brock Biology of Microorganisms, 11th Edn, Pearson Prentice Hall, New Jersey.
12. F. R. Blattner, G. Plunkett 3rd, C. A. Bloch, N. T. Perna, V. Burland, M. Riley, J. Collado-Vides, J. D. Glasner, C. K. Rode, G. F. Mayhew, J. Gregor, N. W. Davis, H. A. Kirkpatrick, M. A. Goeden, D. J. Rose, B. Mau and Y. Shao (1997) The complete genome sequence of *Escherichia coli* K-12. *Science*, **277**, 1453–1474.
13. A. Goffeau, B. G. Barrell, H. Bussey, R. W. Davis, B. Dujon, H. Feldmann, F. Galibert, J. D. Hoheisel, C. Jacq, M. Johnston, E. J. Louis, H. W. Mewes, Y. Murakami, P. Philippsen, H. Tettelin and S. G. Oliver (1996) Life with 6000 Genes. *Science*, **274**, 563–567.
14. H. J. Pel et al. (2007) Genome sequencing and analysis of the versatile cell factory *Aspergillus niger* CBS 513.88 *Nature Biotechnol.*, **25**, 221–231.

15. K. Ishihara, H. Hamada, T. Hirata and N. Nakajima (2003) Biotransformation using plant-cultured cells. *J. Mol. Catal. B-Enzym.*, **23**, 145–170.
16. M. Weis, E.-K. Lim, N. C. Bruce and D. Bowles (2006) Regioselective glucosylation of aromatic compounds: screening of a recombinant glycosyltransferase library to identify biocatalysts. *Angew. Chem. Int. Ed.*, **45**, 3534–3537.
17. G. M. Petsko and D. Ringe (2004) Protein Structure and Function, Wiley-Blackwell, Chichester.
18. T. D. H. Bugg (2004) Introduction to Enzyme and Coenzyme Chemistry, 2nd Edn, Blackwell Publishing, Oxford.
19. D. Kato, S. Mitsuda and H. Ohta (2003) Microbial deracemisation of α, α-substituted carboxylic acids: substrate specificity and mechanistic investigation. *J. Org. Chem.*, **68**, 7234–7242.
20. Y. Oguchi, S. Masada, T, Kondo, K. Terasaka and H. Mizukami (2007) Purification and characterisation of UDP-glucose: curcumin glucoside 1,6-glucosyltransferase from *Catharanthus roseus* cell suspension cultures. *Plant Cell Physiol.*, **48**, 1635–1643.
21. A. R. Fersht (1999) Structure and Mechanism in Protein Science: A Guide to Enzyme Catalysis and Protein Folding, 3rd Edn, W.H. Freeman, San Francisco.

Chapter 2
An Overview of Biocatalyst Sources and Web-Based Information

2.1 Introduction

One of the issues surrounding the use of biocatalysts in synthetic organic chemistry is that of availability of the catalysts themselves. In the post-genomic age, the idea of what constitutes a biocatalyst has changed, and hence the 'sources' of biocatalyst are becoming increasingly diverse and evolving all the time. Essentially, the raw material for a biocatalyst will be either a *micro-organism*, a commercially prepared *enzyme*, or genetic material, in the form of either the *genomic DNA* from the organism of interest, or perhaps the *gene* encoding the enzyme provided already cloned within a convenient 'expression vector'.

As both the interest in preparative biocatalysis and the use of biocatalytic methods have grown, many more sources for the provision of biocatalysts in these many forms have been established. These range from commercial organisations to academic groups and research institutes and provide a range of biological material of the types listed above. Strains of micro-organisms can be obtained from a variety of 'culture collections' – libraries of organisms that are curated to catalogue and preserve microbial diversity and to provide a source of industrial organisms for either the production of bioactives or for catalysts. Sometimes, these organisms may be provided free of charge, but it is more usual to pay a fee. In contrast, isolated enzymes are provided almost exclusively by companies, and in addition to the large established enzyme suppliers, new companies are constantly emerging that will provide enzyme samples either for small scale testing or in bulk, for large-scale application. However, it is recent developments in the area of genomics that have caused a huge increase in the global reservoir of available biocatalysts, in that the gene sequences encoding millions of new enzymes have been revealed. Many of the relevant 'genomic DNAs' or samples of the entire genetic complement of an organism, are available from a mixture of commercial sources and genome research institutes. Such DNA will serve as the basis for cloning experiments leading to recombinant biocatalysts in groups that have the requisite facilities and expertise. 'Single' genes that encode enzymes for use by groups that do not perform cloning in-house are most often obtained through collaboration with other groups, but may now also be obtained from custom gene synthesis companies. We will deal with the three major sources above in turn.

2.2 Microbial Culture Collections

A large number of microbial culture collections exists around the world, based at a mixture of research institutes, universities, government bodies and private companies. These resources will usually provide an organism in a preserved state, sometimes as a freeze-dried preparation or on a slope of agar (Figure 2.1), and will also usually supply directions as to the resuscitation, maintenance and growth of the microbial strain in question.

It is not intended to provide an exhaustive list of sources of micro-organisms here, but to list the culture collections that are mostly encountered by those reading the literature of microbial biotransformation, rather than the literature of secondary metabolite production, ecology or taxonomy. A more extensive list is provided on the website of the Centre of Excellence for Biocatalysis, Biotransformations and Biocatalytic Manufacture (http://www.coebio3.org/Links/Culture%20Collections.htm). For workers in the UK, comprehensive information on national culture collection resources is also provided on the website

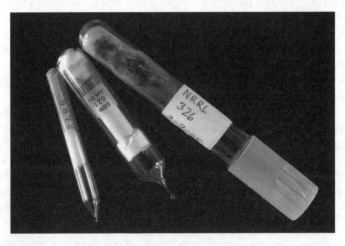

Figure 2.1   Formats for preserved organisms available from culture collections

of the United Kingdom National Culture Collection (UKNCC; http://www.ukncc.co.uk/).

2.2.1 NCIMB

National Culture of Industrial and Marine Bacteria
(http://www.ncimb.com)
NCIMB Ltd, Ferguson Building, Craibstone Estate, Bucksburn, Aberdeen AB21 9YA, UK
Established in the 1950s, NCIMB is the UK's primary collection of bacteria, and holds over 7500 strains. The catalogue can be searched online and provides cultures of bacteria in freeze-dried ampoules on request for a fee. In addition to engaging in other activities, NCIMB also offers a strain identification service, which can be used to identify a strain of interesting biocatalyst that has been derived, perhaps from an enrichment isolation investigation (see below).

2.2.2 NCYC

National Collection of Yeast Cultures
(http://www.ncyc.co.uk)
NCYC, Institute of Food Research, Norwich Research Park, Colney, Norwich NR4 7UA, UK
As part of the Institute of Food Research, Norwich, UK, the NCYC manages over 3000 yeast strains, which are available to both industrial and academic researchers (at a lower cost to the latter). Much of the work of the NCYC is involved with the curation and study of yeasts for the brewing industry, but is also a source of live or freeze-dried strains of yeasts and relevant genomic DNAs, including those whose genomes have been sequenced, and will also provide a yeast strain identification service.

2.2.3 NCAUR (NRRL)

National Center For Agricultural Utilization Research (formerly the Northern Regional Research Laboratory)
(http://nrrl.ncaur.usda.)
NCAUR, at Peoria, Illinois, is one of the US Department of Agriculture's research laboratories, and has for years housed the NRRL culture collection, whose strains still bear the NRRL prefix. With the growing interest in biocatalysts coming from the chemicals and renewable sector, NCAUR now has a dedicated research team in Bioproducts and Biocatalysis Research. The culture collection is managed by the Microbial Genomics and Bioprocessing Research Unit and can be searched online. Freeze-dried cultures are provided to academics at no charge,

unless they were added to the collection after 1983, in which case a small charge is made.

2.2.4 ATCC

American Type Culture Collection
(http://www.lgcpromochem-atcc.com/)
UK: LGC Promochem, Queens Road, Teddington, Middlesex TW11 0LY, UK
Now in partnership with LGC Promochem, the ATCC has for years been one of the primary sources of biotransformations organisms and currently houses approximately 18 000 strains of bacteria and 27 000 of yeasts and fungi. ATCC provides freeze-dried cultures of organisms at a cost (reduced for academics) and has also recently become a source of genomic DNA samples for gene amplification.

2.2.5 CBS

Centraalbureau voor Schimmelcultures, PO Box 85167, 3508 AD Utrecht, The Netherlands
(http://www.cbs.knaw.nl)
An Institute of The Royal Netherlands Academy of Arts and Sciences, CBS provides freeze-dried and agar-slope cultures of actinomycetes, but is primarily a collection of yeasts and filamentous fungi, of which they have 6500 and 37000 strains, respectively.

2.2.6 DSMZ

Deutsche Sammlung von Mikroorganismen und Zellkulturen GmbH, Inhoffen-straße 7, B38124 Braunschweig, Germany
(http://www.dsmz.de/)
The DSMZ is the primary microbial culture collection in Germany, curating over 13 000 cultures including bacteria, yeasts and filamentous fungi.

2.2.7 IFO

Institute for Fermentation, Culture Collection of Micro-organisms, 17–85, Juso-honmachi, 2-chome, Yodogawa-ku, Osaka 532–8686, Japan
Supervision of strains held by the IFO has recently been transferred to the NBRC [Biological Resource Center, National Institute of Technology and Evaluation (NITE), 2-5-8, Kazusakamatari, Kisarazu-shi, Chiba Pref., 292–0812, Japan (http://www.bio.nite.go.jp/pamphlet/e/nbrc-e.html)].
The accession numbers from IFO strains have been transferred to NBRC strains, e.g. IFO 3338 is now NBRC 3338.

2.3 Obtaining Organisms from Other Research Groups

It is worth remembering that most peer-reviewed international journals require that strains and genes described in reports accepted for publication can be made available on request (after consultation and exchange of necessary material transfer agreements or similar arrangements). Hence, where an organism or gene has been isolated by a group and deposited with a culture collection or in an international gene sequence database, direct contact with that research group may be the easiest and most economical route to obtaining a microbial culture.

2.4 Selective Enrichments

The last option in obtaining micro-organisms for use as biocatalysts is the most complex in that it can in some cases constitute a research project in itself. Selective enrichment is the process whereby an organism is isolated from the environment for its ability to use a substrate of interest as a sole source of energy or nutrients. There is an excellent guide to the techniques and considerations involved in selective enrichment in the online microbiology textbook at http://www.bact.wisc.edu:16080/Microtextbook/. As an example of selective enrichment, consider that one may be interested in discovering an enzyme that is capable of hydrolysing the acetate esters of aromatic secondary alcohols such as *sec*-phenyl ethanol. The strategy of selective enrichments (Figure 2.2) would dictate that ideally, one would identify a natural environment in which there was an abundance of esters of aromatic secondary alcohols. It would be thought likely that, given the vast number of bacteria that inhabit even a gram of soil material (millions to billions), one organism within that sample would have evolved the ability to use the relevant acetate esters as a source of carbon for growth, perhaps exploiting an esterase that would furnish the simple carbon source acetate from the substrate. A sample of soil is then taken from the relevant environment, and is grown on a rich medium to encourage the growth of as many organisms as possible. Samples of this culture are then removed and inoculated into a much simpler medium which contains the substrate of interest, in this case an aromatic secondary alcohol, as 'sole carbon source'. Any organism that is able to use this carbon source should thrive on the new medium, and a series of experiments to isolate the relevant strain using selective culture on Petri dishes (see Chapters 3 and 4) is performed in order to obtain a pure culture of that strain. The isolated organism may then be assayed specifically for the transformation in question and studies may begin on the isolation of an enzyme from within the organism that is specific for the transformation of interest.

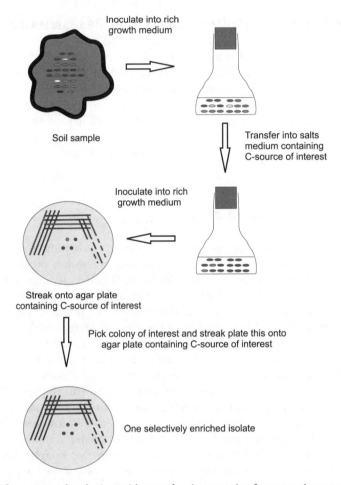

Inoculate into rich
growth medium

Soil sample

Transfer into salts
medium containing
C-source of interest

Inoculate into rich
growth medium

Streak onto agar plate
containing C-source of interest

Pick colony of interest and streak plate this onto
agar plate containing C-source of interest

One selectively enriched isolate

Figure 2.2   A strategy for selective enrichment of a micro-organism from an environmental sample

2.5 Metagenomics

The field of metagenomics is related to that of enrichment selection in that it exploits material from the environment in an effort to source a useful biocatalyst. However, the distinct advantage of metagenomics is that it circumvents the need to culture wild-type organisms from the environment, which is an essential part of enrichment selection. Whilst 1 g of soil may be thought to contain millions of different strains of micro-organisms, it is estimated that a very small proportion of these, perhaps less than 1%, may actually be 'culturable' *in vitro* and thus allow their use as whole-cell biocatalysts or as sources of single-organism genomic DNA.

However, if some information about the gene sequence of the enzyme of interest is known, modern techniques in molecular biology may allow the isolation of genes from these 'unculturable' organisms [1]. This may employ the gene amplification technique known as the polymerase chain reaction (PCR – described in detail in Chapter 7) and will rely to some extent on the similarity of the genes sought and known gene sequences. The advantage of PCR is that it is theoretically able to amplify large amounts of the gene required from a single copy of the gene in any sample so, in theory, any of the genes that exist even in a small soil sample may be accessed (Figure 2.3).

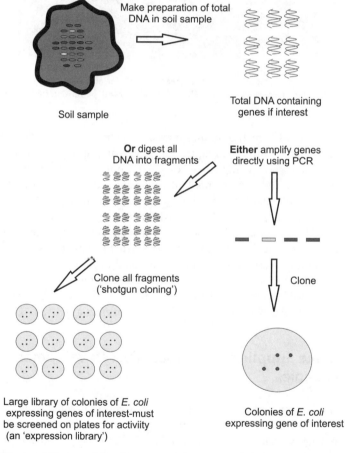

Figure 2.3 A possible strategy for a 'metagenomics' experiment designed to isolate a gene encoding a useful biocatalyst

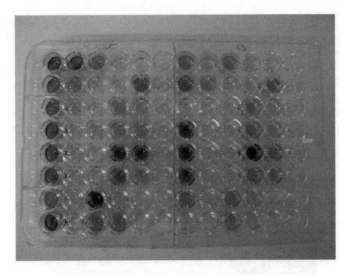

Figure 2.4 A high-throughput colorimetric screening assay for an oxidase activity in multi-well plate format (photograph courtesy of Ingenza Ltd) Reproduced with permission from Ingenza Ltd.

Other strategies for metagenomics may be dependent on so-called 'shotgun cloning' procedures. In this technique, rather than attempting to isolate a single gene directly, the whole complement of DNA from an environmental sample is digested and the DNA fragments that result are *ligated* into plasmid vectors (Chapter 7) that are suitable for the cloning and expression of genes. The resultant 'library' of gene sequences is then transformed into a laboratory expression bacterium such as *E. coli*. What results is a large number of colonies of *E. coli* on agar plates which must then be screened for activity using a suitable high-throughput assay, such as a colour change on the agar plate for the organism or in a multi-well format for the cell extracts derived from them (Figure 2.4). It may also be possible to identify the gene of interest from such a library by using radiolabelled partial genetic sequences of desired enzymes as *probes*.

2.6 Enzyme Suppliers and Biocatalyst Development Companies

There are now many sources available for direct access to isolated enzymes. These are almost exclusively dedicated enzyme companies or chemical companies that feature enzymes as part of their portfolio, and have recognised the increased interest in biocatalytic solutions as a profitable area. It may of course be possible to obtain amounts of isolated enzyme from academic groups, but, considering the amount of effort and expense required to generate the enzyme, this would almost certainly form part of a considerable collaborative effort.

Some of the activity in commercial enzymes and coenzymes stems from biochemical companies such as Sigma-Aldrich-Fluka who have been producing these biochemical reagents for non-biocatalysis related projects involving enzyme assays, in which those enzymes are required in very small amounts. Some of these are very interesting applied biocatalysts in their own right, but their expense precludes their use in synthetic applications. Of the enzymes that are available, hydrolases, being amongst the simplest and more robust, are the most accessible and popular. Enzymes such as proteases, lipases and esterases are of course produced in bulk for applications in the washing powder, textiles and materials industries and can be purchased relatively cheaply. Until recently, the supply of oxidoreductase enzymes was poor, but older companies have been expanding their portfolio, and new companies have been established to meet the growing demand for more 'exotic' enzyme classes.

It is often possible to purchase a screening kit of, for example, lipases or keto-reductases (alcohol dehydrogenases) which would provide sufficient enzyme for preliminary analytical-scale experiments (Figure 2.5).

The researcher could then, on identifying the catalyst of interest, investigate a bulk purchase for a scaled-up investigation. A good place to start a search for a supplier of an enzyme biocatalyst is the 'Enzyme Directory' (http://www.enzymedirectory.com/), which provides an extensive search list of enzymes and

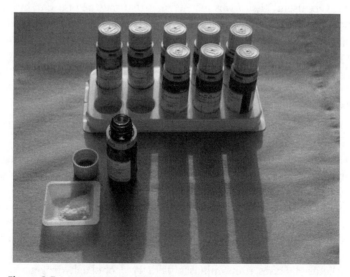

Figure 2.5   A commercially available kit of lipase enzymes available from Fluka

applications. We supply below a list of some companies that provides enzyme preparations. Many of these companies supply enzymes for bulk processing, but the following list is restricted to those with a targeted interest in applied biocatalysis for fine chemicals research and manufacture:

- *Novozymes*

http://www.novozymes.com

Novozymes is one of the world's largest enzyme suppliers, and does not only deal in enzymes for applied biocatalysis for fine chemical synthesis, but also bulk enzymes for materials processing and anticipated domestic application. However, it provides some of the world's most used biocatalysts, notably lipase enzymes, including both 'Novozym 435" (an immobilised preparation of lipase B from *Candida antarctica*) and the protease subtilisin A.

- *Genencor*

http://www.genencor.com

Now a subsidiary of Danisco, Genencor is the world's second-largest manufacturer of industrial enzymes, so a list of enzyme suppliers without the name would be incomplete. Genencor mostly deals in enzymes for bulk materials processing, rather than for fine chemicals applications, but has entered into partnership with e.g. Chiralvision to supply hydrolase enzymes for fine chemical applications.

- *Sigma-Aldrich-Fluka (SAFC)*

http://www.sigmaaldrich.com/

SAFC sells very useful kits of various enzymes that may be of use in biocatalysis, including a 'lipase basic kit' or 'lipase extension kit' of milligram quantities of nine or ten lipases that may be screened against a reaction of choice on a small-scale prior to scale-up. Also available is a range of carbon-carbon bond forming lyases such as aldolases and some oxidoreductases including alcohol dehydrogenases, oxidases and enzymes for nicotinamide cofactor recycling such as glucose and formate dehydrogenases.

- *Codexis*

http://www.codexis.com/

One of a new generation of companies that have exploited the revolution in protein evolution techniques for enzyme improvement, Codexis has recently introduced a range of Codex biocatalyst panels, designed to enable researchers to rapidly screen enzyme families that have been generated using gene shuffling or 'molecular breeding approaches'. Codexis recently acquired Biocatalytics Inc., a company dedicated to the provision of biocatalysts for the preparation of fine chemicals. The company hence acquired an impressive catalogue of enzymes from all classes, including ketoreductases, transaminases, primary alcohol dehydrogenases and the usual lipases. It sells many of its enzymes in kit form for screening, and is unusual in supplying a kit of human cytochromes P450 that might be useful in screening the behaviour of drug molecules *in vivo*. Codexis provides an informative catalogue that may be downloaded from the website. They will also work with clients to identify and develop biocatalytic processes.

- *Amano*

http://www.amano-enzyme.co.jp/

Amano lipases, such as Amano AK, are some of the most frequently encountered in the biotransformations literature. Founded in 1899, the company provides enzymes for a wide range of applications, and its website now features a dedicated section on enzymes for biotransformations including lipases, proteases and acylases.

- *Nzomics*

http://www.nzomics.com/

Nzomics is a new start-up biotechnology company in the north-east of England that supplies a growing range of biocatalysts including nitrile-transforming enzymes, lipases, glycosidases and glycosyltransferases.

- *Chiralvision*

http://www.chiralvision.com/

Chiralvision is a new enzyme company in the Netherlands that is supplying enzymes for biocatalytic applications primarily. In addition to selling 'generic' lipase CAL-B, one of the most widely used of all biotransformation enzymes, the company also sells a kit of proteases in collaboration with Genencor, and a range of immobilised enzymes including lipases and proteases.

2.6.1 Companies providing biocatalysis services

In addition to commercial organisations that sell lipases, there are many other biotechnology companies that have emerged in the past few years that sell biotransformations expertise, either in the identification of biocatalysts, their improvement by protein engineering and evolution techniques, or in scale-up and fermentation engineering. Some of these are:

- *Verenium*

http://www.verenium.com/

Verenium, formerly Diversa are a biofuels company that also uses proprietary directed evolution techniques to develop enzymes for industrial use, in fine chemicals production, materials processing and other industrial settings.

- *Bioverdant*

http://www.bioverdant.com/

Bioverdant is a new company operating out of San Diego, USA with expertise in chemistry, molecular biology and genomics, offering services in biotransformations as applied to pharmaceutical synthesis.

- *Ingenza*

http://www.ingenza.com

Ingenza is a company located in Edinburgh, UK and provides biocatalytic solutions to the production of, especially chiral amines and amino acids based on directed evolution technology and expertise in industrial-scale biotransformations.

At this stage, it is also useful to mention that many of the world's largest chemical companies now have a commitment to the use of biocatalysts in industrial processes, including Merck, BASF, DSM, Lonza, GlaxoSmithKline and AstraZeneca. These companies and others have provided excellent examples of the use of biocatalysis as a real solution to the industrial preparation of synthetic intermediates, some of which are described in an excellent review by the group at BASF [2] and a book, 'Industrial Biotransformations', edited by Professor Christian Wandrey and colleagues [3] which, in addition to listing instances of industrial biotransformations, also provides an introduction to the basics of biochemical engineering and optimisation of industrial-scale biotransformations that lie outside the scope of this book.

2.7 Genome Mining for Biocatalysts

The alternative to finding a microbial biocatalyst from the literature, or by enrichment selection, or to using a commercially available isolated enzyme, is to create one's own recombinant biocatalyst (see Chapter 1) using techniques described in Chapter 7. In order to do this, the crucial ingredient is the genetic material needed to perform a gene amplification experiment, the details of which are provided in Chapter 7. This genetic 'template' can exist in a number of forms: either a preparation of the entire complement of genes in an organism (the 'genomic' or 'chromosomal' DNA); or a smaller piece of DNA that has perhaps been provided by a collaborator or a company in the form of the gene of interest contained within a small circular piece of DNA called a *plasmid* that is suitable for the uptake of genes by bacteria, notably the laboratory workhorse cloning strain *E. coli*. In the latter case, the situation is not complex as, the gene having already been identified, no bioinformatics work needs to be done but in the former case of genomic DNA, a program of genome mining and simple bioinformatics analysis is helpful in locating genes that encode the enzymes of interest.

In addition to the genetic material that will serve as a basis for the cloning experiment, it will of course also be necessary to have the gene sequence encoding the enzyme. Many of the gene sequences that encode enzymes in the literature have resulted from laborious classical 'reverse genetics' cloning experiments which will not be described further in this book. The explosion in gene sequence data has come from the advances in high-throughput gene sequencing and the resultant growth in known genome sequences of micro-organisms. The PEDANT genome database (http://pedant.gsf.de/), which provides an exhaustive and up-to-date list of sequenced genomes, and the tools with which to analyse them, currently lists more than 600 genome sequences, with more than 500 prokaryotes, both eubacteria and archaea, and over 50 eukaryotes, including yeasts and filamentous fungi. We will return to this and other genome databases later in the chapter. When the number of genes emerging from metagenomics studies is included, the number of new possible enzyme catalysts is vast indeed. As part of the activity of the J. Craig Venter Institute, an expedition was recently undertaken that collected samples for sequencing from the Pacific Ocean. The gene sequencing program that resulted from this, the 'Sorceror II expedition', reported in reference [4] unearthed the sequences of six million previously unknown genes, potentially doubling the currently known amount of unique proteins. With this vast unexplored reservoir of potential proteins available, we can see that the exploration of such data could prove extremely useful in providing a host of new genes encoding established or unprecedented biocatalytic activities in the future.

Such gene sequences can be accessed from a variety of databases, the most popular of which are **NCBI** (National Centre for Biotechnology Information; GenBank), **EXPASY** (Expert Protein Analysis System), **EBI** (European Bioinformatics Institute), and **UniPROT**, more of which below. The database of genetic sequences, **GenBank**, managed by the US National Institutes of Health (NIH) is a comprehensive database of all gene sequences in the public domain. Collaboration between the NIH, the DNA DataBank of Japan (**DDBJ**), and the European Molecular Biology Laboratory (**EMBL**) ensures that a comprehensive global resource of genomics data is maintained. Each gene sequence (and hence enzyme coding sequence) is accorded an *accession number* for GenBank, a unique identifier which may be used to access information on the gene, the enzyme that it encodes, and any associated data, perhaps on an X-ray structure or publications that have studies on that enzyme.

In Section 2.8 we detail a procedure for acquiring the gene, and hence the amino acid sequence for a biocatalyst, and how to use this information to mine the available genome databases for homologues using some very simple bioinformatics tools.

2.8 Obtaining Amino Acid and Gene Sequence Information on Biocatalysts

It is likely that the pursuit of an interesting new enzyme activity might be prompted by a publication on a related enzyme, found through a web-based

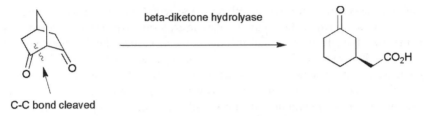

Figure 2.6   Cleavage of a prochiral β-diketone by a lyase enzyme

literature search. The NCBI website (http://www.ncbi.nlm.nih.gov/) is a good place to begin a search for an interesting enzyme activity based on a publication. The NCBI is a literature and data website that enables the interface of the academic literature and associated sequence data, making it easy to isolate the amino acid sequence of the enzyme of interest. For example, a lyase enzyme that catalyses the asymmetric cleavage of bicyclic β-diketones (Figure 2.6) was recently described [5].

On the NCBI website:

- Select >Literature Databases > **PubMed**
  An author search will yield the displayed publication abstract, and a PubMed ID for the article (PMID: 17198383).
- Select >Links
  On the far right-hand side of the screen, select >Protein. This will give a list of polypeptide chains that will enable access to the amino acid sequence of the enzyme. Where a structure is available, these are listed under the accession codes to the Protein Data Bank (pdb). Many of the entries in the example relate to the pdb code 2J5S, and the suffixes A–L refer to the fact that the enzyme in the structural database has twelve identical monomeric units.
- Select >2J5SA, and in the Display window near the top of the page, select 'Genpept'.
  This page will give the amino acid sequence of the enzyme at the bottom of the page:

```
ORIGIN
      1 mgsshhhhhh mtlnqpeyft kyenlhfhrd engilevrmh tngsslvftg kthrefpdaf
     61 ydisrdrdnr vviltgsgda wmaeidfpsl gdvtnprewd ktywegkkvl qnlldievpv
    121 isavngaall hseyilttdi ilasentvfq dmphlnagiv pgdgvhilwp lalglyrgry
    181 flftqeklta qqayelnvvh evlpqsklme raweiartla kqptlnlryt rvaltqrlkr
    241 lvnegigygl alegitatdl rnt
```

- Selecting 'FASTA' in the Display window will give the amino acid sequence in a format that is acceptable to most other enzyme biochemistry servers.

```
>gi|126031004|pdb|2J5S|A Chain A, Structural Of Abdh, A Beta-Diketone
Hydrolase From The Cyanobacterium Anabaena Sp. Pcc 7120 Bound To (S)-3-
Oxocyclohexyl Acetic Acid
MGSSHHHHHHMTLNQPEYFTKYENLHFHRDENGILEVRMHTNGSSLVFTGKTHREFPDAFYDISRDRDNRV
VILTGSGDAWMAEIDFPSLGDVTNPREWDKTYWEGKKVLQNLLDIEVPVISAVNGAALLHSEYILTTDIIL
ASENTVFQDMPHLNAGIVPGDGVHILWPLALGLYRGRYFLFTQEKLTAQQAYELNVVHEVLPQSKLMERAW
EIARTLAKQPTLNLRYTRVALTQRLKRLVNEGIGYGLALEGITATDLRNT
```

The header features the number of the gene within the genome of the organism in which the enzyme is encoded. In FASTA format, the amino acid sequence is prefixed by the header>XXXXX ... ; however, this may of course be simplified for personal archiving, for example:

```
>Lyase
MTLNQPEYFTKYENLHFHRDENGILEVRMHTNGSSLVFTGKTHREFPDAFYDISRDRDNRVVILTGSGDAW
MAEIDFPSLGDVTNPREWDKTYWEGKKVLQNLLDIEVPVISAVNGAALLHSEYILTTDIILASENTVFQDM
PHLNAGIVPGDGVHILWPLALGLYRGRYFLFTQEKLTAQQAYELNVVHEVLPQSKLMERAWEIARTLAKQP
TLNLRYTRVALTQRLKRLVNEGIGYGLALEGITATDLRNT
```

Many online molecular biology tools work with this format, which is most easily saved as a plain text file in Notepad. The use of the Courier New font is deliberate as it allows the precise alignment of sequence residues on different text lines as shown later. Once the sequence of the enzyme has been acquired, it is possible to upload this into dedicated search engines that will find enzymes of related amino acid sequence.

The best way to do this is to access the EXPASY Molecular Biology Server (http://ca.expasy.org/). EXPASY is an excellent multi-purpose molecular biology server that allows for simple manipulations and investigations of gene and amino acid sequence. In the present context, it allows the searching of databases for enzyme homologues and the creation of sequence alignments that permit the observation of conserved regions of amino acid sequence that may, for example, be related to function. All of the functions above and many more that do not fall within the remit of this book can be used with the amino acid sequence in FASTA format highlighted above. In addition, such a search will provide links to the gene sequence encoding the enzyme of interest.

In order to search for homologues of the lyase enzyme described above, a 'BLAST' search may be performed - sometimes also known as 'BLASTing' after the original program [6]. BLAST will search a selection of protein databases that can be chosen from the menu, such as **Swissprot-TREMBL**, and display a list of homologues in order of the sequence identities with the search sequence. It is also possible to perform these searches using gene sequence data.

58

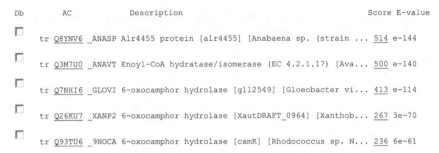

```
Db      AC              Description                                              Score E-value

▯   tr Q8YNV6 _ANASP Alr4455 protein [alr4455] [Anabaena sp. (strain ... 514 e-144

▯   tr Q3M7U0 _ANAVT Enoyl-CoA hydratase/isomerase (EC 4.2.1.17) [Ava... 500 e-140

▯   tr Q7NHI6 _GLOVI 6-oxocamphor hydrolase [gll2549] [Gloeobacter vi... 413 e-114

▯   tr Q26KU7 _XANP2 6-oxocamphor hydrolase [XautDRAFT_0964] [Xanthob... 267 3e-70

▯   tr Q93TU6 _9NOCA 6-oxocamphor hydrolase [camK] [Rhodococcus sp. N... 236 6e-61
```

Figure 2.7   Output of BLAST search. Output of BLAST search (http://www.expasy.org), Gasteiger E., Gattiker A., Hoogland C., Ivanyi I., Appel R.D., Bairoch A. ExPASy: the proteomics server for in-depth protein knowledge and analysis, Nucleic Acids Res. 31, 3784–3788 (2003). Printed with permission from the Swiss Institute of Bioinformatics and Oxford University Press.

To perform a BLAST search with the amino acid sequence of the lyase used as an example above, go to:

> Similarity Searches
> BLAST
> Paste sequence
> Run BLAST

The search will usually take a matter of seconds with output resembling that shown in Figure 2.7.

The first entry will be the enzyme that was used for the search itself. The code number **Q8YNV6** to the left of the title, is the accession number of the enzyme in the UNIPROT database. This number is unique to this particular enzyme sequence and can subsequently be very useful for searching the web for information on this enzyme. Clicking on this number opens a useful page of data that gives multiple links to other databases and information on the enzyme. The gene sequence that encodes the enzyme can be found under 'Cross-References'. Clicking on 'CoDing Sequence' will give a new page, at the base of which will be displayed the gene sequence:

```
SQ    Sequence 762 BP; 223 A; 162 C; 169 G; 208 T; 0 other; 4144027985 CRC32;
      atgactttga atcaacctga atatttcacc aaatacgaaa acctgcactt ccatcgagat        60
      gaaaatggca ttctagaagt gagaatgcac acaaacggta gttcacttgt tttcacaggt       120
      aaaactcatc gagagtttcc cgacgcattc tacgacatta gccgagacag agataaccgt       180
      gttgtcatcc tcacaggtag tggtgatgct tggatggctg aaatcgattt tcctagttta       240
      ggagatgtca ccaatcctcg tgagtgggat aaaacttatt gggaaggtaa aaaggtgctg       300
      caaaacctct tggatattga agtaccagtt atctctgctg tcaatggtgc agcattactt       360
      catagtgaat acattctgac gaccgacatt attcttgcgt ctgaaaacac tgtatttcaa       420
      gatatgcctc atttaaatgc aggtatcgta ccaggtgacg gagtgcatat attatggccg       480
```

```
ttggcacttg gtcttatcg cgggcgctac ttcttgttta ctcaagaaaa gctgacggct    540
caacaagctt atgaacttaa tgttgtccat gaagtactgc cgcaaagcaa actcatggaa    600
cgggcttggg aaattgccag gacacttgcc aaacagccca cattaaactt gcgatacacc    660
cgtgtggcct tgactcaaag actgaagcgg ttagttaatg aagggattgg ctatggtcta    720
gcattagaag gtattaccgc caccgatctc cgcaatactt aa                       762
```

If we go back to the results of the BLAST search, scrolling down the page will show first a list of homologues that have a decreasing sequence identity with the enzyme used as the search model, followed by a series of sequence alignments like the one shown below, for the third 'hit' in the example, a lyase from the bacterium *Gloeobacter*. In the example, the 'Query' is the amino acid sequence entered by the user; the 'Subject' is the new amino acid sequence. The middle line highlights the conserved amino acids in the sequence.

```
trQ7NHI6
Q7NHI6_GLOVI6-oxocamphor hydrolase [gl12549] [Gloeobacter violaceus] 254 AA
align
Score =  413 bits (1061), Expect = e-114
Identities = 202/252 (80%), Positives = 223/252 (88%)

Query:   1  MTLNQPEYFTKYENLHFHRDENGILEVRMHTNGSSLVFTGKTHREFPDAFYDISRDRDNR  60
            M  QPEYF+KYENL F RDE GIL VRMHT G  LVFTGKTHREFDAFYDISRDRDNR
Sbjct:   1  MATVQPEYFSKYENLIFTRDEQGILTVRMHTQGGPLVFTGKTHREFVDAFYDISRDRDNR  60

Query:  61  VVILTGSGDAWMAEIDFPSLGDVTNPREWDKTYWEGKKVLQNLLDIEVPVISAVNGAALL  120
            VVILTG+G+AWM +IDF SLGDVTNPREWDKTYWEGKKVLQNLLDIEVPVI+AVNG ALL
Sbjct:  61  VVILTGTGNAWMDQIDFASLGDVTNPREWDKTYWEGKKVLQNLLDIEVPVITAVNGPALL  120

Query: 121  HSEYILTTDIILASENTVFQDMPHLNAGIVPGDGVHILWPLALGLYRGRYFLFTQEKLTA  180
            H+EYILT+DI+LA+++ VFQDMPHLNAGIVPGDGVH+LWPL LG  RGRYFL TQ+KLTA
Sbjct: 121  HTEYILTSDIVLAAQSAVFQDMPHLNAGIVPGDGVHVLWPLVLGPSRGRYFLLTQQKLTA  180

Query: 181  QQAYELNVVHEVLPQSKLMERAWEIARTLAKQPTLNLRYTRVALTQRLKRLVNEGIGYGL  240
            Q+A +L VV+E+LP KLMERA ++A  LA QPTL LRYTRVALTQRLKRLV+EG+GYGL
Sbjct: 181  QEALDLGVVNEMLPDEKLMERAQQLAEQLAHQPTLTLRYTRVALTQRLKRLVSEGLGYGL  240

Query: 241  ALEGITATDLRN  252
            LEGITATDLRN
Sbjct: 241  VLEGITATDLRN  252
```

A quick inspection reveals that this homologue is very similar indeed to the enzyme sequence used as search model ('Identity' of 80%) and might therefore be an interesting new candidate for study. Clicking on the Gene accession number Q7NHI6 raises the UniProtKB/TrEMBL entry Q7NHI6 page, which has a link to the amino acid sequence of this new lyase at the bottom of the page, and also the Coding Sequence, which can then be used in cloning experiments described in Chapter 7.

A second example concerns an enzyme for which the structure has not been determined. The cyclohexanone-1,2-monooxygenase (CHMO) from

Acinetobacter calcoaceticus NCIMB 9871 is an oxidoreductase that catalyses the Baeyer–Villiger reaction and was first described by Trudgill and colleagues in 1976 [7]. The gene encoding CHMO was cloned in 1988 by Walsh and colleagues [8]. A literature search for CHMO in the PubMed database reveals a number of publications including the gene sequencing paper:

Chen YC, Peoples OP, Walsh CT.
Acinetobacter cyclohexanone monooxygenase: gene cloning and sequence determination.
J Bacteriol. 1988 Feb;170(2):781-9. PMID: 3338974

Clicking on this link will bring up the abstract of the paper. Clicking on the 'Links' button at the far right-hand side of the page, will bring up a menu that includes 'Protein'. Clicking on this link will bring up, in this case, a series of proteins that are related in some way to CHMO, either because they are also monooxygenases, or perhaps they are other enzymes contained within *Acinetobacter*.

BAA86293
Cyclohexanone 1,2-monooxygenase [Acinetobacter sp. NCIMB9871]
gi|6277322 | dbj|BAA86293.1 | [6277322]

Clicking on the gene accession number BAA86293 brings up a page at the bottom of which can be found the amino acid sequence:

```
ORIGIN
        1 msqkmdfdai vigggfggly avkklrdele lkvqafdkat dvagtwywnr ypgaltdtet
       61 hlycyswdke llqsleikkk yvqgpdvrky lqqvaekhdl kksyqfntav qsahyneada
      121 lwevtteygd kytarflita lgllsapnlp nikginqfkg elhhtsrwpd dvsfegkrvg
      181 vigtgstgvq vitavaplak hltvfqrsaq ysvpigndpl seedvkkikd nydkiwdgvw
      241 nsalafglne stvpamsvsa eerkavfeka wqtgggfrfm fetfgdiatn meanieaqnf
      301 ikgkiaeivk dpaiaqklmp qdlyakrplc dsgyyntfnr dnvrledvka npiveiteng
      361 vklengdfve ldmlicatgf davdgnyvrm diqgknglam kdywkegpss ymgvtvnnyp
      421 nmfmvlgpng pftnlppsie sqvewisdti qytvennves ieatkeaeeq wtqtcaniae
      481 mtlfpkaqsw ifganipgkk ntvyfylggl keyrsalanc knhayegfdi qlqrsdikqp
      541 ana
```

As with the lyase example, >'Display' >FASTA will then give a page with the amino acid sequence of CHMO in FASTA format for personal archiving.

```
>gi|6277322|dbj|BAA86293.1| cyclohexanone 1,2-monooxygenase [Acinetobac-
ter sp. NCIMB9871]
MSQKMDFDAIVIGGGFGGLYAVKKLRDELELKVQAFDKATDVAGTWYWNRYPGALTD TETHLYCYSWDKE
LLQSLEIKKKYVQGPDVRKYLQQVAEKHDLKKSYQFNTAVQSAHYNEADALWEVTTE YGDKYTARFLITA
LGLLSAPNLPNIKGINQFKGELHHTSRWPDDVSFEGKRVGVIGTGSTGVQVITAVAP LAKHLTVFQRSAQ
YSVPIGNDPLSEEDVKKIKDNYDKIWDGVWNSALAFGLNESTVPAMSVSAEERKAVF EKAWQTGGGFRFM
FETFGDIATNMEANIEAQNFIKGKIAEIVKDPAIAQKLMPQDLYAKRPLCDSGYYNT FNRDNVRLEDVKA
NPIVEITENGVKLENGDFVELDMLICATGFDAVDGNYVRMDIQGKNGLAMKDYWKEG PSSYMGVTVNNYP
NMFMVLGPNGPFTNLPPSIESQVEWISDTIQYTVENNVESIEATKEAEEQWTQTCAN IAEMTLFPKAQSW
IFGANIPGKKNTVYFYLGGLKEYRSALANCKNHAYEGFDIQLQRSDIKQPANA
```

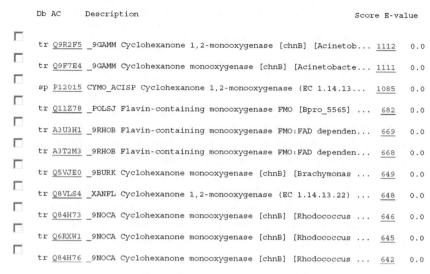

```
 Db AC       Description                                                    Score E-value

□ tr Q9R2F5 _9GAMM Cyclohexanone 1,2-monooxygenase [chnB] [Acinetob...      1112   0.0
□ tr Q9F7E4 _9GAMM Cyclohexanone monooxygenase [chnB] [Acinetobacte...      1111   0.0
□ sp P12015 CYMO_ACISP Cyclohexanone 1,2-monooxygenase (EC 1.14.13...       1085   0.0
□ tr Q11Z78 _POLSJ Flavin-containing monooxygenase FMO [Bpro_5565] ...       682   0.0
□ tr A3U3H1 _9RHOB Flavin-containing monooxygenase FMO:FAD dependen...       669   0.0
□ tr A3T2M3 _9RHOB Flavin-containing monooxygenase FMO:FAD dependen...       668   0.0
□ tr Q5VJE0 _9BURK Cyclohexanone monooxygenase [chnB] [Brachymonas ...       649   0.0
□ tr Q8VLS4 _XANFL Cyclohexanone 1,2-monooxygenase (EC 1.14.13.22) ...       648   0.0
□ tr Q84H73 _9NOCA Cyclohexanone monooxygenase [chnB] [Rhodococcus ...       646   0.0
□ tr Q6RXW1 _9NOCA Cyclohexanone monooxygenase [chnB] [Rhodococcus ...       645   0.0
□ tr Q84H76 _9NOCA Cyclohexanone monooxygenase [chnB] [Rhodococcus ...       642   0.0
```

Figure 2.8   Output of BLAST search

On the EXPASY site, a BLAST search using this sequence brings up many more examples of related enzymes than was observed with the lyase example, reflecting the amount of new enzymes in this class that have been discovered as a result of genome sequencing (Figure 2.8).

Clicking on the UniProt number Q9R2F5 will bring up the appropriate page for CHMO and on this will be found the Coding sequence as before:

```
atgtcacaaa aaatggattt tgatgctatc gtgattggtg gtggtttttgg cggactttat      60
gcagtcaaaa aattaagaga cgagctcgaa cttaaggttc aggcttttga taaagccacg     120
gatgttgcag gtacttggta ctggaaccgt tacccaggtg cattgacgga tacagaaacc     180
cacctctact gctattcttg ggataaagaa ttactacaat cgctagaaat caagaaaaaa     240
tatgtgcaag gccctgatgt acgcaagtat ttacagcaag tggctgaaaa gcatgattta     300
aagaagagct atcaattcaa taccgcggtt caatcggctc attacaacga agcagatgcc     360
ttgtgggaag tcaccactga atatggtgat aagtacacgg cgcgtttcct catcactgct     420
ttaggcttat tgtctgcgcc taacttgcca aacatcaaag gcattaatca gtttaaaggt     480
gagctgcatc ataccagccg ctggccagat gacgtaagtt ttgaaggtaa acgtgtcggc     540
gtgattggta cgggttccac cggtgttcag gttattacgg ctgtggcacc tctggctaaa     600
cacctcactg tcttccagcg ttctgcacaa tacagcgttc caattggcaa tgatccactg     660
tctgaagaag atgttaaaaa gatcaaagac aattatgaca aaatttggga tggtgtatgg     720
aattcagccc ttgcctttgg cctgaatgaa agcacagtgc cagcaatgag cgtatcagct     780
gaagaacgca aggcagtttt tgaaaaggca tggcaaacag gtggcggttt ccgtttcatg     840
tttgaaactt tcggtgatat tgccaccaat atggaagcca atatcgaagc gcaaaatttc     900
attaagggta aaattgctga aatcgtcaaa gatccagcca ttgcacagaa gcttatgcca     960
caggatttgt atgcaaaacg tccgttgtgt gacagtggtt actacaacac ctttaaccgt    1020
gacaatgtcc gtttagaaga tgtgaaagcc aatccgattg ttgaaattac cgaaaacggt    1080
gtgaaactcg aaaatggcga tttcgttgaa ttagacatgc tgatatgtgc cacaggtttt    1140
gatgccgtcg atggcaacta tgtgcgcatg gacattcaag gtaaaaacgg cttggccatg    1200
aaagactact ggaaagaagg tccgtcgagc tatatgggtg tcaccgtaaa taactatcca    1260
```

```
aacatgttca tggtgcttgg accgaatggc ccgtttacca acctgccgcc atcaattgaa      1320
tcacaggtgg aatggatcag tgataccatt caatacacgg ttgaaaacaa tgttgaatcc      1380
attgaagcga caaaagaagc ggaagaacaa tggactcaaa cttgcgccaa tattgcggaa      1440
atgaccttat tccctaaagc gcaatcctgg atttttggtg cgaatatccc gggcaagaaa      1500
aacacggttt acttctatct cggtggttta aaagaatatc gcagtgcgct agccaactgc      1560
aaaaaccatg cctatgaagg ttttgatatt caattacaac gttcagatat caagcaacct      1620
gccaatgcct aa                                                         1632
```

This second example reveals the opportunities, but also the challenges of biocatalyst discovery in the post-genomics era. Searches of the type above may reveal a bewildering array of enzymes that may have the desired function, and it will be necessary to either choose one target based on more specific criteria (perhaps the strain of source organism or stretches of amino acid sequence that may give clues to substrate specificity) or to adopt high-throughput approaches to cloning, expression and evaluation as described in Chapters 7 and 8. The identification of more specific criteria may be aided by studying the alignment of the search sequence with the set of homologous sequences (see below).

2.8.1 Amino acid sequence alignments

The alignments that result from BLAST searching are often partial, indicating the identity only between stretches of relevant sequence in the search sequence and the result. More information is to be gained from performing full length amino acid sequence alignments. These can be generated when one is in possession of a number of related sequences and would like to examine which amino acid residues are *conserved* between a family of sequences, that is, the ones that occur in an equivalent position in the polypeptide chains, perhaps as this may give a clue as to the identity of amino acid residues that are important in substrate binding or in catalysing a certain kind of reaction. There are several different programs available for running alignments, including TCoffee [9] and ClustalW [10], both of which are accessible through the EXPASY server.

To perform a simple Clustal W alignment using the EXPASY server:

>Under 'Tools and Software Packages': 'Proteomics and sequence analysis tools' Select 'Alignment'
>Under 'Alignment'- 'Multiple' Select Clustal W at [EBI]
>In the box at the base of the interactive screen, two or more sequences are pasted in FASTA format, which includes the >XXXX prefix at the top of the sequence. In a first example we will return to the lyase used in the description of BLAST. In this case, two lyase enzymes, a β-diketone hydrolyase from *Anabaena* sp. PCC 7120 (>ABDH) and 6-oxo camphor hydrolase from *Rhodococcus* sp. NCIMB 9784 (>OCH) will be aligned.
>Paste the sequences into the box
>Press 'Run'

The length of time taken to retrieve the alignment may vary on the number of input sequences. For the two sequences used in the first example, the results are presented in the following format:

```
CLUSTAL W (1.83) multiple sequence alignment

ABDH    -MTLNQP--EYFTKYENLHFHRDENGILEVRMHTNGSSLVFTGKTHREFPDAFYDISRDR 57
OCH     MKQLATPFQEYSQKYENIRLERD-GGVLLVTVHTEGKSLVWTSTAHDELAYCFHDIACDR 59
        *   *  **  ****:::.** .*:* * :**:*.***:*...: *:. .*:**: **

ABDH    DNRVVILTGSGDAWMAEIDFPSLGDVTNPREWDKTYWEGKKVLQNLLDIEVPVISAVNGA 117
OCH     ENKVVILTGTGPSFCNEIDFTSF-NLGTPHDWDEIIFEGQRLLNN-LSIEVPVIAAVNGP 117
        :*:******:* :: ****.*: :: .*::**: :**:::*:* *.******:****.

ABDH    ALLHSEYILTTDIILASENTVFQDMPHLNAGIVPGDGVHILWPLALGLYRGRYFLFTQEK 177
OCH     VTNHPEIPVMSDIVLAAESATFQDGPHFPSGIVPGDGAHVVWPHVLGSNRGRYFLLTGQE 177
        . *.*  : :**:**:*.:.*** **: :*******.*::** .** ******:* ::

ABDH    LTAQQAYELNVVHEVLPQSKLMERAWEIARTLAKQPTLNLRYTRVALTQRLKRLVNEGIG 237
OCH     LDARTALDYGAVNEVLSEQELLPRAWELARGIAEKPLLARRYARKVLTRQLRRVMEADLS 237
        * *:  *  :  ..*:***.:.:*: ****:** :*::* *   **:* .**::*:*::: .:.

ABDH    YGLALEGITATDLRNT--- 253
OCH     LGLAHEALAAIDLGMESEQ 256
        *** *.::* **
```

Residues which are absolutely conserved between the sequences are shown by asterisks; those with very close similarity (such as valine and leucine – both hydrophobic aliphatic side-chains) by two dots, and less, but still similar (such as aspartate and serine – both polar) residues by one dot. The alignment reveals that residues that are implicated in substrate binding and catalysis [5] including Histidine 122 (H122) and Histidine 145 (H145, both in bold) are conserved in both of these isofunctional biocatalysts, as we may expect.

As a second example, let us return to the cyclohexanone 1,2-monooxygenase. Here, we have used the ClustalW program to perform an alignment of CHMO from *Acinetobacter* (UniProt number Q9R2F5) with a previously not studied 'putative' monooxygenase Q84H88 from *Arthrobacter* sp. BP2. The following alignment results:

```
Q9R2F5  --------------------------------------------------MSQK---MDF 7
Q84H88  MSTRSWPGGPPSWHRSSTSSRPGTGNNPATLRSHTIHYFVPCIRTTKEFAMTAQNTFQTV 60
                                                          *: :    .

Q9R2F5  DAIVIGGGFGGLYAVKKLRDELELKVQAFDKATDVAGTWYWNRYPGALTDTETHLYCYSW 67
Q84H88  DAVVIGAGFGGIYAVHKLHNEQGLTVVGFDKADGPGGTWYWNRYPGALSDTESHVYRFSF 120
        **:***.****:***:**::*   *.* .**** . .************:***:*:* :*:

Q9R2F5  DKELLQSLEIKKKYVQGPDVRKYLQQVAEKHDLKKSYQFNTAVQSAHYNEADALWEVTTE 127
Q84H88  DKGLLQDGTWKHTYITQPEILEYLEDVVDRFDLRRHFRFGTEVKSATYLEDEGLWEVTTG 180
        ** ***.   *:.:*  *:: :**::*.::.**::  ::*.* *:** * *  :.****** 
```

```
Q9R2F5    YGDKYTARFLITALGLLSAPNLPNIKGINQFKGELHHTSRWPDDVSFEGKRVGVIGTGST 187
Q84H88    GGAVYRAKYVINAVGLLSAINFPNLPGIDTFEGETIHTAAWPQGKSLAGRRVGVIGTGST 240
          *   * *::*.*:***** *;**: **: *;** **: **:. *; *:**********

Q9R2F5    GVQVITAVAPLAKHLTVFQRSAQYSVPIGNDPLSEEDVKKIKDNYDKIWDGVWNSALAFG 247
Q84H88    GQQVITALAPEVEHLTVFVRTPQYSVPVGKRPVTTQQIDEIKADYDNIWAQVKRSGVAFG 300
          * *****;** .;***** *:.*****;*: *;: :::.;** ;**;** * .*.;***

Q9R2F5    LNESTVPAMSVSAEERKAVFEKAWQTGGGFRFMFETFGDIATNMEANIEAQNFIKGKIAE 307
Q84H88    FEESTVPAMSVTEEERRQVYEKAWEYGGGFRFMFETFSDIATDEEANETAASFIRNKIVE 360
          ::*********: ***: *;****: *********** .****: *** * .**:.**.*

Q9R2F5    IVKDPAIAQKLMPQDLYAKRPLCDSGYYNTFNRDNVRLEDVKANPIVEITENGVKLENGD 367
Q84H88    TIKDPETARKLTPTGLFARRPLCDDGYFQVFNRPNVEAVAIKENPIREVTAKGVVTEDGV 420
          :*** *:** * .*:*:*****.**::.*** **.  :* *** *;* ;** *:*

Q9R2F5    FVELDMLICATGFDAVDGNYVRMDIQGKNGLAMKDYWKEGPSSYMGVTVNNYPNMFMVLG 427
Q84H88    LHELDVIVFATGFDAVDGNYRRMEISGRDGVNINDHWDGQPTSYLGVSTAKFPNWFMVLG 480
          : ***::: ********** **;*.*::*: ::*:*. *;**;**:. ;:** *****

Q9R2F5    PNGPFTNLPPSIESQVEWISDTIQYTVENNVESIEATKEAEEQWTQTCANIAEMTLFPKA 487
Q84H88    PNGPFTNLPPSIETQVEWISDTVAYAEENGIRAIEPTPEAEAEWTETCTQIANMTVFTKV 540
          *************;********: *; **;.;;**.* *** ;**;**;;**;**;*.*.

Q9R2F5    QSWIFGANIPGKKNTVYFYLGGLKEYRSALANCKNHAYEGFDIQLQRSDIKQPANA 543
Q84H88    DSWIFGANVPGKKPSVLFYLGGLGNYRGVLDDVTANGYRGFELKSEAAVAA----- 591
          :*******:**** ;* ****** ;**..* : . ;.*.**::: : :
```

Two conserved areas of sequence or 'motif' are highlighted in bold. The first, on the second line, is representative of a **GXGXXG** motif involved in the binding of the ADP moiety of the coenzyme FAD, essential for the activity of this class of monoxygenases. The second, on the fourth line, shows a **FXGXXHXXXWP** motif thought to be characteristic of enzymes that catalyse the Baeyer–Villiger reaction [11].

The presence of such motifs, if supported by experimental evidence from the literature, may give some confidence that the new gene, if cloned and expressed, will have the desired activity. Some caution should be exercised, however. First, the gene may not be suitable for cloning and soluble expression in *E. coli* (see Chapter 7). Secondly, attempts to assign function to an enzyme from primary structural information only can come unstuck; often the 'annotation' of possible function of an amino acid sequence may be based on limited sequence identity and there are certainly cases where even the conservation of amino acid residues that have been reported in the literature to be catalytically relevant, are not functional in the new homologue, which may catalyse a different chemical reaction entirely. Even when predicted function is more certain, for example with the cytochrome P450 family of oxidases, whose identity is comparatively easy to predict from sequence, it is at this time very difficult to predict *substrate specificity* of an enzyme from within a group of related sequences. Some clue may be obtained from what is known as the *genomic context* of the gene that encodes the enzyme of interest, i.e. the location of the gene within the genome. If the gene is located adjacent to a number of other genes involved in, say, the biosynthesis of methionine, this may give some clues to the types of substrates on which the enzyme acts naturally. It is easy to examine the genomic context of a gene using genome sequence databases of the type described below.

2.8.2 Genome sequence databases

It is also possible to search sequenced genomes directly for biocatalyst sequences using servers such as **PEDANT** [12] (http://pedant.gsf.de/), which is run by the Munich Information Centre for Protein Sequences (MIPS). This website lists all the available sequenced genomes alphabetically or by kingdom (eubacteria, archaea, etc.) and allows searching based on amino acid sequence, text-based searches using the names of enzymes and also direct observation of the order in which genes are laid down on the chromosome – the *genomic context* alluded to above.

One may select a genome of choice from an alphabetical list. We will choose the genome of *Nostoc* (*Anabaena*) PCC7120 as an example. First we select 'Bacteria' from the genome menu at the top of the page, and from the alphabetical list select the genome of the relevant bacterium.

Clicking on the name of this bacterium raises a page with a list of genes encoded within the genome ordered by 'Contig. Start' – essentially the number of the gene within the genome from a predetermined starting point. The top menu provides a search option under which both *sequence*-based and *text*-based searching would be possible. One could therefore, search for homologues of the lyase example used above by selecting sequence search and entering the amino acid sequence of the lyase in FASTA format. The search in this instance only finds two hits: the search sequence itself, and an enzyme of related sequence, but different function, napthoate synthase.

Alternatively, a 'text-based search' of the genome allows the option of entering the name of an enzyme 'lyase' or 'cytochrome P 450" and discovering homologues in this way. The disadvantage here is that sequences which have not had functions assigned to them - 'hypothetical proteins' as they are sometimes called – will not feature in the results.

To examine the genomic context of a gene, let us consider the genome of the well-studied *Pseudomonas putida* KT2440, an organism, which amongst many other biocatalytic capabilities, is known to degrade aromatic compounds. The enzymes involved in the pathways of aromatic catabolism can be very useful biocatalysts, catalysing reactions such as dioxygenation and C-C bond hydrolysis. Very often, the genes encoding a whole enzymatic pathway of catabolism are *clustered*: that is to say, feature next to each other in a genome sequence, and are under the same mechanism of genetic control, being perhaps expressed

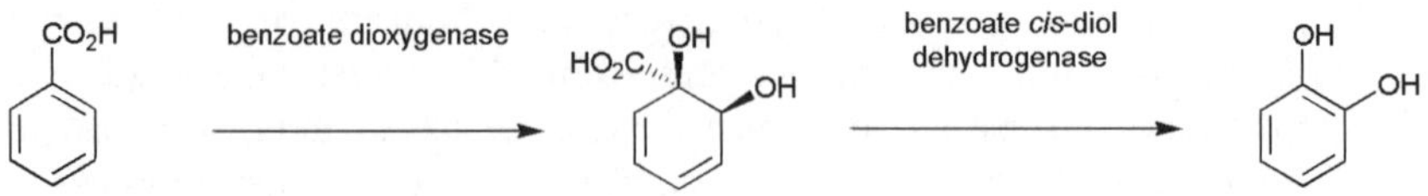

Figure 2.9 Initial steps in the degradation of benzoate in *Pseudomonas*, showing the cascade of enzymatic reactions

in response to the presence of an aromatic compound in the environment (see inducible expression in Chapter 4). When *P. putida* is exposed to some aromatic compounds, such as benzoate, a pathway of enzymes is induced in order to catabolise that substrate. The pathway might first involve dioxygenation of benzoate (by *benzoate dioxygenase*), followed by reduction of the oxygenated intermediate (by a *cis-diol dehydrogenase*) (Figure 2.9).

If we search the genome of *P. putida* KT2440 on the PEDANT database using the sequence of the *cis*-diol dehydrogenase (gene number gi_26989883) and then select from the top of the page >Viewers >DNA viewer, a page results that shows the gene, represented by a red arrow, within the genome of the organism. Genes that lie next to the selected gene can be viewed by zooming out. It is found that the genes encoding the components of the benzoate dioxygenase are to be found next to the dehydrogenase. There are also other genes close by that are very obviously associated with the metabolism of aromatics, such as 'benzoate transport protein' and 'benzoate-specific porin'. Whilst such an approach is far from certain to give information about the possible function or substrate specificity of a previously unstudied enzyme, it may be of limited use when the enzymes are associated with genes involved in either the biosynthesis, or degradation of certain compounds.

One of the other major online genome resources is **TIGR** (The Institute for Genomic Research) (http://www.tigr.org/) now operated through the J. Craig Venter Institute (JCVI). The website of the JCVI, established in October 2006, now accommodates the web-based resource of TIGR, which includes the 'Comprehensive Microbial Resource' – a service encompassing, among other tools, a comprehensive means of searching completed and unfinished genomes for enzyme sequences of interest. Genomes are listed by organism name and may be searched by BLASTing using the amino acid sequence, or by enzyme name or gene accession number. There are also useful links to genome sequence publications and a host of molecular biology tools.

2.9 Obtaining DNA Templates for Cloning

Once an interesting enzyme homologue has been identified, one may wish to obtain a DNA template for cloning experiments (a beginner's guide to a simple cloning experiment can be found in Chapter 7). There are various avenues for obtaining DNA for these studies.

Genomic DNA, or indeed the expression plasmid that has been used for the production of recombinant protein, may be obtained from the corresponding author of the relevant publication – indeed it is often a condition of publication that such material should be made available to academic researchers on request, as previously stated. As well as research groups, various institutes around the world may provide genomic DNA either free-of-charge, or for a small fee, to academic investigators, either 'no strings attached' or as part of a collaboration. Genomic DNA templates may also sometimes be purchased from culture collections such as the **ATCC** and **NCYC**. It is probable that the genomic DNA of an organism that has been sequenced (and for which the genomic sequence is publicly accessible) will be available from one or more sources.

It is of course also possible to obtain samples of genomic DNA from small culture of the chosen organism. There are various simple protocols for this, and like many contemporary techniques in molecular biology, may be accomplished using commercial kits from companies supplying reagents for molecular biology such as Sigma, Qiagen or Promega. The preparation of genomic DNA samples from live bacteria may be absolutely necessary in some cases, but is of course undesirable when the organism is difficult to grow as it is perhaps thermophilic, or is a pathogen. The techniques associated with performing a simple cloning experiment using a genomic DNA template are detailed in Chapter 7.

2.10 Custom Gene synthesis

Until very recently, the cloning of a gene for a biocatalysis experiment was dependent on the availability of the genetic material from the organism producing the enzyme of interest. However, even when genetic sequence information was in the public domain, it could sometimes be difficult to acquire the necessary material from either companies or other academic groups. The need for genomic DNA, or a sample of the organism, is being obviated by the emergence of affordable gene synthesis. A number of companies, including NextGen, Blue Heron and GeneArt and DNA 2.0, will, on being provided with the sequence of interest, perform a custom-made gene synthesis service that, in addition to providing that gene within a user-friendly plasmid, will customise various aspects to ease cloning into different plasmid vectors ('sub-cloning') and expression in laboratory strains such a *E. coli* (see Chapter 7). The disadvantage is the expense. At the time of writing, gene synthesis companies charge in the range of $0.5–0.8 'per base-pair' i.e. $1.50–2.40 per amino acid in the resultant protein, so a synthetic gene encoding an average-sized enzyme of perhaps 300 amino acids would cost in the region of $450–750. It is certain that the costs of gene synthesis will decrease in the future, however, as techniques improve.

2.11 Other Web Resources for Finding Information on Biocatalysts

2.11.1 EBI

The European Bioinformatics Institute
(http://www.ebi.ac.uk/)
Operated by the European Molecular Biology Laboratory in Cambridgeshire, the EBI is one of the world's leading institutues whose mission is to 'collect, store and curate' the huge amount of biological information that is now arising from genome sequencing and associated high-throughput endeavours such as proteomics, structural genomics and microarray analysis. It provides a number of useful search engines in its own right that might be of use to biocatalysis researchers, including access to the RCSB database of protein structures, EMBLbank (gene sequences) and UniProt (protein sequences).

2.11.2 RCSB PDB

Research Collaboration for Structural Bioinformatics Protein Data Bank
http://www.rcsb.org/pdb/home/home.do
A repository of X-ray and NMR crystal structures of proteins and enzymes that is becoming increasingly easy to use as a result of the provision of freeware for the visualisation of protein structures in three dimensions. Although an extension to the work described in this book, researchers may certainly be interested in the three-dimensional structure of biocatalysts when trying to rationalise observed selectivity or to provide a basis for rational protein engineering. Again, the structure database here or at EBI may be searched using a mixture of text-based approaches or using amino acid sequence to obtain the four-character deposition code for the structure (e.g. 1o8u). Some software for viewing these structures is available within the RCSB site, but coordinate files (xxxx.pdb) can be downloaded for personal archiving and viewing using programs such as Coot, Quanta or O.

2.11.3 CoEBio3

Centre of Excellence for Biocatalysis, Biotransformations and Biomanufacture
http://www.coebio3.manchester.ac.uk/
   The Centre was established in 2005 at the University of Manchester, UK. The Centre is a collaborative effort between academics at the Universities of Manchester, York, Strathclyde and Heriot-Watt and industrial affiliates representing more than fifteen companies. The website offers information on enzyme sources,

procedures and relevant literature, and offers to members an abstracts service that provides regular updates on the literature on applied biocatalysis in organic chemistry once a month.

2.11.4 University of Minnesota biocatalysis and biodegradation database

http://umbbd.msi.umn.edu/

This website is an excellent resource that provides information on microbial biocatalytic reactions and also the pathways by which microbes degrade certain organic substrates. The website can be searched using organism and compound names and also provides other useful links to those interested in identifying a possible biocatalyst for an intended process.

2.11.5 CBB

The University of Iowa Center for Biocatalysis and Bioprocessing
http://www.uiowa.edu/~biocat/

Established in 1983, the CBB is one of the world's leading centres for Biocatalysis research. The CBB carries out research funded by both public bodies and industry and runs academic programmes for graduates. It has also recently begun to perform contract fermentation reactions for industry.

2.11.6 Synopsys biocatalysis database

http://www.accelrys.com/products/chem_databases/databases/biocatalysis.html

Now administered by Accelrys, the Biocatalysis database uses information gathered by Professor J. B. Jones (University of Toronto) to provide a database of information on biocatalysed reactions that can be searched by compound substructure, micro-organism name, publication author and other strands.

2.12 Conclusion

The availability of biocatalysts and the information that can be used to acquire them is now greater than it has ever been, as a result of high-throughput techniques that reveal the content of genomes and the development of tools that can be used to access the reservoir of new genes. It is certain that, in addition to providing a whole host of new candidate enzymes for existing reactions that may offer new opportunities in selectivity, new enzymatic activities will continue to be described that catalyse chemical transformations as yet unprecedented for biological systems.

References

1. P. Lorenz, K. Liebeton, F. Niehaus and J. Eck (2002) Screening for novel enzymes for biocatalytic processes: accessing the metagenome as a resource of novel functional sequence space. *Curr. Opin. Biotechnol.*, **13**, 572–577.
2. M. Breuer, K. Ditrich, T. Habicher, B. Hauer, M. Keßeler, R. Stürmer and T. Zelinski (2004) Industrial methods for the production of optically active intermediates. *Angew. Chem. Int. Ed.*, **43**, 788–824.
3. A. Liese, K. Seelbach and C. Wandrey (Eds) (2006) *Industrial Biotransformations*, Wiley-VCH, Weinheim.
4. S. Joseph et al. (2007) The Sorcerer II Global Ocean Sampling Expedition: expanding the universe of protein families. *PLoS Biol.*, **5**, e16.
5. J. Bennett, J. L. Whittingham, A. M. Brzozowski, P. M. Leonard and G. Grogan (2007) Structural characterisation of a β-diketone hydrolase from the cyanobacterium *Anabaena* sp. PCC 7120, in native and product-bound forms – a Coenzyme-A-independent member of the Crotonase suprafamily. *Biochemistry*, **46**, 137–144.
6. S. F. Altschul, T. L. Madden, A. A. Schäffer, J. Zhang, Z. Zhang, W. Miller and D. J. Lipman (1997) Gapped BLAST and PSI-BLAST: a new generation of protein database search programs. *Nucleic Acids Res.*, **25**, 3389–3402.
7. N. A. Donoghue, D. B. Norris and P. W. Trudgill (1976) The purification and properties of cyclohexanone oxygenase from *Nocardia globerula* CL1 and *Acinetobacter* NCIB 9871. *Eur. J. Biochem,.* **63**, 175–192.
8. Y-C. Chen, O. P. Peoples and C. T. Walsh (1988) Acinetobacter cyclohexanone monooxygenase: gene cloning and sequence determination. *J Bacteriol.*, **170**, 781–789.
9. C. Notredame, J. Higgins and J. Heringa (2000) T-Coffee. A novel method for multiple sequence alignments. *J. Mol. Biol.*, **302**, 205–217.
10. D. Higgins, J. Thompson, T. Gibson, J. D. Thompson, D. G. Higgins and T. J. Gibson (1994) CLUSTAL W: improving the sensitivity of progressive multiple sequence alignment through sequence weighting, position-specific gap penalties and weight matrix choice. *Nucleic Acids Res.*, **22**, 4673–4680.
11. M. W. Fraaije, N. M. Kamerbeek, W. J. H. van Berkel and D. B. Janssen (2002) Identification of a Baeyer-Villiger monooxygenase sequence motif. *FEBS Lett.*, **518**, 43–47.
12. D. Frishman, M. Mokrejs, D. Kosykh, G. Kastenmüller, G. Kolesov, I. Zubrzycki, C. Gruber, B. Geier, A. Kaps, K. Albermann, A. Volz, C. Wagner, M. Fellenberg, K. Heumann and H. W. Mewes (2003) The PEDANT genomes database. *Nucleic Acids Res.*, **31**, 201–207.

Chapter 3
Setting up a Laboratory for Biotransformations

3.1 Introduction

Biotransformations in the laboratory are certainly not more complicated to carry out than traditional chemical synthetic reactions, but the equipment used and the techniques may of course be somewhat unfamiliar to the uninitiated. It is comparatively straightforward for the worker in an adequately equipped synthetic organic laboratory to use off-the-shelf powdered enzymes of the sort that are dealt with in Chapter 4. The use of such enzymes does not of itself constitute any health risk as such, apart from that which might arise from any individual sensitivity to contact with the protein powder acting as an allergen. Therefore, the use of sterile equipment (*aseptic technique*), detailed below, and other precautions and equipment associated with the use of microbes are not required for enzymatic reactions using powdered enzymes. Experiments using bought enzymes may thus be performed in a standard organic chemistry laboratory, even if some considerations, such as the use of biological buffers as reaction media (see Chapter 5), will be unfamiliar. However, any use of either 'wild-type' (natural) strains of micro-organisms or genetically modified strains for biotransformations will require either the setting-up of a small microbiological facility within the research laboratory or access to appropriate facilities and equipment through collaboration with a microbiology/biochemistry group.

In this chapter, we will address considerations involved in setting up and running a basic microbiology facility for conducting whole-cell microbial bio-transformations, including equipment, reagents and basic safe practice in the handling and disposal of micro-organisms. We also describe additional equip-ment for the handling of proteins and DNA, which will be relevant to protocols described in Chapters 6 and 7.

Practical Biotransformations: A Beginner's Guide   © 2009   Gideon Grogan

3.2 Microbiological Containment

It should be emphasised that the vast majority of organisms used in biotransformation reactions pose little or no risk to the worker under ordinary laboratory conditions and everyday exposure. However, when working with micro-organisms, the issue of containment, or the control of contamination hazards, inevitably arises and the worker new to biotransformations should be absolutely certain before commencing work that the handling of any micro-organism takes place in a suitably equipped laboratory that is appropriately licensed and capable of both handling micro-organisms and disposing of them once they have been used.

The UK government provides detailed guidelines on the handling of micro-organisms hazardous to human health under the auspices of the Advisory Committee on Dangerous Pathogens (ACDP, http://www.hse.gov.uk/aboutus/meetings/committees/acdp/) [1]. The ACDP defines containment as 'the way in which biological agents are managed in the laboratory environment so as to prevent, or control the exposure of laboratory workers, other people and the outside environment to the agent(s) in question.' Considerations of containment are usually divided into 'primary' and 'secondary' levels, being, respectively, those issues directly affecting the worker and colleagues in the relevant laboratory and those which may impact outside the laboratory. Primary containment may be addressed by observing appropriate microbiological safety practices within the laboratory; secondary containment may involve considerations such as laboratory design, but also issues related to the safe and appropriate disposal of waste.

The ACDP has divided organisms into four Hazard Groups, namely HG 1, 2, 3 and 4, requiring, respectively, laboratories of physical containment levels CL1, CL2, CL3 and CL4. These classifications are made on the bases of criteria associated with pathogenicity, the possibility of transmission and the existence (or not) of effective prophylaxis or treatment. HG 3 (CL3) and HG 4 (CL4) are reserved for high-risk pathogens and are not relevant to this discussion. The overwhelming majority of organisms used for biotransformations are likely to be HG 1, which are 'unlikely to cause human disease'. HG 1 organisms may be used in an appropriate laboratory facility where standard *aseptic technique* (see Section 3.8) may be applied. Occasionally, the use of an organism that is HG 2 may be reported in the literature (notably, certain strains of *Aspergillus, Pseudomonas, Acinetobacter*) In the words of the ACDP, 'these organisms can cause human disease and may be hazard [sic] to employees; it is unlikely to spread to the community and there is usually effective prophylaxis or treatment available.' Such strains may also pose additional risks to any immunocompromised personnel working in or near the laboratory. The use of HG 2 organisms in CL2 facilities raises several implications for the upgrading of laboratory facilities and practices that would be beyond the ordinary facility provided by the average organic laboratory. In any case, HG 2 organisms are generally avoided for biotransformations these days, as there is no prospect of scaling up reactions to an industrial level using organisms that might pose any problems with respect to pathogenicity. We will therefore restrict our discussion to the use of ACDP HG 1 wild-type organisms, and address the installation of equipment and handling techniques for a CL1 containment laboratory only.

Some simple requirements for the containment of HG 1 micro-organisms are summarised below. These guidelines have been adapted from those currently listed at the University of York (http://www.york.ac.uk/depts/biol/web/safety/saf_bsmas/index.htm), but the reader is, in all cases, referred to the relevant government information supplied by ACDP [1] for authoritative guidance.

3.2.1 Clothing and general safety in the microbiology laboratory

Safety spectacles and a labcoat should of course be worn in any laboratory environment, but the type of labcoat worn for microbiological work should be a side-fastening one (Figure 3.1), rather than a more standard labcoat. It is not considered necessary per se to wear gloves for microbial work – the ACDP

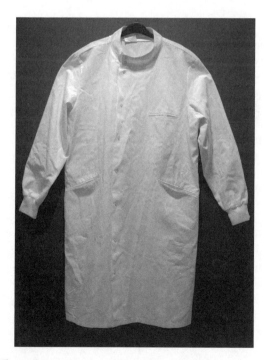

Figure 3.1 A labcoat suitable for microbiological experiments

HG 1 organisms, or recombinant strains of *E. coli* expressing biocatalyst genes to be used are by definition extremely unlikely to cause any kind of infection, but hands should be washed using a biocidal soap following all microbiological manipulations.

3.2.2 General considerations for a CL1 laboratory*

- The laboratory should be easy to clean. Bench surfaces should be impervious to water and resistant to acids, alkalis, solvents and disinfectants.
- If the laboratory is mechanically ventilated, it is preferable to maintain an inward airflow while work is in progress by extracting room air to the atmosphere.
- The laboratory must contain a wash-basin or sink that can be used for hand washing, preferably situated at the entrance to the laboratory and of the kind that has elbow-operated taps.
- The laboratory door should be closed when work is in progress.
- Laboratory coats (see Clothing and General Safety below) or gowns should be worn in the laboratory and removed when leaving the laboratory suite.
- Personal protective equipment, including protective clothing, must be:
  - stored in a well defined place;
  - checked and cleaned at suitable intervals;
  - when discovered to be defective, repaired or replaced before further use.
- Personal protective equipment which may be contaminated by biological agents must be:
  - removed on leaving the working area;
  - kept apart from uncontaminated clothing;
  - decontaminated and cleaned or, if necessary, destroyed.
- Eating, chewing, drinking, taking medication, smoking, storing of food and applying cosmetics must not take place in the laboratory.
- Mouth pipetting must not take place.
- Hands must be disinfected or washed immediately when contamination is suspected, after handling viable materials and also before leaving the laboratory.
- All procedures must be performed so as to minimise the production of aerosols of bacteria or other biological material
- Effective disinfectants (see below) should be available for immediate use in the event of spillage.
- Bench tops should be cleaned after use.

---

* Reprinted with permission from David Nelson, University of York.

- Used laboratory glassware and other materials awaiting disinfection must be stored in a safe manner. Pipettes, if placed in disinfectant, must be totally immersed.
- All waste material which is not incinerated should be disposed of safely by the appropriate means
- Materials for disposal must be transported in robust and leak-proof containers without spillage.
- All accidents and incidents should be immediately reported to and recorded by the person responsible for the work or other delegated person.
- The effectiveness of the autoclave used for sterilization (see Section 3.4) must be monitored at regular intervals to ensure that all material is being effectively sterilised.

The implementation of these simple rules should help to ensure that microbiological techniques using HG 1 wild-type micro-organisms are carried out safely. The containment of HG 1 wild-type organisms should, therefore, be well within the scope of an ordinary synthetic chemistry laboratory, assuming that the necessary equipment is accessible and provided of course that some dedicated space can be made available and that the simplest of containment measures can be put into place. Any use of genetically modified organisms will require more careful planning, however, as detailed below.

3.3 On Containment Issues and Genetically Modified Organisms (GMOs)[†]

It is becoming common practice for genes encoding useful biocatalysts to be expressed in workhorse laboratory strains of easy-to-grow bacteria and yeasts, the most notable being *E. coli*, leading to the recombinant or designer biocatalysts already alluded to in earlier chapters. The manipulation and application of these designer biocatalysts is straightforward once the genetic manipulations detailed in Chapter 7 have been completed – this is the advantage of such catalysts – but their containment poses serious consideration for laboratories and institutions, as they must be appropriately licensed for the handling of recombinant micro-organisms before such material can be received into the laboratory and stored and handled therein. The UK Health and Safety Executive (HSE) again provides clear advice on the contained use of GMOs through the Scientific Advisory Committee on Genetic Modification (Contained Use) [SACGM(CU);

[†] Biological Agents. Managing the Risks in Laboratories and Healthcare Premises 2005. Advisory Committee on Dangerous Pathogens. Health and Safety Executive. © Crown copyright material is reproduced with the permission of the Controller of HMSO and Queen's Printer for Scotland.

http://www.hse.gov.uk/aboutus/meetings/committees/sacgmcu/index.htm] [2]. The Committee defines genetic modification (GM) thus: 'GM occurs where the genetic material of an organism (either DNA or RNA) is altered by use of a method that does not occur in nature and the modification can be replicated and/or transferred to other cells or organisms.' The use of strains of *E. coli* in which genes from other organism have been heterologously expressed clearly constitutes an instance of GM in all cases therefore, and the regulations must therefore be strictly adhered to. These are available from the document on 'Genetically Modified Organisms (Contained Use) Regulations 2000' (http://www.opsi.gov.uk/si/si2000/20002831.htm) [2].

Under these regulations, there are six main duties of the institution proposing to commence such work. These are taken from reference [2] and include:

- *notification* that such activities will be initiated;
- a thorough *risk assessment* of all activities to be performed prior to their commencement;
- the establishment of a *genetic modification safety committee* at the site;
- the *maintenance of detailed records*;
- the classification of all intended activities and their regular review (see below);
- the implementation of all required containment measures.

The ACDP HG classification does not apply with respect to the use of GMOs; rather, an additional classification system is adopted, consisting of containment levels for activities of Class 1, 2, 3 or 4, in increasing order of risk. Again, given the intended scaled-up applications of biocatalytic processes, only low risk, Class 1 GM experiments defined by the GMO regulations as presenting 'no or negligible risk' are relevant to our discussion, particularly those that involve recombinant strains of *E. coli* as the host strain; After initial notification of the use of premises, further Class 1 activities do not require separate notification to the HSE. CL1 is appropriate for Class 1 activities 'to protect human health and the environment', and will cover the majority of simple cloning experiments that involve the heterologous expression of genes encoding biocatalytic enzymes into *E. coli*, unless the products of the gene themselves or the vectors and strains generated are inherently hazardous. Class 2 GM activities would require specific notification to the HSE in each case.

The GMO Regulations established in 2000 classify the 'contained use' of GMOs as any activity in which: 'a) organisms are genetically modified or b) GMOs are cultured, stored, transported, destroyed, disposed or used in any other way for which physical, chemical or biological barriers, or any combination of barriers, are used to limit their contact with, and to provide a high level of protection for, humans and the environment' [2]. Crucially, this means that it is not only genetic modification activities, for which licensing and approval must be sought, but even just the *storage* or *use* of GM material (GMM) such as a recombinant plasmid or recombinant strain of *E. coli*. Essentially, once permission to commence has been acquired and the relevant structures and risk assessments are in place, good microbiological practice should be sufficient to control the risks associated with the use of recombinant biocatalysts in the laboratory. There are some specific infrastructure requirements detailed in the document, many of which are common to those required for ACDP CL1 level work, including the use of:

- 'work surfaces that are impervious to water, resistant to acids, alkalis, solvents, disinfectants and decontamination agents and easy to clean';
- 'suitable protective clothing';
- specified disinfection procedures (required where and to what extent the risk assessment shows they are required);
- efficient control of vectors (e.g. rodents or insects) that could disseminate GMMs;
- inactivation of GMMs in contaminated material and waste (by validated means);
- safe storage of GMMs.

There is also a specific section on waste disposal that exceeds the requirements of the normal use of ACDP HG 1 organisms which states that, prior to disposal: 'any risks to humans and the environment associated with any GMO must be removed by use of validated inactivation methods. Inactivation refers to the complete or partial destruction of GMMs so as to ensure that any contact between the GMMs and humans or the environment is limited to provide a high level of protection to both humans and the environment.' General methods of organism deactivation and validation methods that would be suitable are covered later in this chapter.

3.4 Equipment for Handling Micro-Organisms

The handling of micro-organisms may well be unfamiliar to those with a training in synthetic organic chemistry, and there are many associated standard protocols that should become second-nature after a few manipulations. A list of equipment follows that would be considered essential for the safe handling of Class 1 micro-organisms or GM strains of *E. coli* or *S. cerivisiae* in a laboratory – this will introduce some terms, such as *asepsis* or techniques, such as *spread plating*, which will be described in detail in subsequent sections. The equipment listed in this section is also required, and suitable for the handling of recombinant strains of *E. coli*, if additional, relevant containment facilities and documentation are in place.

- *Autoclave.* An autoclave is effectively a large 'pressure cooker' that is used for sterilisation of microbiology equipment, including growth media and small pieces of apparatus (Figure 3.2). An instrument that can accommodate one or multiple 2 L Erlenmeyer flasks is ideal. It is imperative that glassware to be used for microbial culture is pre-sterilised to ensure the integrity of the organism grown in the flasks. Autoclaves will typically allow the contents to be heated to a temperature and pressure of 121 °C and 15 p.s.i., respectively. Autoclaves are routinely used for the sterilisation of media, glassware and plasticware when not provided sterile by the supplier.
- *Disinfectant (Virkon, Chloros in powdered form from supplier).* Virkon, widely available form many suppliers, is a general disinfectant provided as a pink powder that can be made up as a 1% solution (w/v) in water for the purposes of both disinfection of liquids contaminated by micro-organisms and the relevant glassware prior to cleaning. Chloros is a pink liquid disinfectant that can be made up as a dilute solution for an equivalent purpose. It is

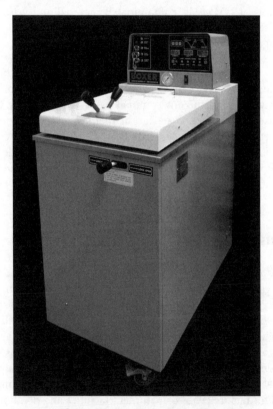

Figure 3.2 A laboratory autoclave

also useful to maintain a receptacle on the microbiology bench containing a solution of disinfectant for the routine disposal of, for example, plastic pipette tips or capped plastic tubes (see below) that have been contaminated with micro-organisms.

- *Wash-bottle containing 70% (v/v) ethanol in water.* A 70% ethanol solution is routinely used as a bacteriostatic agent in microbiology. It is most commonly used to wipe workbenches both before and after microbiological manipulations, and also to sterilise, for example, centrifuge rotors that have been used to harvest micro-organisms and also as part of aseptic technique in the culturing of micro-organisms on agar plates.
- *Nichrome wire loops.* It is possible to use either pre-sterilised packaged plastic loops for the handling of micro-organisms or a metal loop made from nichrome alloy – the latter can be sterilised in a Bunsen flame during aseptic procedures and may of course be re-used any number of times (Figure 3.3).
- *Orbital shaker with temperature control.* A variety of orbital shakers is available that allows the culturing of micro-organisms in Erlenmeyer flasks of volumes from a few millilitres to, commonly, 2 L. The size of experiment will of course dictate the size of orbital shaker required. Given that most large shakers are equipped with platforms that will take flasks of various sizes, it may only prove necessary to invest in one large chest-type shaker that could serve a range of flask sizes, but it is also useful to have a small orbital shaker for smaller reactions that may require different temperatures (Figure 3.4).
- *Bunsen burner or portable camping gas burner.* If the laboratory is supplied with a natural gas supply, then a simple Bunsen burner will suffice for aseptic microbial manipulations. However, if this is not the case, then a portable propane gas burner taking cylinders of 500 mL capacity is very useful and serves exactly the same purpose (Figure 3.5).
- *Plastic Petri dishes.* Petri dishes of 9 cm diameter are usually supplied pre-sterilised in packets of 10 or 20 and can be used directly from the packet (Figure 3.6).

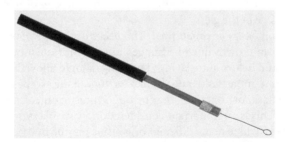

Figure 3.3 A nichrome loop for the inoculation of solid and liquid media with organisms

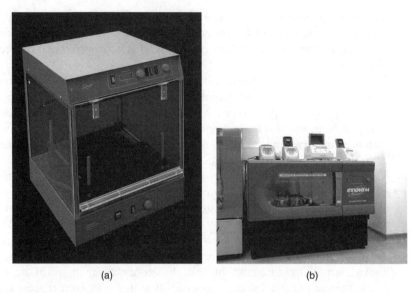

(a)   (b)

Figure 3.4   Orbital shakers suitable for the growth of micro-organisms and incubation of biotransformation reactions

- *Erlenmeyer flasks (50 mL, 250 mL, 1000 mL or 2000 mL)*. Erlenmeyer (conical)
  flasks are routinely used for microbial culture. It is recommended that
  volumes of growth medium would not exceed 50% of the flask volume (e.g.
  1 L in a 2 L Erlenmeyer flask) and the working volume is often lower than
  that, given considerations of the swirl volume in flasks when they are in an
  orbital shaker, and the necessary levels of effective mixing and oxygenation
  that may be required for optimal organism growth. Erlenmeyer flasks may
  have either small indentations at their base or vertically up the sides, giving
  rise to projections ('baffles') inside the flask (Figure 3.7). The value of these
  baffles is somewhat case-specific – they may improve oxygenation or mixing
  but can also lead to excessive frothing or the adhesion of micro-organisms
  to the surface of the glass.
- *Silicon foam bungs or cotton wool*. The necks of the Erlenmeyer flasks used
  for organism growth must be bunged after the formulation of the growth
  medium and before autoclaving. The bungs should allow the transfer of air
  into and out of the flask, but should restrict both the escape of the organism
  from the flask, or contamination by opportunistic organisms from outside
  the flask. This is an important consideration, particularly when working in
  a laboratory where many different organisms may be in use.
- *Sterile disposable syringe barrels, needles and plastic filters*. There are some
  medium additives that are not suitable for autoclaving, or which cannot

Figure 3.5 A portable burner suitable for use in a biocatalysis laboratory

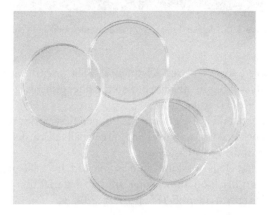

Figure 3.6 Sterile disposable plastic Petri dishes

Figure 3.7 'Baffled' Erlenmeyer flasks for microbiological culture

be autoclaved together (see Chapter 4). The aseptic addition of these supplements will require the use of a sterile filter attached to a sterile syringe barrel and needle (Figure 3.8).

- *Aluminium foil.* Foil is used to cap the bunged flasks prior to autoclaving, to stop the bungs soaking up water. They are usually discarded after the aseptic inoculation of the growth medium.
- *Autoclave tape.* This masking tape is marked by heat-pressure sensitive lines that turn black on autoclaving, so that material that has already been autoclaved is easy to identify (Figure 3.9).

Figure 3.8 Disposable plastic filters for sterile additions to growth media

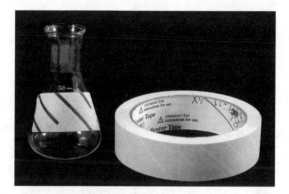

Figure 3.9 Autoclave tape, showing black stripes that develop on heating

- *Plastic pipettes and pipette filler.* Plastic pipettes (Figure 3.10), of 10, 25 and 50 mL volumes, which can be purchased in sealed bags to ensure sterility, are useful for the aseptic transfer of buffers and growth media between flasks. Mouth pipetting, as with all laboratory processes, is of course strongly discouraged.
- *Centrifuge and centrifuge tubes.* Different types of centrifugation will be described throughout the practical chapters, but ideally, a larger machine that has the capacity to spin six tubes of 250–500 mL and also smaller volumes of 25–50 mL would be required for experiments at a scale above tens of millilitres. A *microcentrifuge*, a benchtop instrument that holds tubes of up to 2 mL capacity, is also recommended (Figures 3.11 and 3.12).
- *Small UV spectrophotometer.* The growth of bacteria is usually best monitored using the absorbance at a certain wavelength – for example, the growth of *E. coli* is routinely monitored by measuring the increase in turbidity or

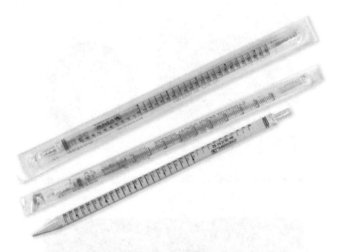

Figure 3.10   Disposable plastic pipettes

Figure 3.11   Centrifuge tubes for a larger capacity centrifuge

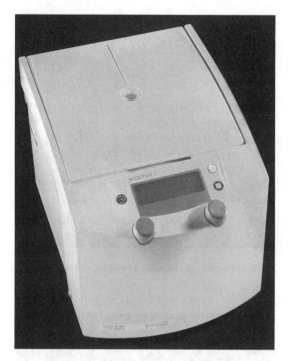

Figure 3.12 A microcentrifuge

'optical density' of the cell suspension at 600 nm. This is usually done by transferring 1 mL of the growing culture, using aseptic technique, into a 1 mL plastic cuvette, and reading against a blank of the non-inoculated medium (Figure 3.13).

- *1.5 mL or 2 mL capped plastic tubes.* Small plastic reagent tubes with attached click-on caps are the standard laboratory receptacle for microbiology and molecular biology and are most commonly used in 1.5 mL tapered or 2 mL formats (Figure 3.14).
- *Whirlimixer.* This is a benchtop device for rapid mixing, consisting of a rotating rubber crucible or pad on a shaking motor (Figure 3.15). It is used for mixing samples in Falcon tubes and capped plastic tubes.
- *Microlitre automated pipettes.* The accurate dispensing of sub-millilitre volumes of materials routinely encountered in microbiology and molecular biology requires the use of automated volume adjustable pipettes which are capable of dispensing volumes of as low as 1 µL and up to 5000 µL or 10000 µL. Most laboratories will have automated pipettes of 1, 2, 20, 100, 200, 1000 and 5000 µL volumes. The pipettes are used routinely with

Figure 3.13   A small benchtop spectrophotometer

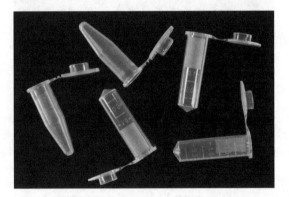

Figure 3.14   A range of capped plastic tubes

disposable plastic tips which may be sterilised by autoclaving before use in plastic boxes that allow the tips to be easily picked up by the pipette without contamination (Figure 3.16).

- *pH meter.* The accurate monitoring of pH is crucial in the manipulation of enzymes, as these may be denatured in pHs that are unfavourable. Many of the techniques involved with handling proteins take place in buffered solutions (see below), which must be adjusted to the pH of choice using a pH meter.

Figure 3.15 A whirlimixer

- *Light microscope, slides and cover slips.* A simple light microscope will be found within most microbiology laboratories, but in practice is not essential for everyday biocatalysis work, as, of course, many of the strains of organism that are being used are 'type' cultures that are defined by the culture collection that has provided them. In most cases, contamination on agar plates or in culture may be quite obvious – perhaps due to a change in morphology or colour of a culture – in the case of recombinant strains of *E. coli*, where the use of antibiotic markers should preclude the chances of contamination, the different recombinant strains expressing foreign genes will not in practice be observable by light microscopy anyway.
- *Microwave oven.* It may be convenient to purchase a simple microwave oven for use in the laboratory. Once agar is prepared by sterilisation, it can be stored in reagent bottles, but will, of course, set whilst in storage. The agar is readily melted in a microwave oven. The oven will also be useful for the mixing of agarose–buffer mixtures for DNA analysis gels.

The extent of equipment purchased will of course vary, dependent on the type of microbes to be used, their applications and also the existing resources of the relevant laboratory, but the list above should equip the laboratory with sufficient apparatus to begin some simple experiments.

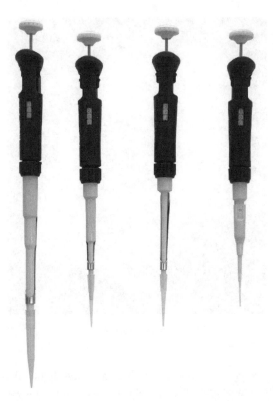

Figure 3.16   A range of automated microlitre pipettes

3.5 Techniques and Terms in Microbiology – Sterility, Asepsis and Aseptic Technique

Important concepts for the handling of micro-organisms, whether wild-type or recombinant, are: *sterility* – ensuring that growth media and tools for handling micro-organisms are sterile; and *asepsis* – ensuring that the organism of choice is the only one that contaminates the equipment and growth medium being used. Maintenance of sterility and asepsis in, respectively, medium that has not been used, and that which has been inoculated with organism, are two major challenges in the laboratory, but should be straightforward if good *aseptic technique* is practised.

3.5.1 Aseptic technique

Aseptic technique refers to a general approach to microbiology that ensures sterility and/or asepsis when appropriate during medium preparation, organism growth and, in this particular case, biotransformation experiments. There are various simple precautions that can be taken in the first instance, which would be considered good laboratory practice in any synthetic organic laboratory, including clean hands, tying long hair back and wearing a suitable labcoat.

- Both before beginning work and after any microbiological manipulations have been completed, the bench area (to be) used should be swabbed, either with a solution of a commercial disinfectant such as Virkon or Chloros, or with a solution of 70% ethanol, a common bacteriostatic agent used in the laboratory for sterilisation purposes.
- The growth flask, and any materials that are to come into contact with the interior of the growth flask, such as plastic pipette tips, and indeed any subsequent vessels of transfer tools that will come into contact with the pure organism must be sterilised by autoclaving.
- A Bunsen burner, or portable camping gas burner is indispensable for ensuring sterility and asepsis in a laboratory environment except where access to a microbiological safety cabinet is provided. The latter is an expensive installation not accessible to most synthetic organic chemistry laboratories, so we will proceed with the consideration of a burner. The presence of a naked flame being incompatible with the standard organic laboratory, it is almost certain therefore that these microbiological manipulations need to be carried out in a dedicated or separate area, or in one where no volatile solvents are present.

The burner is used to 'flame' glassware and associated transfer tools (such as metal loops, pipettes, sample bottles) when these are exposed to the open air, and as such, to reservoirs of contamination. For example, when removing a sample from a growing culture of a micro-organism, the bung is removed and the neck of the open culture flask passed through the Bunsen flame before sampling with a sterile pipette. After the sample has been removed, the neck of the flask is once again flamed before the bung is replaced.

There are many instances of the requirement for good aseptic technique contained within the experimental chapters of this book, and each will be detailed as appropriate.

3.6 Disposal of Viable Microbial Waste and Disinfection of Reusable Equipment

The application of sound aseptic technique should ensure the integrity of a microbial culture even where sophisticated facilities such as microbiological safety cabinets are not available. However, a consideration that is at least as important as asepsis is the disposal of waste materials contaminated by micro-organisms after their use. These types of waste usually fall into the following categories:

- *Spent liquid waste.* This may include spent growth medium that has accumulated from centrifuge runs once an organism has been harvested, or buffers that have been used to wash cells. Often, liquid waste of this nature can accumulate in large volumes.
- *Organism waste.* Organism material itself – cell pellets or fungal cell mats; old agar plates with organism present.
- *Reusable glassware and plasticware.* Growth flasks, centrifuge tubes, beakers and storage vessels.
- *Disposable miscellaneous solid waste.* Pipette tips, capped plastic tubes, organism-contaminated tissues, gloves.

The key consideration is that the waste to be discharged should not be 'viable' or contain material that would result in microbial growth if inoculated into a suitable medium. A straightforward mechanism for testing the viability of such waste might therefore be to, using aseptic technique, streak a suitable agar plate with the contaminating material (see Chapter 4). If growth ensues, then the waste material is still viable.

The two most frequently used methods of microbial deactivation are autoclaving and 'chemical poisoning' by the use of disinfectants. Autoclaving spent liquid medium under the same condition as used for sterilising new media should deactivate the biological material within. The liquid may then be safely discharged to the environment if shown to be non-viable. Solid organismal waste is also routinely treated by autoclaving. Most laboratories have a supply of autoclavable waste bags, into which both solid organism waste, such as old agar plates, is deposited, along with the miscellaneous contaminated solid waste, such as used pipette tips and capped plastic tubes. The bags of solid waste may then be autoclaved and then removed for incineration.

Reusable glassware and plasticware may also be autoclaved for decontamination, but in the case of the latter, some plasticware may not be suitable for autoclaving – some centrifuge tubes are weakened by autoclaving for example, and must be deactivated by the use of disinfectants. Again, the test of a successful decontamination of such a vessel is whether any part of the disinfectant solution is viable after incubation in the vessel. Common practice is to soak a centrifuge tube in 1% Virkon or Chloros for a minimum of 2 h prior to the discharge of the material, once it has been confirmed that such treatment would render the material non-viable.

One extra consideration is the disposal of the aqueous waste that results from the work-up of a microbial biotransformation reaction that has been extracted into solvent. In such cases, it is usual that the spent organism has been removed by either filtration or centrifugation prior to the extraction of the reaction supernatant in a separating funnel. The combined aqueous phases should, of course, be tested for viability, but it is unlikely that the viable organism will have survived the extraction washes. If non-viability has been established, the aqueous phases should be combined and discharged to the appropriate waste receptacle for the solvent used for extraction (halogenated, ether, water-miscible, etc.).

3.7 Equipment for Enzymology and Molecular Biology

In addition to the equipment and general consumables that form part of the everyday requirements of microbiological work, there are pieces of equipment that are necessary for the techniques to be described in Chapters 6 and 7 that relate to the study of proteins isolated from organisms, or the genes that encode them. It should be emphasised that these are simply not necessary if the group is only interested in wild-type whole-cell biotransformations or biotransformations using recombinant biocatalysts that have been provided by a collaborator. Examples will be divided into equipment specific for the studies of *enzymes* (proteins) and *genes* (DNA).

3.7.1 Enzymes (Proteins)

You will need access to:

- *A method of cell disruption.* In order to obtain the soluble fraction of cell material that will contain the biocatalyst of interest, it will be necessary to break open the cell membrane or cell wall usually using a physical method of cell disruption. This could include either an ultrasonicator (Figure 3.17) or pressure-cell equipment, such as a French press (Figure 3.18, see Chapter 6) or glass-bead mill.
- *Mini-protean or (similar) gel electrophoresis kit with gel plates, combs and moulds for pouring gels.* Gel electrophoresis, detailed in Chapter 6, is a chromatographic technique that allows the analysis of protein mixtures and can be used to check quickly for the presence of a protein (although this has to be verified by other means) and/or its purity.
- *Power packs for running gel equipment.* Power packs can be suitable for running both agarose gel electrophoresis for DNA analysis (see Chapter 7) and sodium dodecyl sulfate polyacrylamide gel electrophoresis (SDS-PAGE) for the analysis of proteins (Chapter 6). Packs that deliver a maximum voltage of 300 V are sufficiently large; SDS-PAGE is often run at 200 V and simple agarose gel electrophoresis at 100 V.

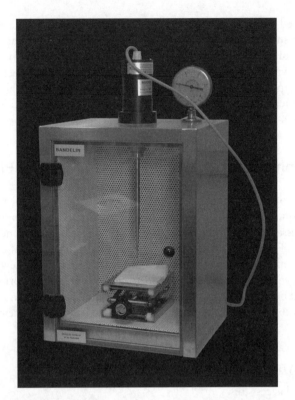

Figure 3.17   An ultrasonicator for the disruption of bacterial cells

- *Bench peristaltic pump for running affinity columns.* It is not absolutely necessary to purchase an expensive fast protein liquid chromatography (FPLC; see Chapter 6) machine for simple isolation of 'tagged' proteins, as in some cases these can be isolated using a nickel-agarose affinity column (below), run using a simple benchtop peristaltic pump that will deliver eluant at $5-10\,\text{mL}\,\text{min}^{-1}$, complemented by manual sample collection in an analogous manner to a flash silica column. These can be purchased from a number of relevant laboratory suppliers.

- *Nickel-agarose affinity columns.* For the simple purification of proteins from recombinant strains of *E. coli* that have been engineered with hexa-histidine tags that aid purification, small packed columns of an agarose polymer are used. These often come in volume sizes of either 1 mL or 5 mL and are suitable for the purification of protein from laboratory-scale shake-flask cell-culture volumes of 100 mL to a few litres.

- *Cuvettes for measuring protein concentration and assaying enzyme activity using UV spectrophotometry.* Many enzymatic reactions can be assayed using UV

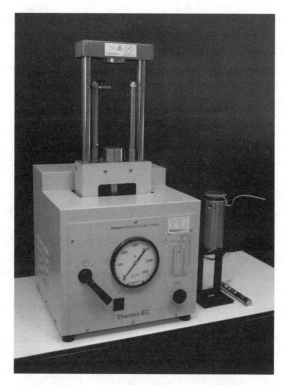

Figure 3.18   A French press for the disruption of bacterial cells

spectrophotometry as detailed in Chapter 6, perhaps by direct observation of the (dis)appearance of (substrate) product or the oxidation/reduction of cofactors such as nicotinamides. Many of these assays may require quartz cuvettes. Estimations of protein concentration using one of many established methods are also routinely performed in disposable plastic cuvettes in a UV spectrophotometer.

3.7.2 Genes (DNA)

You will need access to:

- *PCR machine or 'thermal cycler'.* Thermal cyclers come in various forms and degrees of complexity that allow one to exploit the various and evolving techniques in PCR (polymerase chain reaction – see Chapter 7). A simple small machine that is able to hold 16 reactions, has a heated lid, and that is able to deliver the four major segments of the PCR cycle (i.e. DNA melting at

95 °C; DNA annealing at 35–65 °C; DNA-polymerase-catalysed extension at 72 °C; cooling at 4 °C) is perfectly adequate for simple experiments. Suppliers include Techne, Hybaid, ABI, MJ Research and Perkin-Elmer.

- *Gel tanks, moulds and combs for DNA analysis by agarose gel.* Simple horizontal gel tanks of approximately 50–100 mL volume are required for the assembly and running of agarose gels for the analysis of DNA samples such as those generated by PCR reactions, restriction digests and other cloning experiments. Examples of these techniques are given in Chapter 7.
- *Small power packs for running agarose gels.* See above.
- *UV lamp 'transilluminator' and protective mask/goggles.* A flat-bed UV lamp or 'transilluminator' is necessary for visualising the DNA fragments in agarose gels resulting from e.g. PCR reactions. This will also be necessary for visualising DNA whilst cutting it from the gel in DNA isolation procedures such as gel extraction (see Chapter 7). Such open UV lamps must be used with great caution, and appropriate face-masks or goggles should be worn. In practice, such UV lamps are now often purchased as part of 'gel-doc' systems, in which the lamps are enclosed in cabinets that have a camera inside them for the photographing of gels.
- *Molecular biology kits (PCR clean-up; plasmid mini/midi prep).* As will be seen in Chapter 7, many of the simple DNA manipulation experiments are simply performed using molecular biology kits purchased from a number of suppliers (e.g. Qiagen, Stratagene, Sigma, New England Biolabs). These contain pre-mixed bottles of reagents for the relevant procedures and a ready supply of these is a common feature of those laboratories working with creating recombinant biocatalysis.
- *Water-bath.* Whilst water-baths are of course common features of chemistry laboratories, the purchase of a heating-block type water-bath, which specifically allows the incubation of capped plastic tubes, is extremely useful when it comes to, for example, the routine procedure of 'heat-shocking' cells during transformation protocols (see Chapter 7).

Most other protocols, once transformation of the recombinant plasmid has been achieved, can be performed using the equipment purchased for general microbiological manipulation.

3.8 General Reagents and Chemicals in a Biotransformations Laboratory

In addition to the general collection of organic chemicals that would be found in a standard synthetic organic laboratory, a facility that also performs biotransformations will possess a collection of general chemicals that is standard for a laboratory engaged in microbiological culture and enzyme assay. These include ingredients for *growth media* and *buffers*.

3.8.1 Growth media

The growth requirements of organisms are many and varied, and there is a wide range of formulations of growth medium – the liquid phase in which the micro-organisms of interest will be grown. Some specific recipes for the media required for the growth of micro-organisms are detailed in Appendix 4. The choice of medium for the growth of different micro-organisms used in biocatalysis is discussed in Chapter 4. The list of general reagents is likely to include fairly large supplies (1–2 kg) of:

- monobasic and dibasic phosphate salts [either potassium or sodium (cheaper)]
- ammonium sulfate
- ammonium chloride
- sodium chloride
- magnesium sulfate
- simple carbon sources for organisms such as glucose, sodium acetate, sucrose or glycerol.

Growth media may also contain a number of metal salts such as *ferrous sulfate* and *manganese sulfate* as constituents for a trace element solution that is added to some medium formulations. In addition to inorganic salts, media often contain variable amounts of powdered hydrolysates of biological material, most commonly *yeast extract, tryptone* and *peptone* (all available from Sigma or Oxoid), which serve as a source of vitamins and sometimes nitrogen and carbon in a growth medium. There are also speciality biologically derived medium formulations that can be bought ready-made that are often used for fungal growth such as 'corn-steep solids' and 'Czapek-Dox medium'. Every biotransformations laboratory will also have a supply of *agar* – a freeze dried preparation of the polysaccharide that forms the basis of the solid media found in Petri dishes.

The availability of these chemicals should ensure that the laboratory is equipped to make up standard basic salts medium for the growth of wild-type organisms and also the standard growth media used for the growth of recombinant strains of *E. coli*.

3.8.2 Buffers

Biological material, whether this be in the form of cells of organisms that have been harvested from the growth media, or crude protein extracts that have resulted from the disruption of these cells, or purified protein themselves, will be handled in buffered solutions, or those which are resistant to small changes in pH that result from the addition of small amounts of acid or alkali. These will provide the best protection against denaturation or inactivation as a result of changes in pH and can also maintain the optimum pH at which washed cells of micro-organisms and enzymes catalyse their reactions.

It is common therefore to find a range of powdered buffers within biocatalysis laboratories that buffers effectively in the region of pH 4–10. Some of these buffers are used in organic chemistry, such as a mixture of monobasic and dibasic phosphate salts, but many are said to be 'biological buffers' in that they have been selected for their performance in biological systems. These include:

- Tris (trishydroxymethylaminomethane)
- PIPES [piperazine-1,4-bis(2-ethanesulfonic acid)]
- HEPES [4-(2-hydroxyethyl)piperazine-1-ethanesulfonic acid]
- MES (morpholinoethanesulfonate).

Each of these is used typically at concentrations of 10–100 mM and usually adjusted to the required pH using bench-strength hydrochloric acid or sodium hydroxide. A more extensive list can be found in Appendix 5.

There are also many chemicals that are commonly added to buffers that are usually included in an effort to enhance the stability or activity of proteins. These are described in more detail in Chapter 6, but they include solvents (most commonly glycerol), detergents (such as Triton X-100 or Tween 20), ethylenediaminetetraacetic acid (EDTA), which can chelate divalent metal ions that may be deleterious to enzyme activity and β-mercaptoethanol or dithiothreitol, which are thiol reagents used to stabilise enzymes. Specific inhibitors of proteases are also often added, to protect the proteins from being hydrolysed by endogenous proteases once cell walls have broken. The formulation of buffers for whole-cell reactions, isolated enzyme biotransformations and protein purifications is addressed in more detail in the relevant chapters.

3.9 Conclusion

The use of commercially available enzymes for biotransformations by synthetic organic chemistry laboratories is certainly well within their technical abilities and facilities, requiring little more than the purchase of some buffer salts and, perhaps, a small temperature-controlled orbital shaker. It will be apparent from this chapter though, that establishing a laboratory that is equipped for small microbiological manipulations of the type used in biotransformations should not present a significant obstacle, the major expense probably being a centrifuge of sufficient capacity to remove cells from their growth media. It is certainly the case that many synthetic chemistry facilities now have more ready access to biochemistry and microbiology-friendly apparatus with the increasing interest in bioorganic chemistry and chemical biology in university departments, or between relevant departments in companies, to render the initiation of such studies of biocatalysis more feasible than in previous years.

References

1. 'Biological agents: managing the risks in laboratories and healthcare premises', UK Health and Safety Executive and available at http://www.advisorybodies.doh.gov.uk/acdp/publications.htm.
2. 'Genetically Modified Organisms (Contained Use) Regulations 2000', available at http://www.hse.gov.uk/aboutus/meetings/sacgmcu/index.htm.

Chapter 4
A Beginner's Guide to Preparative Whole-Cell Microbial Transformations

4.1 Introduction

The origins of modern biotransformations lie in the use of organisms such as baker's yeast for reduction reactions and the filamentous fungi that have been exploited for the industrial production of penicillins. Whole-cell biotransformations are still extremely important in a research or industrial context, if only in their new form, as recombinant strains of easy-to-handle organisms that overproduce enzymes of interest that are to be found in more esoteric strains. The tools and skills necessary for both growing small amounts of microbial cell cultures and exploiting them in biotransformation reactions are described in this chapter. This will include details of organism maintenance and storage; growth in shake flasks and biotransformation reactions using a number of illustrative examples, including those employing bacteria, fungi and yeasts and existing recombinant strains of *E. coli*.

4.2 Storage, Maintenance and Growth of Microorganism

It is important that once a microbial culture has been obtained from a culture collection or collaborator, steps are taken to ensure the *sustainability* of that microbial strain within the laboratory. It is not uncommon that strains stored, for example, on agar in Petri dishes, lose the growth characteristics or enzymatic activities for which they have been purchased over time, so it is important that the integrity of the fresh, original culture is maintained for as long as the organism is required. The method of storage and maintenance of the culture of

Practical Biotransformations: A Beginner's Guide © 2009 Gideon Grogan

micro-organism being used as a source of biocatalyst is thus of major importance to the reproducibility of the biotransformation under investigation over the lifetime of a project.

4.2.1 Maintenance of Organisms on Solid Media – Agar plates or Slopes

Solid agar is by far the most commonly used medium for the storage of micro-organisms in the short to medium term (weeks to a few months). Agar, which can be obtained from simple chemicals or biochemical suppliers, is added at 1–2% (w/v) concentration to a medium of known formulation prior to autoclaving. After autoclaving and partial cooling, the mixture is poured out into either plastic Petri dishes (usually packaged sterile) or into glass universals which are then rested at an angle (Figure 4.1) so that the agar sets at a slope, giving a larger and accessible surface area for the deposition and removal of biological material. Agar sets at approximately 46 °C, so the pouring must be done before the agar sets, or it may be easily re-melted. Agar plates do dessicate, even if left in the fridge and should really only be used as a source of organism for inoculation for a period of 2–3 weeks. Regular housekeeping of organisms kept on agar plates should be practised, to ensure the disposal of unwanted biological material. Fresh plates (2–3 weeks old) prepared from agar slopes, glycerol stocks or freeze-dried samples (below) should be used ideally to ensure the use of fresh organisms in any experiment. Agar slopes, which are kept under screw-caps at 4°C should provide a reliable source of culture for the short to medium term.

Figure 4.1   Agar slopes for medium- to long-term maintenance of microbial cultures

4.2.2 Glycerol stocks

Glycerol stocks are a convenient way of preserving material derived from original cultures for long periods of time. After resuscitating the organism in liquid culture (see below) from the format provided by the supplier, an amount of this culture is mixed with one equivalent of sterile glycerol in an capped plastic tube – this should then be frozen at −20°C, or ideally, −80°C, and should remain viable for months or years.

4.2.3 Freeze-dried cultures

Freeze-drying can be an excellent method of preserving microbial cultures in a dormant state for long periods of time (years). After resuscitating the organism from the culture collection in liquid culture, an amount of this is transferred to an open ended glass vial and plugged with sterile cotton wool. The culture is lyophilised and then the other end of the glass vial is sealed in a Bunsen flame. These freeze-dried ampoules (Figure 4.2) are the format in which many culture collections preserve and distribute their material.

4.2.4 Growth of Organisms in Liquid Medium

The growth of organisms on a preparative scale suitable for biotransformations is most usually done in liquid culture in shake flasks or fermentors. The liquid medium can be made from a variety of simple chemicals depending on the organism used, and in some cases, the enzymatic activity that one wants to elicit. The formulation of liquid media is addressed later in the chapter. In practice, as described below, a small amount of biomass from a solid culture, or a small pre-grown liquid culture is used as the *inoculum* for establishing the growth of the organism in the flask. The subsequent rate of growth of the organism, or indeed,

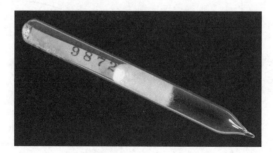

Figure 4.2   Freeze-dried ampoule culture of micro-organisms for long-term storage

whether there is any growth at all, will be dependent on a number of factors such as temperature, nutrient availability and oxygenation as achieved in shake flasks through mixing whilst incubated on the orbital shaker or in the fermentor.

The progress of growth of an organism in liquid culture can be monitored either by *spectrophotometry* (for bacteria) or by the extraction of samples followed by lyophilisation or dessication in an oven to remove water and hence estimate biomass by *dry mass*. For bacteria such as *E. coli*, the accumulation of biomass is routinely measured by extracting a small volume of the growing culture and reading the absorbance at 600 nm. If this reading is performed at intervals, and the absorbance of the samples is plotted against time, a *growth curve* results, which, in shake flasks, is usually sigmoidal in shape (Figure 4.3).

It will be seen that, in the first phase of growth, the *lag phase*, there is not a rapid increase in optical density as the number of bacterial cells remains roughly constant. This phase corresponds to the organism making intracellular preparation for rapid cell division, including the synthesis of RNA and proteins. The second phase, the *exponential* or *logarithmic phase*, describes that section of the growth curve where the number of cells doubles for every 'generation time' i.e. if the cell population at any one time is 1×10^6, this will increase to 2×10^6, and then to 4×10^6, each in an equivalent time. This unit of time is also known as the 'doubling time' and for *E. coli* is approximately 20 min. The exponential growth continues until the nutrients in the growth medium have been exhausted. Exponential growth then ceases, giving rise to a third phase, the *stationary phase*. This third phase usually signals that nutrients of one type or another have become limiting. A decrease in optical density may then be seen, which may be attributed to some *lysis* or breaking of cells in what is sometimes called the *death phase*. It is usually desirable to harvest cells for biotransformation in the late exponential or stationary phase, therefore maximising the amount of biomass (and hence

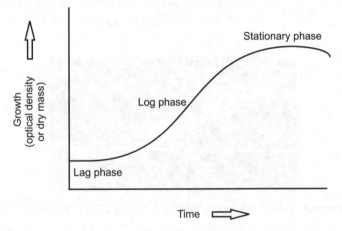

Figure 4.3   A typical growth curve for micro-organisms grown in shake-flask culture

biocatalyst) although it should be emphasised that the expression of some genes that encode proteins is only associated with one or other stages of growth, and it may be informative to assay the biocatalytic activity of an organism at intervals throughout the growth curve by taking samples at different points, in order to ascertain the optimum time for cell harvesting.

4.3 General microbiological methods

4.3.1 Preparation of agar plates for bacterial growth on Luria-Bertani (LB) agar

The preparation of solid agar plates is a routine activity in a microbial biotrans-formations laboratory. Whilst we provide here a step-by-step guide to preparing particularly agar plates made using LB broth mixture, it is largely correct to say that most liquid growth media can be formulated as solid media if 1% agar is added to the mixture containing the medium constituents, autoclaved and then poured into Petri dishes to form agar plates. The preparation of agar plates also provides the first example of *aseptic technique*, described in Chapter 3.

1. Prepare the area to be used by swabbing with 70% ethanol or a 1% solution of disinfectant.
2. Into a 500 mL screw cap bottle, add 12.5 g of LB broth mix (see Appendix 4) and 2.5 g of agar. Add 250 mL of deionised water and cap *loosely*. (NB Do not overtighten caps prior to autoclaving, as the build-up of pressure within the autoclave can cause bottles to break.)
3. Autoclave at 15 p.s.i. and 121°C for 20 min. Remove the agar from the autoclave and allow this to cool. The agar will set at approximately 46°C, but may be poured when it is hand-hot. Should the agar set, it can be easily re-melted either in the autoclave or in a microwave oven but, again, do not overtighten the cap on the bottle.
4. Light the Bunsen burner and turn the Bunsen airhole to make a blue flame. Unscrew the cap of the bottle containing the agar and gently flame the top of the bottle (Figure 4.4). Then pour approximately 20 mL of the molten agar into each plastic Petri dish. Do *not* cover at this stage as the condensation from the steam will obscure the view through the lid of the dish, and also accumulate unwanted fluid in the finished culture.
5. Leave plates to cool, and store at 4°C.

4.3.2 Reviving a culture from a collection

In the usual course of events, one will have identified, from the literature, an interesting transformation that is catalysed by an organism, either a yeast,

(a)   (b)

(c)   (d)

Figure 4.4   Aseptic technique used in the preparation of agar plates. The bottle containing the molten agar is flamed prior to pouring the agar into the plate, which is then left to set either at room temperature or in the refrigerator

bacterium or filamentous fungus. It is usually a simple matter to order the organism from one of the sources described in Chapter 2, and is usually provided in one of the forms discussed below. Many of the techniques, such as spread plating, will be common to each method.

4.3.3 Reviving from a freeze dried ampoule

As described above, an amount of the organism derived from liquid culture has been deposited in a sterile glass ampoule, plugged with cotton wool, the sample lyophilised and then sealed (Figure 4.1). Instructions are often provided for the resuscitation of the dormant organism, but a general guide follows: use aseptic technique throughout.

1. Prepare the area to be used by swabbing with 70% ethanol or a 1% solution of disinfectant. Light the Bunsen burner or gas.
2. Score the ampoule using a diamond cutter or similar, at the top. A small plastic cap device is usually provided, which is then placed over the ampoule covering the score. The cap may then be broken, with care, to expose the organism. Flame the top of the open vial.
3. Mobilise the biological material with either a small amount of sterile water or growth medium (the nature of which is usually provided with the microbial culture). A small amount of this can be added to the ampoule using a sterile Pasteur pipette and the contents transferred directly onto a Petri dish containing agar of the required formulation.
4. Transfer the pipette contents to a Petri dish containing agar of the required formulation. The material must then be 'plated out' using one of two techniques: *spread plating* or *streaking*.

4.3.4 Spread plating and streaking

There are two commonly used techniques for spreading biomass around an agar plate. The first technique, *spread plating*, involves merely distributing the biomass around the surface of the agar to obtain a lawn that will result in a large amount of biomass on the surface of the plate – this will act as a reservoir of organism for inoculation. The second, *streaking*, is designed to result in the growth of the organism in ordered lines of progressively decreasing numbers of colonies, with the objective that each individual colony, or spot of organism on the plate is representative of growth originating from one cell only. The technique of streaking is especially important when performing enrichment experiments (see Chapter 2), wherein it is necessary to separate out different organisms from each other. Streaking is often implemented, as it gives rise to a more aesthetically pleasing plate, allowing the separation and observation of individual colonies of bacteria.

4.3.4.1 Spread plating

In order to perform spread plating, you will need:

- a Bunsen burner or portable gas;
- automated pipette and sterile tips for dispensing 100 μL;
- a glass plate spreader – a glass tube that has been moulded into a handle with a flat triangle at the base (see Figure 4.5) – it is also possible to mould a glass pipette for this purpose, but the open end of the pipette should be sealed first;
- a covered beaker containing 70% ethanol.

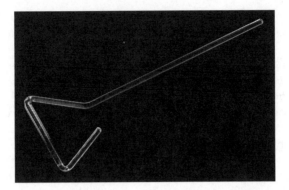

Figure 4.5   A glass plate spreader

1. Pipette, using either a sterile plastic pipette tip or a sterilised glass Pasteur pipette, approximately 100 µL of the cell suspension onto the surface of the agar.
2. Dip the plate spreader in the 70% ethanol and then carefully insert the wet spreader into the Bunsen flame. A flame will appear on the spreader, which can be waved out.
3. Apply the spreader to the agar firmly, but not with such force that the surface of the agar breaks. Spread the culture to the limits of the plate, but make sure that you do not concentrate a liquid mass at the plate edge. Once uniformly spread, place a lid on the plate and invert it (Figure 4.6).
4. Label the plate around the circumference of the plate base once inverted – this allows you to see the content of the plate clearly during incubation – and store at a temperature appropriate for the organism used.

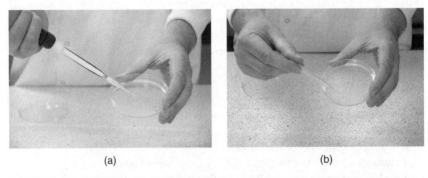

(a)   (b)

Figure 4.6   Preparing a bacterial culture on solid medium using the spread-plating technique. A small amount of liquid culture is transferred aseptically from a growth flask to an agar plate (a). The liquid is then moved around the plate using the glass spreader (b), which has been sterilised by flaming

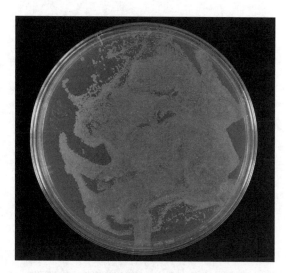

Figure 4.7   An agar plate of bacteria prepared using the spread-plating technique

The storage temperature will be different depending upon which organism is being cultivated. Most terrestrial bacteria (strains of *Pseudomonas, Arthrobacter, Bacillus, Rhodococcus, Streptomyces,* etc.) should be cultivated at 30°C. Recombinant strains of *E. coli* should be incubated at 37°C. The plates should be left for a sufficient time so that an even lawn of bacteria has grown on the surface of the plate (Figure 4.7). This may take from overnight (in the case of *E. coli*), to a couple of days (*Pseudomonas, Arthrobacter*) to a week (*Rhodococcus, Streptomyces*).

4.3.4.2 Streaking

For streaking a plate, you will need:

- a Bunsen burner or portable gas;
- automated pipette and sterile tips for dispensing 100 μL;
- a nichrome alloy wire loop.

1. Flame the nichrome loop in the hottest part of the Bunsen flame until it glows red.
2. Cool the loop on the agar at the edge of the plate, and dip the loop into the culture suspension. At one edge of the plate, make a line of culture with the loop as shown in Figure 4.8. Flame the loop again, and then drag four or five lines of culture from the original line over to the other side of the plate. Flame the loop again, and drag four or five lines from the last set over to a new area of the plate. Repeat until you have the original line plus four sets

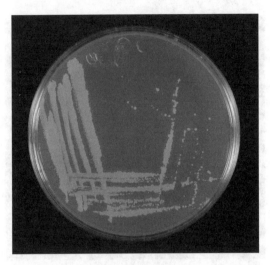

Figure 4.8   An agar plate of bacteria prepared using the streaking technique

of sub-lines around the edge of the plate. Then, flame the loop one last time and draw a last line into the centre of the plate from the last set of lines.
3. Replace the lid of the Petri dish, invert, and label and incubate as before. When the cells grow you should see lines of organism growth, growing less dense as you progress around the plate, and then individual colonies as a result of the last set of lines and the last line into the centre (Figure 4.8).

In most cases, if the culture received from the collection is viable, and the instructions with respect to the correct medium have been followed, successful growth should be observed. However, occasionally, the organism will fail to establish on solid medium and in this case it may have been useful to establish a small liquid culture from the received sample in the ampoule as an insurance measure, as detailed below.

4.3.5 Growth in liquid culture

In order to prepare a small liquid culture suitable for reviving an organism from an ampoule culture:

1. Prepare in a 100 mL Erlenmeyer flask, 25 mL of a suitable growth medium. In the current example, this would mean LB broth (see Appendix 4). Bung the flask with cotton wool or a silicon bung and place aluminium foil over the top. Autoclave the flask and allow to cool.
2. Once the contents of the flask are cool, remove the foil. Light the Bunsen burner and adjust the flame to blue. Remove the bung from the culture flask

and, holding the flask by the base, pass the open neck of the flask through the Bunsen flame.

3. Using a sterile Pasteur pipette, transfer some of the suspended contents of the ampoule into the flask. Pass the open neck of the flask through the flame again and replace the bung.

4. Place the flask in an orbital shaker and set the shaker to 150 r.p.m. The temperature setting, and the time of growth will vary according to the organism. As with the solid medium above, *E. coli* grows very quickly, usually overnight, but some other organisms will take longer to become established. When a significant amount of biomass has become established within the flask, as indicated by an appreciable level of turbidity, this biomass may be used to prepare glycerol stocks of the organism or agar plates and slopes.

4.3.6 Reviving a culture from solid media (agar slope or plate) into liquid medium

In these instances the organism might be provided on an agar slope in a glass or plastic test tube, plugged with cotton wool or a plastic cap (Figure 4.9) or on an agar plate.

1. If reviving from a slope, remove the lid or plug of cotton wool and flame the aperture of the tube. If using a plate, remove the lid.

2. Using a sterile nichrome or plastic loop, scrape some of the biomass from the slope or plate [Figure 4.10(b)]

3. Transfer the biomass on the loop to a sterile Erlenmeyer flask containing a suitable growth medium [Figure 4.10(d)] and grow the organism in an orbital shaker. You could additionally transfer some of the biomass to a Petri dish containing agar of the required formulation, for storage on solid media, and store in an incubator at the appropriate temperature.

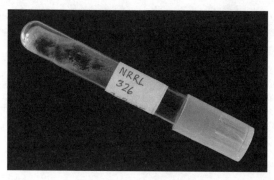

Figure 4.9   An agar slope from a culture collection, supporting the growth of a filamentous fungus

(a)

(b)

(c)

(d)

Figure 4.10 Inoculation of liquid medium using aseptic technique. The nichrome loop is flame-sterilised (a) then used to pick up a small amount of biomass from the plate (b). The bung is removed from the neck of the growth flask, which is then flamed (c). The loop is then agitated in the liquid medium to release the inoculum (d)

4.3.7 Formulations of Growth Media for Micro-Organisms

Micro-organisms will grow on a variety of solid and liquid media of diverse chemical complexity. When considering on which medium to grow a bacterium, fungus or yeast, it is useful first to consider the possible effects of that growth medium on the production of the enzyme of interest. In this context, it is useful at this stage to distinguish between *inducible* and *constitutive* expression.

Expression is a term that strictly describes the activation of a gene to give the mRNAs or proteins that constitute the products of the transcription-translation process (Chapter 7). However, it is often used trivially to describe the production of protein (enzyme) by an organism, so the term 'protein expression', whilst strictly incorrect, is widely used. *Inducible* expression describes the situation wherein an organism, for example a species of *Rhodococcus*, will only produce the enzyme of interest if grown on a medium containing a compound that specifically elicits that enzyme activity. In bacteria, the genes that encode enzymatic pathways for the metabolism of a certain compound are often arranged next to each other in what is known as an *operon* (Chapter 7). The expression of any of the genes within this operon may come under the control of the same *promoter*, a sequence of DNA to which RNA polymerase can bind to begin the process of transcription that leads ultimately to the production of the enzyme of interest. The activity of this promoter may be directly affected by the composition of the growth medium, and may respond directly to the addition of the chemical, the degradation of which is encoded by the pathway.

As an example, it may be that in order to produce an oxidase enzyme that hydroxylates the hydrocarbon octane, octane must be added to the growth medium either as a supplement, or in some cases as sole carbon source (SCC), a medium in which this inducing substrate is the only source of carbon that the organism may access for growth. The octane interacts indirectly (usually via a protein) with the promoter sequence and the expression of all the octane metabolising enzymes, including the oxidase, is switched on, and the resultant cells contain large amounts of the required oxidase. If induction of this kind is necessary to elicit the production of an enzyme of interest, this can create problems for preparing sufficient amounts of enzyme for preparative biocatalysis, as the inducers may be toxic, expensive (or, in the case of octane, flammable). However, genes that are expressed in response to specific induction often result in large amounts of the relevant protein, which has good consequences both for activity and for the attempted purification of the enzyme later on.

Constitutive expression describes a situation in which the organism will produce the enzyme of interest when grown on a medium which does not specifically elicit that activity. Constitutively expressed enzymes will be produced when the organism is grown on a medium such as LB broth or M9 medium supplemented with a non-specific carbon source such as acetate or glucose (although it is possible that these may act as inducers in their own right). Enzymes derived from organisms grown in such a way lend themselves very well to large-scale growth for biotransformations as their growth requirements are cheaper, non-toxic and therefore less complicated overall, but proteins resulting from constitutive expression may be present in lower amounts than those which result from inducible systems.

Generally therefore, the types of medium that are used to grow micro-organisms are divided into two types, dubbed *complex* or *basic*. Confusingly, it is the *complex* media, which are rich media made from sugars, salts and yeast and animal extracts that are easiest to prepare, and most often used for the resuscitation and maintenance of micro-organisms, as in this stage of experiments, it is not usually necessary to elicit the enzymatic activity for biotransformation – the objective is merely to keep the organism alive and to propagate it. *Basic* media, conversely are user-defined mixtures made from simple salts and a separate carbon source, and hence allow much greater control over the growth characteristics of an organism in a particular run.

4.3.8 Water in microbiological media

The range of chemical compositions of tap water found across the UK and the world makes it unsuitable for use in defined media of a reproducible nature. Hence, all of the media and buffer recipes referred to in this and later chapters make use of deionised or distilled water as a basis.

4.3.9 Complex media for Bacterial Growth

Probably the most widely used complex media are Nutrient broth obtainable from large suppliers such as Sigma and Oxoid, and LB broth, a simply defined medium containing tryptone (an acid hydrolysate of animal protein), yeast extract and sodium chloride, made up in distilled water. These are suitable for the maintenance of many of the bacterial species used for biotransformation, including *Pseudomonas, Arthrobacter, Rhodococcus, E. coli* and others. Variations of LB include Terrific broth (see Appendix 4), which is richer, and can give more biomass. Companies that provide strains of *E. coli* for cloning experiments also provide more specialised complex media such as SOC broth.

4.3.10 Basic Media for Bacterial Growth

Basic media usually consist of a mixture of salts and trace element additives that are, by their nature, more complex to prepare, but being more defined, allow greater control over the medium constituents and hence the growth of the organism. Owing to the more complex mixture of ingredients, it may be that some constituents of these media need to be prepared and autoclaved separately. For example, for media that contain magnesium salts and phosphates, the phosphate salts must be autoclaved in the absence of magnesium to prevent the formation of insoluble magnesium phosphate salts. One of the most popular basic media is known as M9 (see Appendix 4). M9 contains a mixture of phosphate salts, sodium chloride and ammonium chloride. Calcium chloride, magnesium sulfate and a source of vitamins are added subsequent to autoclaving. The last of these may be trace amounts of yeast extract, tryptone or thiamine. Glucose or glycerol may also be used as non-inducing carbon sources, although additional carbon and/or nitrogen sources may be added on top of the basal salts medium in order to elicit inducible enzyme activities as described above.

4.3.11 Media for growth of filamentous fungi and yeasts

Filamentous fungi such as *Aspergillus, Mucor* and *Rhizopus* often require somewhat different growth requirements and standard media from bacteria. Common solid media for the resuscitation of fungi include potato dextrose agar and malt extract agar. A variety of defined media has also been used in the literature, but a couple of generally good recipes are given below:

- Corn steep medium is a general medium for fungal growth that uses $7.5\,g\,L^{-1}$ of corn steep solids, basically a freeze-dried hydrolysate of corn-processing waste, and $10\,g\,L^{-1}$ of glucose. The pH of the medium is very low prior to correction – this can be adjusted to 5.0 using 2.0 M sodium hydroxide, a slightly acid pH being very favourable for the growth of some filamentous fungi.
- Czapek-Dox medium can be purchased from Oxoid and is a more complex fungal medium which contains sucrose and sodium nitrate as sole nitrogen source and is constituted in water with an approximately neutral pH that can be adjusted to an acidic pH after formulation. This medium is suitable for the growth of a range of fungi in both solid (with the addition of agar) and liquid phases.

There are various medium formulations available for the growth of simple yeasts such as *Saccharomyces cerivisiae* (baker's yeast). These include YPD, a simple formulation containing yeast extract, tryptone and dextrose and yeast nitrogen base, a highly complex medium that allows the defined study of the nutrient requirements of different yeast species.

Having established some simple methods for the growth and maintenance of micro-organisms, we now move to consider examples of whole-cell biotransformations using selected examples from the literature.

4.4 Examples of Whole-Cell Biotransformations using Bacteria

In this section, we provide examples of some growth and biotransformation protocols. We specify whether the biotransformation is *whole cell*, catalysed by *resting or growing cells*, and illustrative of *constitutive* or *inducible* expression.

4.4.1 Hydrolysis of linalyl acetate by *Rhodococcus* sp. DSM 43338

(whole cell, resting cells in buffer, constitutive expression)

Bacteria of the genus *Rhodococcus*, often of a distinctive orange or bright red coloration (Figure 4.11), are described frequently in the biotransformations literature.

As a genus, they have been shown to be equipped with a huge variety of biocatalytic activities which themselves stem from the ability to use a wide variety of natural and xenobiotic compounds as sole carbon source for growth [1]. Consequently, *Rhodococcus* bacteria often emerge from either microbial screens or enrichment selections as the organism of choice for a wide range of enzymatic processes.

Amongst these enzymatic activities are enantioselective esterases. The following example, adapted from methods described in reference [2], details the use of a strain of *Rhodococcus* for the hydrolysis of a tertiary ester, linalyl acetate (Figure 4.12).

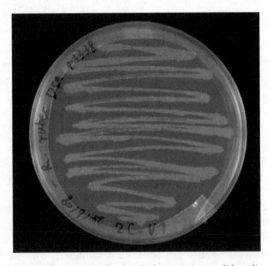

Figure 4.11   The growth of a *Rhodococcus* sp. on solid medium

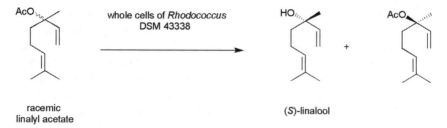

Figure 4.12 Enantioselective hydrolysis of linalyl acetate by *Rhodococcus*

In this example, it is not necessary to add an inducing compound during the growth of the organism, as the esterase activities are expressed by cells grown only on simple LB medium. We will assume that the bacterial culture has been revived and is growing healthily on an agar plate. *Rhodococcus* organisms are slow-growing, in contrast to some Gram-negative organisms such as *E. coli*, with slow doubling times, and it can take up to 7 days for a culture of *Rhodococcus* to establish itself on an agar plate.

Aseptic technique (see Chapter 3) should be applied throughout, and will include the routine flaming of the openings of flasks and tubes and the heating of the nichrome wire loop to a red heat prior to use and after being in contact with the biological material.

Apparatus and reagents

- Established culture of *Rhodococcus* sp. DSM 43338
- 250 mL Erlenmeyer flasks
- Foam bungs and aluminium foil
- Sterile syringe barrels, needles and bacterial filters
- Bunsen burner or laboratory gas
- Orbital shaker at 30°C
- Ingredients for LB growth medium (see Appendix 4)
- Linalyl acetate
- 200 mL 100 mM Phosphate buffer pH 7.0

1. First it will be necessary to make up 100 mL of LB medium. Weigh the ingredients for 100 mL LB medium (see Appendix 4) into a 250 mL Erlenmeyer flask. To this add 100 mL of distilled water. The flask is then plugged either with cotton wool or a foam rubber bung. A square of aluminium foil is then placed over the top of the cotton wool and wrapped around the top and the neck. This is then sealed with a strap of autoclave tape, in order to protect the bung from moisture in the autoclave.

2. The medium is autoclaved at 121°C for 15 min and allowed to cool.
3. Gently scrape a loopful of the organism from the surface of the agar plate bearing the *Rhodococcus* strain using the nichrome wire loop (previously heated to red in the Bunsen flame) and deposit this into the medium as shown in Figure 4.9. The loop can be agitated in the medium if there is difficulty dislodging the biomaterial. Flame the opening and replace the lid of the Falcon tube. This culture should be incubated for 24 h at 30°C or until turbid.
4. The growth can be monitored, if desired, using a spectrophotometer during this period to establish a growth curve for the organism under these conditions. In the case of the *Rhodococcus*, the culture can be turbid and the colour distinctly pink-red in liquid culture as well as on a plate (Figure 4.13).
5. After 24 h, remove the flask from the incubator and, using aseptic technique, pour the culture into a plastic centrifuge tube and centrifuge at 5000 r.p.m for about 15 min. Centrifuging whole cell samples either for longer times or at higher speeds can result in the organism being very difficult to remove

Figure 4.13   Growth of *Rhodococcus* in shake-flask culture

from the sides of the tube. The resulting supernatant is discarded – this is best done by returning this to the growth flask. The spent supernatant should be treated with disinfectant (see Chapter 3) or autoclaved prior to disposal.

6. To the organism in the centrifuge tube, add 10 mL of 100 mM phosphate buffer pH 7.0. Replace the lid of the centrifuge tube and agitate the contents until the pellet at the base of the tube has been resuspended thoroughly. Centrifuge again at 5000 r.p.m for about 15 min. Discard the supernatant and disinfect prior to disposal.

7. Add another 10 mL of 100 mM phosphate buffer pH 7.0 to the cells. Resuspend the contents again and then transfer the cells-in-buffer suspension to a clean 250 mL Erlenmeyer flask.

8. Add neat linalyl acetate to the washed cell suspension to give a final concentration of 10 mM. Incubate the reaction with shaking at 30°C.

9. The reaction can then be monitored using thin layer chromatography (TLC) or gas chromatography (GC). For TLC, the reactions can be spotted directly from the flask, or, if suitable extracted first into an organic solvent. The latter will also serve for GC analysis. To do this, using aseptic technique, remove 0.5 mL of the reaction using a sterile plastic pipette tip into a clean 1.5 mL capped plastic tube. To this, then add 0.5 mL of ethyl acetate or other suitable solvent. Close the tube and mix the contents either by inversion or by agitating on a whirlimixer. The organic and aqueous layers should settle out, but this can be helped by placing the tube into a micro-centrifuge and centrifuging at maximum speed for 2 min. This usually results in a cell pellet, and the clearly resolved aqueous and organic phases, the superior layer of which may be used for either spotting a TLC plate or injecting into the gas chromatograph.

10. Once the reaction has gone to completion, as revealed by TLC/GC, the cells can be removed by centrifugation at 5000 r.p.m. The supernatant can be poured off and extracted into solvent as for a standard organic work-up.

4.4.2 Hydrolysis of an aromatic nitrile by *Rhodococcus* sp. NCIMB 11216

(whole cell, resting cells in buffer, inducible expression)

Among the reactions catalysed by *Rhodococcus*, the hydrolysis of nitriles is the one that has received perhaps the most attention. The Nitto process for the hydrolysis of acrylonitrile to acrylamide is one of the world's largest and most significant commercial biotransformation reactions, and whilst that specific example is perhaps of limited interest to fine chemical synthesis, the enzymes responsible for the conversion of nitriles to either amides or carboxylic acids have been found to display a much wider substrate specificity.

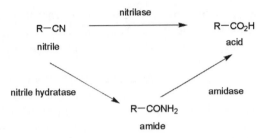

Figure 4.14   Two pathways for nitrile metabolism in bacteria

Nitriles are metabolised essentially by one of two pathways in micro-organisms (Figure 4.14).

They are either converted directly from the nitrile to the acid by enzymes know as nitrilases (E.C.3.5.5.1) or are converted first to the intermediate amide by the action of a nitrile hydratase (E.C. 4.2.1.84), and then to the carboxylic acid by an amidase (E.C. 3.5.1.4) enzyme.

Rhodococcus strains often have both systems operating within their cells – however, it may still be possible to exploit only one of these activities in a whole cell reaction, especially if only one of the enzymes within the cell has a substrate specificity that encompasses the molecule in question.

Techniques

The following method is adapted from reference [3] and uses cells of *Rhodococcus* sp. NCIMB 11216 that have been grown using benzonitrile as an inducer to hydrolyse a selection of aromatic nitriles including α-cyano-*o*-tolunitrile (Figure 4.15).

Until very recently, the growth of a wild-type organism on an inducing substrate was fairly commonplace, but with the advent and simplicity of recombinant systems, it is rarer to find such examples in the literature. We have already seen that the use of recombinant *E. coli* as a designer biocatalyst has removed both considerations of possible wild-type organism pathogenicity for biotransformations, but has also removed the need to grow some of these organisms on 'difficult' (perhaps volatile, toxic, expensive, or even explosive) carbon sources such as hydrocarbons

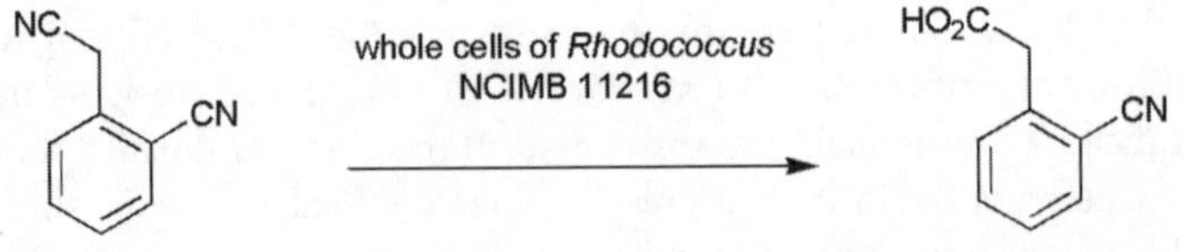

Figure 4.15   Hydrolysis of α-cyano $-o$-tolunitrile by a *Rhodococcus* sp

or terpenes. Induction of activity in an organism used to be essential for eliciting activity, however, and so two examples are included for the sake of completeness, and in the interests of providing the reader with a strategy should repetition of a literature method prove necessary.

The growth techniques differ from that used in the esterase example above, in that the basal salts medium M9 is used for growth and this requires some mixing of constituent components *after* autoclaving. *Aseptic technique* should again be applied throughout.

Apparatus and reagents

- Established culture of *Rhodococcus* sp. NCIMB 11216
- Sterile Falcon tubes
- 250 mL Erlenmeyer flasks
- Foam bungs and aluminium foil
- Sterile syringe barrels, needles and sterile plastic filters
- Bunsen burner or laboratory gas
- Orbital shaker at 30°C
- Ingredients for M9 growth medium (see Appendix 4)
- Sodium benzoate
- Benzonitrile
- 200 mL EPPS-NaOH buffer pH 8.0

1. It is first necessary to prepare a small starter culture of the organism. Transfer 2 mL of LB broth (see Appendix 4), using aseptic technique and either a sterile disposable pipette or an automated pipette with a sterilised disposable tip to a sterile 15 mL or 50 mL Falcon tube.
2. Transfer a loopful of the organism from the agar plate into the LB medium using the wire loop. This culture should be incubated for 24 h at 30°C or until turbid.
3. Whilst the starter culture is incubating, it will be necessary to make up 50 mL of M9 minimal medium (see Appendix 4). Weigh the ingredients for 50 mL M9 medium *except* the ammonium salt, into a 250 mL Erlenmeyer flask. To this add 50 mL of distilled water. Plug with cotton wool, cover with foil and autoclave as for the LB medium in the linalyl acetate hydrolysis example above.
4. Weigh out the other M9 supplements (magnesium sulfate, calcium chloride and thiamine dichloride), dissolved in a minimum (1 mL) of water. Add ammonium chloride (to a final concentration of 5 mM) and sodium benzoate (20 mM). These will act as nitrogen and carbon source for the organism, respectively. Add these supplements to the solution in the Erlenmeyer flask via the syringe and sterile filter, using aseptic technique throughout.

5. Now, transfer the turbid starter culture to the new flask containing the M9 medium. Flame the opening, replace the bung and place the Erlenmeyer flask on the orbital shaker at a speed of 150 r.p.m at 30°C. Incubate the organism for 12 h.

6. Remove the flask from the incubator and, using aseptic technique, add benzonitrile to a final concentration of 20 mM. Replace the culture in the shaker and incubate for a further 5 h.

7. Harvest the culture using centrifugation as in Section 4.3, then, to the organism in the centrifuge tube, add 25 mL of 100 mM EPPS-NaOH buffer pH 8.0. Replace the lid of the centrifuge tube and agitate the contents until the pellet at the base of the tube has been resuspended thoroughly. Centrifuge again at 5000 r.p.m for about 15 min. Discard the supernatant and disinfect prior to disposal.

8. Add another 25 mL of 50 mM Tris-HCl buffer to the cells. Resuspend the contents again and then transfer the cells-in-buffer suspension to a clean 250 mL Erlenmeyer flask.

9. The substrates are added to the washed cell suspension to give a final concentration of 20 mM. If liquid, the nitriles are added using an automated pipette; if solid, the nitriles can be ground up using a pestle and mortar as added as a powder. The compounds that proved to be successful substrates for the enzymatic activity induced by benzonitrile in this study included 4-hydroxybenzonitrile, *p* - tolunitrile and α-cyano-*o*-tolunitrile. Incubate the reactions with shaking at 30°C.

10. The reactions can then be monitored using TLC, GC or HPLC. For HPLC, a 1 mL sample can be taken, and the biomass first removed by centrifugation at maximum speed in a benchtop microcentrifuge to give a clear solution for injection.

11. Once the reaction has gone to completion, the cells are removed by centrifugation and the aqueous supernatant worked up in the normal manner.

4.4.3 Lactonisation of norbornanone by an inducible Baeyer–Villigerase activity in *Pseudomonas putida* ATCC 17453

(whole cell, resting cells in buffer, inducible expression, bacterial)

Strains of *Pseudomonas putida* often possess interesting complements of oxidising enzymes. They have been the source of, amongst other activities, the dioxygenases that give rise to optically pure *cis*-dihydrodiols, from metabolic pathways associated with the breakdown of toluene and other aromatics. They are also known to possess a number of Baeyer–Villiger monooxygenase (BVMO)

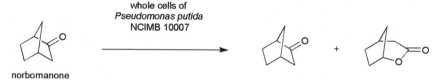

Figure 4.16   Lactonisation of norbornanone by camphor-grown *Pseudomonas putida*

activities, and well-known work in the 1960s by Gunsalus, and later by Trudgill in the 1980s and 1990s, described the nature of these monooxygenase enzymes in some detail. These BVMOs are inducible by growth on the monoterpene ketone camphor and are used to cleave both bicyclic and monocyclic ring systems in the associated catabolic pathway. In the example below, the inducible activities will be exploited for the biotransformation of the bicyclic ketone norbornanone (Figure 4.16). This method has been adapted from reference [4].

The organism will be grown on a basal salts medium that has been supplemented with camphor as sole carbon source, so as to elicit the monooxygenase activity. The recipe for this medium can be found in Appendix 4 (basal salts medium 1). Note that, as with M9, the medium needs to be prepared in two separate flasks for autoclaving and mixed after cooling. Also note that iron, in the form of ferrous sulfate must be added to the growth medium – this is especially important as the metabolic pathway for the degradation of camphor is dependent on the induction of cytochrome P450, an enzyme containing iron. The ferrous sulfate must also not be autoclaved, as this will oxidise the Fe^{2+} to Fe^{3+}, which cannot be used by the organism. This must therefore be added using a sterile filter (see above).

Apparatus and reagents

- Established culture of *Pseudomonas putida* ATCC 17453 (NCIMB 10007)
- 250 mL Erlenmeyer flasks
- Foam bungs and aluminium foil
- Sterile syringe barrels, needles and bacterial filters
- Bunsen burner or laboratory gas
- Orbital shaker at 30°C
- Ingredients for basal salts medium 1 (see Appendix 4)
- (1*R*)-(+)-Camphor
- Norbornanone
- 200 mL Potassium phosphate buffer pH 7.0

Use *aseptic technique* throughout, for all medium mixing, inoculation and sampling steps.

1. Make up 50 mL of component 1 in a 250 mL Erlenmeyer flask. Bung with cotton wool, cover with foil and autoclave. Also make up a concentrated solution of component 2 solution – such that 1 mL of the solution would be the correct addition to 50 mL - i.e. weigh out the amount required for 1 L and dissolve this in 20 mL of distilled water in a 50 mL Erlenmeyer flask. Bung the flask and protect with foil and autoclave. Finally, make up a 20x solution of ferrous sulfate in water i.e. 1 g in 20 mL, but do not autoclave this.

2. When all the medium components have been sterilised and are cooled, light a Bunsen burner and remove the foil and bung from the flask containing component 1. Open the flask containing component 2, flame the neck of this flask and remove 1 mL of component 2 solution from it using a sterile plastic pipette tip. Transfer this to the flask containing component 1, flame the neck of the flasks and replace both bungs.

3. Using a 1 mL plastic syringe, draw 1 mL of the 20x ferrous sulfate solution into the barrel and place a 0.2 µm filter on the end of the syringe. Open the flask containing the mixture of components 1 and 2, flame the neck of the flask and inject the content of the syringe into the medium. Flame the flask and replace the bung.

4. Clean a spatula with 70% ethanol. Weigh out 50 mg of camphor into a clean, sterile glass vial or plastic-capped tube and transfer this to the flask of medium, flaming the neck of the flask as you proceed. Replace the bung of the flask. The medium is now complete and you can proceed to inoculate it.

5. Transfer a loopful of organism from the agar plate into the culture flask using a nichrome loop. Place the culture in a rotary shaker at 150 r.p.m. and 30°C for 24 h. After this time, a turbid culture of the *Pseudomonas* should have resulted that appears grey-green. Some solid camphor crystals may persist in the flask, but these will be removed during centrifugation and washing.

6. Harvest the cells by centrifugation at 5000 r.p.m. and wash once using buffer as described previously. Resuspend the cells in 10 mL of the buffer and pour into a clean 100 mL Erlenmeyer flask.

7. Dissolve 10 mg of the substrate, norbornanone, into 200 µL of ethanol, and add the ethanolic solution to the cells. Monitor the reaction as described previously by extraction, and then analysis by TLC or GC until conversion of the ketone to the lactone is complete, or has reached a plateau.

8. The cells can be removed by centrifugation at 5000 r.p.m and the aqueous supernatant saturated with sodium chloride, followed by extraction into ethyl acetate. Dry the combined organic extracts using magnesium sulfate, remove the solvent and isolate the lactone using flash chromatography.

4.4.4 Hydroxylation of 1, 4-cineole by a *Streptomyces* sp

(whole cell, growing cells in medium, constitutive expression, bacterial)

Organisms of the genus *Streptomyces* contain some of the most fascinating enzymes, many of which are associated with the construction of the elaborate architectures that constitute the arrays of secondary metabolites produced, notably the polyketides. They are hence rich in enzymes such as oxidoreductases, and cytochromes P450, which are often used to elaborate secondary metabolites such as the P450 EryF that hydroxylates the erythromycin precursor 6-deoxyerthyronolide B. It has been shown, for example, that *Streptomyces coelicolor* possesses at least twenty cytochromes P450 that might find an application in biocatalytic synthesis or the production of drug metabolites.

Streptomyces spp. are sporulating organisms and can be grown on a variety of solid or liquid media, but for purposes of biotransformation it is often best to grow them on solid media that encourage the formation of these spores. One such solid medium employs, somewhat bizarrely, tomato purée and oatmeal, but is very effective at inducing sporulation in the organisms. This makes the organisms easy to scrape for inoculation into liquid media, such that they are suitable for biotransformation reactions. We will describe a procedure for the hydroxylation of the monoterpene compound 1,4-cineole (Figure 4.17) by *Streptomyces*, although these reactions have also often been shown to be useful for the synthesis of small amounts of drug metabolites for toxicology studies.

The advantage of such whole-cell biotransformations using growing cells is that the growth and biotransformation procedures are very simple. The disadvantages, as detailed in Chapter 3, are that there are many cytochromes P450 in the organism, and as such a complex mixture of products may result in the biotransformation mixture. The yields of these reactions may be low, but often only work at a relatively small scale and consequently it has been usual to run many small-scale reactions in parallel as shown below.

We will assume that the *Streptomyces* strain has been revived from a freeze-dried ampoule by growth on oatmeal-tomato purée agar (see Appendix 4) and has been grown for 7–10 days, resulting in a white, dusty morphology on the agar plate.

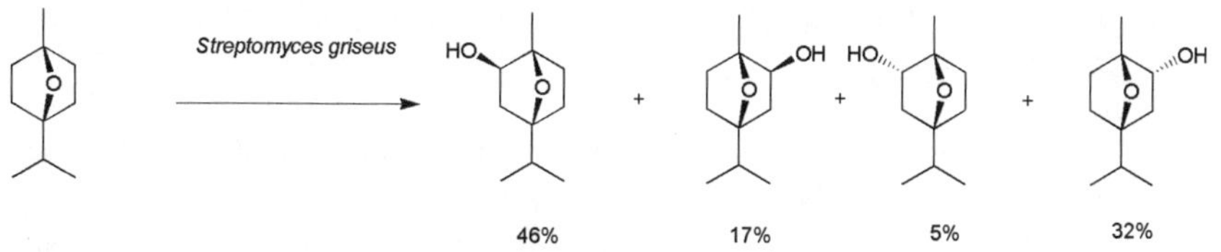

Figure 4.17   Hydroxylation of 1,4-cineole by a *Streptomyces* sp

The following procedure is adapted from references [5] and [6] and would be suitable for screening a range of *Streptomyces* species for different reaction profiles. Use *aseptic technique* throughout.

Apparatus and reagents

- Established culture of *Streptomyces griseus* ATCC 10137
- 250 mL Erlenmeyer flasks
- Foam bungs and aluminium foil
- Bunsen burner or laboratory gas
- Orbital shaker at 30°C
- Ingredients for soybean meal-glucose culture medium
- 1,4-Cineole

1. Prepare (multiples of) 25 mL of soybean meal-glucose medium in a 250 mL Erlenmeyer flask. The reduced volume of medium in the flask can sometimes improve the yield of oxidative reactions, either through better mixing or oxygenation of the medium or a combination of both. Bung the flask with cotton wool, cover with foil and sterilise by autoclaving.
2. Light the Bunsen burner and adjust to the blue flame. Remove the bung from the cooled flask and flame the open neck. Using a nichrome loop, scrape a small amount of biomass from the dusty white mycelium on the *Streptomyces* plate and agitate the loop in the medium until the biomass has been dislodged into the liquid. Flame the neck of the flask and replace the bung.
3. Place the flask in an orbital shaker at 150 r.p.m. and 28°C and grow the organism for 72 h. *Streptomyces* can be rather slow-growing, and 72 h may be required for the organism to attain a suitable level of biomass for bioconversion, and also to begin to synthesise the enzymes involved in secondary metabolism that will be exploited in the hydroxylation reaction.
4. To each 25 mL culture, add 25 mg of 1,4-cineole. It is *crucial* that sound aseptic technique is applied at both stages of substrate addition and sampling in growing cell biotransformations, as the organism is still in growth medium, which is susceptible to infection from opportunistic micro-organisms.
5. Take samples for analysis by either TLC or GC at intervals over a period of about 3 days. If the reaction is going to work, you should certainly see some transformation after 1–2 days. After this time, there is a danger that any reactions that might be observed may be due either to other organisms having infected the flask or possibly a different behaviour of the organism under anaerobic conditions that have arisen as a result of the protracted growth time.

6. Remove the *Streptomyces* mycelium by either filtration using a Büchner funnel or centrifugation at 5000 r.p.m. The aqueous supernatant should then be saturated with sodium chloride and then extract the metabolites into ethyl acetate or suitable solvent, dry, evaporate to dryness and purify metabolites by flash chromatography or preparative HPLC.

In the literature example cited, this strain of *Streptomyces* gives four different products of hydroxylation, but screening of different strains using the general method applied above may well lead to organisms of different selectivity and productivity.

4.5 Biotransformation by filamentous fungi and yeasts

Yeast is a well-established and user-friendly biocatalyst, demonstrated by the fact that the baker's yeast purchased from a local supermarket will probably be an excellent catalyst for, for example, the stereoselective reduction of carbonyl groups. Biotransformations by fungi, however, are perceived as being more complicated than those using bacteria. Rather than growing as simple smears on agar plates, the growth of fungi is observed to be more plant-like, with spreading aerial hyphae and spores resulting a in a wide variety of morphologies in both solid and liquid media and, possibly, more capricious behaviour in respect of the reproducibility biotransformations. It is often seen that filamentous fungi lose the activity for which they have been selected if maintained on agar plates for too long for example. The previous points about maintaining the identity and integrity of the culture to be used for a project are to be emphasised here therefore, through the use of glycerol stocks and the use of fresh plates for inoculation.

4.5.1 Maintenance of filamentous fungi on agar

Filamentous fungi can be revived on solid agar in much the same way as for bacteria in examples described above. The morphologies of various fungal species are shown in Figure 4.18. A couple of good standard solid media for the resuscitation of filamentous fungi are made from potato dextrose agar (PDA), or agar added to the corn steep solids or Czapek-Dox media listed in Appendix 4. The plates should be stored at a temperature somewhat lower than 30°C (perhaps 25°C), as fungal plates tend to dessicate easily. Once grown to an appreciable level of biomass, say after 10 days, they can be stored in the refrigerator at 4°C, with Parafilm wrapped around the edge, joining the base and the lid of the plate.

One of the challenges in using filamentous fungi for biotransformation arises in the next step, cultivation in liquid medium, as fungus in liquid culture may not grow in homogeneous suspension, such as planktonic bacteria, and can form

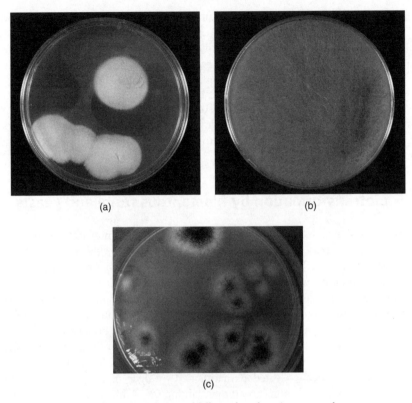

(a) (b)

(c)

Figure 4.18 Morphology of different fungal species on agar plates

dense cell masses that introduce complications such as poor surface area–volume ratio for catalysis and mass transfer into the cells themselves. The problems are largely caused by the large amount of polysaccharide cell wall that surrounds a fungal cell. The problems of cell aggregation can be overcome with careful inoculation protocols that disperse the inoculum finely throughout the growth medium. Having said this, some enzymes are secreted extracellularly, as fungi are often saprophytic and need to secrete enzymes onto growth substrates such as wood, and biotransformations, for example by the fungus *Caldariomyces fumago* which contains an active haloperoxidase, are catalysed by enzymes that have been secreted into the growth medium.

4.5.2 Inoculation of filamentous fungi into liquid medium from agar plates

The transfer of fungal mycelium into liquid medium for the purposes of growth for biotransformation provides challenges, depending on the morphology of the fungus. For example, the basidiomycete *Beauveria bassiana* grows on solid medium such as potato dextrose agar, as a dusty white mat, that can be scraped with a nichrome loop in much the same way as a bacterium. Other organisms grow as feathery white mycelia that bear spores – these may be harvested for inoculation – spores being small provide small nucleations for organismal growth and should prevent the clumping of organisms to form the large masses that one is trying to avoid. However, some fungi grow as mats that are difficult to scrape, with the result that, if used for inoculating liquid medium, the organism grows as a clump of biomass. Hence, some method of macerating the fungal tissue from the agar prior to inoculation is often necessary. There are a number of methods of doing this, with various degrees of sophistication, although a tissue homogeniser is usually the best way.

A two-part tissue homogeniser can be purchased from most purveyors of biochemical laboratory equipment – this consists of a plunger with a plastic end and a tapered glass tube the base of which will allow close contact of the plunger with the glass (Figure 4.19). The two parts of the homogeniser can be sterilised separately, by wrapping both in foil and autoclaving and pieced together using aseptic technique when cool. Mechanical tissue homogenisers are also available that operate using rotating blades like a food blender – these are of course more effective, but also are more expensive.

In order to use the manual homogeniser:

1. Using a sterile scalpel blade, cut a small (1 cm) square of agar from the plate on which the fungus is growing.
2. Remove the plunger from the homogeniser and, after flaming the top of the tube, insert the square of agar into the bottom of the tube.
3. Add 1–5 mL of sterile water to the tube and replace the plunger. Push the plunger in and out of the tube until an even suspension results in the tube.
4. Then 1 mL of this suspension can be transferred aseptically to 50 mL of liquid growth medium as a suitable inoculum.

Some examples of whole-cell biotransformations using wild-type fungi are now addressed, starting with the reaction for which they have perhaps become most well-known, that of the hydroxylation of non-activated carbon atoms.

4.5.3 Hydroxylation of N-CBz-piperidine by growing cells of *Beauveria bassiana* ATCC 7159

(whole cell, growing cells in medium, constitutive expression, fungal)

Beauveria bassiana ATCC 1759 (previously known as *Beauveria sulfurescens* and *Sporotrichum sulfurescens*) has a long history of use as a biocatalyst in the academic arena since the late 1960s when Johnson and co-workers at Upjohn first realised

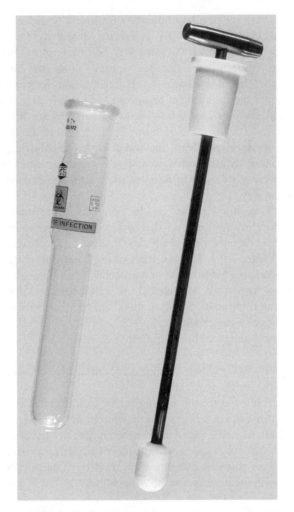

Figure 4.19   A tissue homogenizer for preparing fungal inocula

its potential, particularly as a hydroxylation catalyst [7]. Although no genome of a *Beauveria* species has yet been completed, it appears that a number of cytochromes P450 are genetically encoded in the organism that will catalyse the hydroxylation of a range of substrates. In many cases, these substrates require some aromatic character – this may help to secure the substrate in the cell membranes wherein the relevant enzymes probably reside. The following is an example of such a reaction, the hydroxylation of an N-CBz piperidine (Figure 4.20), with methods adapted from reference [8].

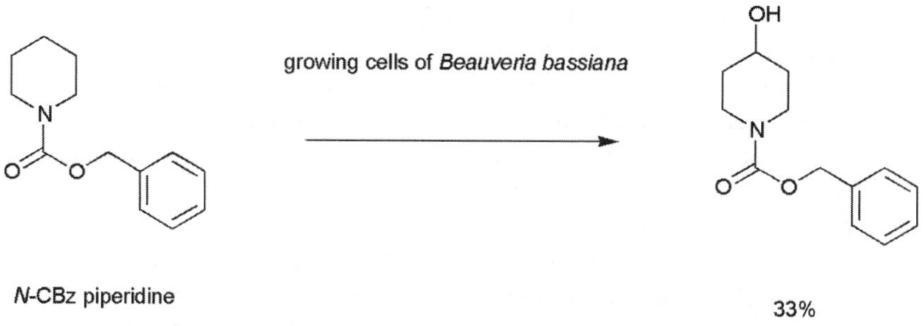

growing cells of *Beauveria bassiana*

N-CBz piperidine

33%

Figure 4.20   Hydroxylation of an N-CBz piperidine by *Beauveria bassiana*

It is assumed that the organism from the ATCC has been revived onto agar plates that are best made in this case from potato dextrose agar (see Appendix 4). The organism will take up to a week to establish and will present as dusty white streaks on the plate.

Apparatus and reagents

- Established culture of *Beauveria bassiana* ATCC 7159
- 250 mL Erlenmeyer flasks
- Foam bungs and aluminium foil
- Bunsen burner or laboratory gas
- Orbital shaker at 30°C
- Ingredients for corn-steep liquor medium
- *N*-CBz protected pyrrolidine

1. Prepare 4 × 250 mL Erlenmeyer flasks containing corn-steep liquor medium (see Appendix 4) and adjust to pH 4.9 using 2 M sodium hydroxide. Bung the flasks with cotton wool or silicon bungs, cover with foil and sterilise by autoclaving. The medium will be somewhat particulate even after autoclaving, but this insoluble material often disappears during the biotransformation reaction.
2. Transfer a loopful of the white fungal biomass from the agar plate to the corn-steep medium. Agitate the loop holding the fungal biomass into the sterile medium until the inoculum has been dislodged. Place the fungus in an orbital shaker at 28°C at 150 r.p.m and grow for 72 h.
3. After 72 h growth, the fungus should appear as a thick, off-white soup – sometimes small spherical particles are also visible. It may be that fungus has deposited on the wall of the flask at the level of the swirl, but this is normal. To each flask, add the substrate in a small amount (0.5. mL) of ethanol to a final concentration of 0.1 mg mL^{-1} and leave to incubate at 28°C. As with

hydroxylation reactions by *Streptomyces* it can be common to grow many small identical reactions in parallel.

4. Extract 0.5 mL samples at intervals. This can be done by using 1 mL Gilson pipette tips, the end of which has been removed with a pair of scissors – this prevents clogging of the pipette tip with the fungal biomass. Analysis can be performed by extraction into ethyl acetate and monitoring by TLC as described for bacterial reactions.

5. When the reaction has stabilised, or reached completion as judged by analysis, the fungal biomass should be removed by either centrifugation or filtration. The aqueous supernatant should then be saturated with sodium chloride and extracted with three times the volume of ethyl acetate. If an emulsion forms during extraction, this may be resolved by the addition of small amounts of either acetone or methanol to the separating funnel, followed by swirling.

6. The solvent is removed *in vacuo* and the residue subjected to standard flash chromatography to isolate the product.

In the study cited, a 45% yield of the monohydroxylated product alone was obtained, but, as with *Streptomyces*-catalysed hydroxylations, it may be that different hydroxylated regioisomers will result when using filamentous fungi such as *Beauveria, Cunninghamella, Mucor* or *Rhozopus* as biocatalysts.

4.5.4 Enantioselective hydrolysis of styrene oxide by washed, resting cells of *Aspergillus niger*

(whole cell, growing cells in medium, constitutive expression, fungal)

The field of enzymatic epoxide hydrolysis has gained in popularity as a facile method of gaining access to enantiopure epoxides and diols. Recent studies largely focus on the use of a purified enzyme from the fungus *Aspergillus niger* that has been shown to possess both a wide substrate tolerance and excellent enantioselectivity with respect to a range of racemic epoxide substrates. The utility of the organism was first demonstrated with the hydrolysis of styrene oxide by Furstoss and co-workers in 1993 (Figure 4.21). The general method of using washed cell preparations of *Aspergillus niger* to carry out hydrolysis reactions provides a good illustration of the general method of washed cell fungal reactions, for uncomplicated processes such as epoxide or ester hydrolysis. The following method is adapted from reference [9].

We will assume that the organism has been revived on potato dextrose agar plates. The organism should have the appearance of a white mycelium turning black and crusting on storage.

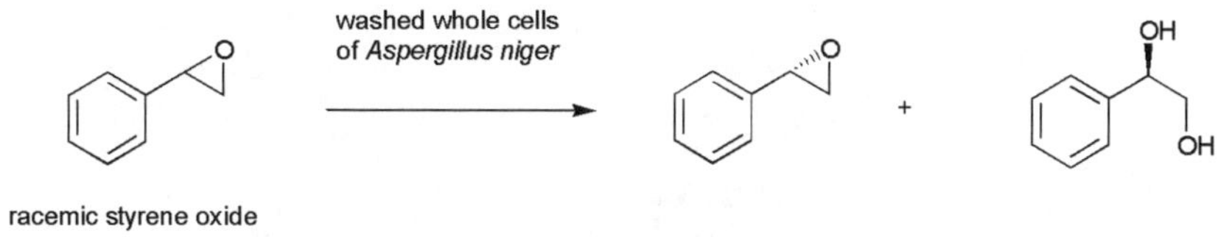

Figure 4.21   Hydrolysis of styrene oxide by *Aspergillus niger*

Apparatus and reagents

- Established culture of *Aspergillus niger*
- 250 mL Erlenmeyer flasks
- Two-part tissue homogeniser (sterilised)
- Foam bungs and aluminium foil
- Bunsen burner or laboratory gas
- Orbital shaker at 30°C
- Ingredients for corn-steep liquor medium
- Styrene oxide
- 100 mM Potassium phosphate buffer pH 7.0

(NB It should be noted that styrene oxide is a carcinogen and should be handled and disposed of with care.)

1. Prepare 50 mL of corn-steep solids medium (see Appendix 4) in a 250 mL Erlenmeyer flask. Bung the flask with cotton wool or a silicon foam bung, cover with foam and autoclave. Allow the flask to cool.
2. Using the tissue homogeniser described earlier, prepare a sterile amount of water that contains a homogenised square of agar from the plate bearing the *Aspergillus niger*. Remove the bung from the culture flask and flame the open neck. Using a sterile plastic pipette that has had the tip removed, transfer approximately 1 mL of the homogenate to the culture flask. Flame the neck of the flask and replace the bung.
3. Place the flask in a rotary shaker at 150 r.p.m. and 25°C. Grow the organism for 48–72 h, or until a large amount of cream-turning-black pellets are observed floating in the flask (Figure 4.22). When an appreciable amount of biomass has been accumulated, the cells can be recovered by filtration using a Büchner funnel. Pour the fungal culture into the Büchner funnel and pull the liquid through until a dry mat of fungus results. Wash the mat with 3 × 50 mL of sterile 100 mM phosphate buffer pH 7.0, or until the filtrate runs clean.

Figure 4.22   Morphology of *Aspergillus niger* in liquid culture. The cream-coloured pellets begin to turn black on further incubation

4. Resuspend the fungal mat in 50 mL of sterile 50 mM phosphate buffer pH 7.0 in a clean 250 mL Erlenmeyer flask and replace in the orbital shaker. Leave to resuspend for 30 min – an even suspension of fungal pellets should result. Dissolve the substrate (100 mg) in 0.5 mL of ethanol and transfer to the reaction flask. Replace the flask in the shaker and at intervals, withdraw samples as described previously for extraction and analysis. The rates of hydrolysis are fast – much faster than for oxidation reactions such as hydroxylation–so reactions must be monitored carefully, particularly where a desired optical resolution necessitates stopping the reaction at approximately 50% conversion.
5. When the reaction has reached the desired extent level of conversion, as measured by GC or TLC, remove the fungus by filtration using a Büchner funnel – this removal of the catalyst will also stop the reaction of course – and extract the supernatant with ethyl acetate. The diol and residual epoxide may then be easily separated by flash chromatography.

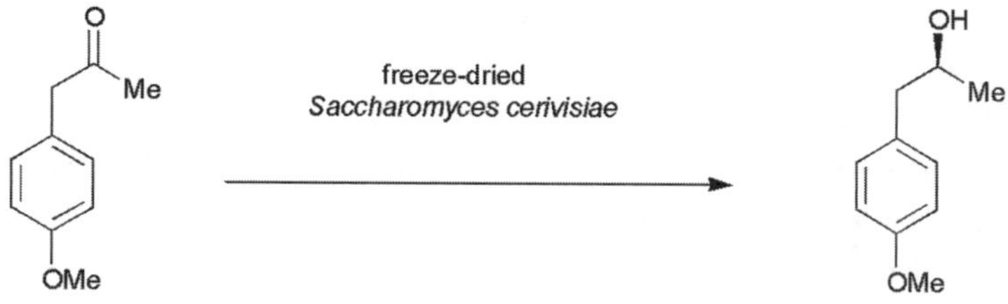

1-(4-methoxyphenyl)propan-2-one   (S)-

Figure 4.23 Reduction of 1-(4-methoxyphenyl)propan-2-one by freeze-dried *Saccharomyces cerivisiae*

In the study cited, this technique was extended to use two organisms, *Aspergillus niger* and *Beauveria bassiana*, each displaying the opposite *regioselectivity* for the ring opening of styrene oxide to give an *enantioconvergent* biotransformation in which an 89% yield of near enantiopure diol was obtained. The gene encoding the epoxide hydrolase from *Aspergillus* has been cloned and expressed in *E. coli*, so now in practise the *Aspergillus* strain is not the biocatalyst of choice for this transformation.

4.5.5 Reduction of a carbonyl group using *Saccharomyces cerivisiae* (baker's yeast)

(whole cell, growing cells in medium, constitutive expression, yeast)

Strains of yeast, notably *Saccharomyces cerivisiae*, but also *Candida* and *Geotrichum*, have been used for many years as biocatalysts, being easy to use and particularly active for asymmetric reductive transformations. Once the genome of *S. cervivisiae* had been determined, it was then shown that the genome encoded many different keto-reductase enzymes, some with different or even opposing selectivities [10]. Given the detailed description of these activities that now exists, and the tools available for expressing predominantly a single keto-reductase in a recombinant expression strain [10], it might be unusual for wild-type *S. cerivisiae* to be used as a biocatalyst for fine chemicals production. However, in the interests of providing a guide to the use of readily accessible biocatalysts, we include below a protocol for a simple yeast-catalysed reduction of a chiral secondary alcohol. The reduction described is the asymmetric reduction of 1-(4-methoxyphenyl)propan-2-one by a freeze-dried commercial preparation of *S. cerivisiae* (Figure 4.23).

It is necessary to first incubate the yeast in a solution of sucrose in order for it to be in the required fermentative state, as described below. This example is adapted from reference [11].

Use *aseptic technique* throughout.

Apparatus and reagents

- Freeze-dried baker's yeast (obtained from supplier e.g. Sigma)
- 250 mL Erlenmeyer flasks
- Foam bungs
- Bunsen burner or laboratory gas
- Orbital shaker at 30°C
- Water
- Sucrose

1. To a 250 mL sterile Erlenmeyer flask, add sucrose (7 g), 50 mL of distilled water and 14 g of a freeze-dried preparation of baker's yeast. Bung with cotton wool and agitate for 30 min at 30°C in an orbital shaker.
2. Remove the yeast suspension from the incubator and add 1.2 mmol of the ketone substrate. Monitor the reaction using extraction and TLC as described for the bacterial whole cell biotransformation above.
3. The reaction may take a few days to reach completion. At daily intervals, add a further dose of catalyst – 7 g of baker's yeast in a suspension of sterile water (22.5 mL) to which has been added 3.5 g sucrose.
4. When the reaction has reached completion as adjudged by TLC, pour the reaction suspension through a pad of celite to remove the catalyst. Extract the reaction mixture into ethyl acetate or diethyl ether, dry the combined extracts and remove the solvent carefully *in vacuo*. Purify the alcohol product using flash chromatography.

4.6 Whole-cell Biotransformations by recombinant strains of *E. coli*

In the present day, it is often the case that the 'designer' biocatalysts described above, in which the gene from one organism has been heterologously expressed, usually in the common laboratory host *E. coli*, is the biocatalyst of choice for a required transformation. Either the recombinant strains of *E. coli* or the expression plasmids into which the gene of interest has been ligated are usually available from the authors of relevant publications – indeed it is most often a condition of publication that these materials are made freely available to academic workers. The following, therefore gives brief instructions on the handling of such recombinant biocatalysts, either acquired as strains of *E.coli* or as plasmids which are used to 'transform' commercial strains of *E. coli* in the laboratory. More details on the background to recombinant biocatalysts, and how to construct them, are given in Chapters 2 and 7.

4.6.1 Commercial expression plasmids and competent cells

It is likely that you will have been provided with an expression plasmid into which has been ligated the gene encoding the enzyme of choice. The plasmid will also feature one or more 'antibiotic resistance markers' that encode, for instance a β-lactamase which will confer resistance to β-lactam antibiotics such as ampicillin when the transformed strain of *E. coli* is plated out onto solid agar containing the antibiotic. The mechanism of induction of gene expression, leading to production of the enzyme in growing cells of *E. coli* will be described in more detail in Chapter 7. For the techniques detailed here, it is enough to know that appropriate antibiotics must be included in both solid and liquid growth medium for the recombinant strain of *E. coli* used, and that an inducer, most usually isopropyl-β-D-thiogalactopyranoside (IPTG), will probably be required to induce gene expression and protein production.

The *E. coli* cells that will be used to perform the biotransformation need to be of an 'expression strain', that is, they need to be able to take up the plasmid that contains the gene of interest (during 'transformation') make multiple copies of that plasmid, and to possess the necessary cellular machinery to induce expression of the required gene under controlled growth conditions. Examples of these expression strains include the common laboratory hosts BL21 (DE3) and B834 (DE3). These expression strains are to be distinguished from simple 'cloning strains' that are used merely for the replication and synthesis of plasmid material. The cells in question need to be 'competent' – that is, they need to be equipped with the ability to take up the recombinant plasmid. Competent cells can be bought from commercial suppliers, but can also be made in the laboratory using the procedure below.

4.6.2 Preparation of competent cells for cloning and expression

In order to begin making competent *E. coli* cells for use in the laboratory it is necessary of course to have the required strain available and growing on an LB-agar plate. The procedure will involve partial growth of the organism, followed by harvesting in a *sterile* centrifuge tube and resuspension and washing in special buffers containing either calcium chloride or rubidium chloride that confer competence on the cells.

Apparatus and reagents

- Agar plate with established culture of required expression strain of *E. coli* (BL21, B834, Rosetta, etc. – see below for more details)
- 500 mL Erlenmeyer flask

- Sterile centrifuge tube
- Foam bungs
- Bunsen burner or laboratory gas
- Orbital shaker at 30°C
- Sterile 1.5 mL capped plastic tubes
- Ice bath
- Sterile Psi broth (see Appendix 4)
- Sterile TfB1 buffer (see Appendix 5)
- Sterile TfB2 buffer (see Appendix 5)

Use *aseptic technique* throughout.

1. Inoculate 100 mL of sterile Psi broth (see Appendix 4) in a 500 mL Erlenmeyer flask with a sample of the *E. coli* strain using a sterile nichrome loop.
2. Grow the culture on an orbital shaker at 37°C for 3 h 45 min.
3. Remove the culture from the incubator and put the flask on ice for 45 min.
4. Centrifuge the culture in a sterile centrifuge tube for 5 min at a speed of 4000 r.p.m. Pour off the supernatant.
5. Resuspend the cells in 30 mL sterile TfB1 medium and then incubate the tube on ice for 15 min.
6. Centrifuge the culture in a sterile centrifuge tube for 5 min at a speed of 4000 r.p.m. Pour off the buffer supernatant.
7. Resuspend the cells in 6 mL sterile TfB2 medium and then incubate the tube on ice for 15 min.
8. Using an automated 200 μL pipette fitted with a sterile tip, transfer the cell suspension in 200 μL aliquots each to a sterile 1.5 mL capped plastic tube.

These aliquots of competent cells can either be stored at −20°C, or to best ensure their performance over a longer period, each tube can be snap-frozen in liquid nitrogen and stored at −80°C.

4.6.3 Transformation of competent *E. coli* strains with plasmids

If it is an expression plasmid that has been provided by the researcher or collaborator, it is most often the case that the conditions for transformation (or the process during which a strain of *E. coli* will take up the expression plasmid) will be provided, but the following provides a general guide to transformation protocols. These details also suggest the identity and required final concentration of the antibiotic to be used that will ensure growth of only transformed clones of *E. coli* that have taken up the plasmid bearing the gene of interest.

Apparatus and reagents

- Sterile plastic Petri dishes
- Ice bath
- 200 µL aliquot competent cells
- Automated pipettes
- Bunsen burner or laboratory gas
- Plate spreader
- Sterile 1.5 mL capped plastic tubes
- Sterile LB agar
- Antibiotic relevant to expression plasmid
- Water bath at 45°C

Use *aseptic technique* throughout.

1. Prepare 100 mL of LB agar (Appendix 4), sterilise by autoclaving and allow to cool.
2. Calculate the amount of the recommended antibiotic (usually ampicillin, kanamycin or carbenicillin, or sometimes more than one) that would need to be added to the agar to achieve the final concentration. The final concentration will usually be in the region of $30-100$ µg µL^{-1}. In practise it is best to make up 1000x concentration stocks and to freeze them in appropriately sized aliquots.
3. When the bottle has cooled sufficiently for it to be held in the hand, add the antibiotic to the agar, mix by swirling and pour approximately 20 mL of the agar into one of each of five sterile plastic Petri dishes.
4. Thaw a 200 µL aliquot of competent cells on ice and split into two 100 µL fractions in sterile 1.5 mL capped plastic tubes. To these small aliquots, add 1 µL of the solution of plasmid that has been adjusted to a concentration of $50-100$ ng µL^{-1}. Mix with the end of a plastic pipette tip, then leave on ice for a further 30 min.
5. The cell suspension is 'heat shocked' in a water bath at 42°C for 45 s, then transferred to ice for a further 2 min.
6. Add 1 mL of sterile LB medium and grow the cells in an orbital shaker for 1 h at 37°C. After 1 h the suspension should appear slightly cloudy. Pipette 100 µL of the suspension onto the now solid LB-ampicillin agar plates and use the spread plating technique (Section 4.3) to spread the liquid culture around the surface of the agar. Leave the plates to incubate at 37°C overnight.
7. On the following day, you should observe a number of colonies (<10spor>1000 depending on the 'transformation efficiency'; Figure 4.24).

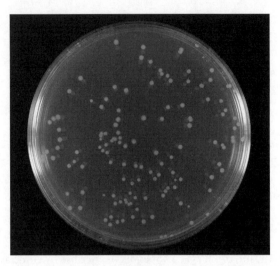

Figure 4.24   Colonies of *E. coli* resulting from transformation with a plasmid

The number of colonies obtained is an indication of the 'transformation efficiency' (see below) of the strains used and will vary for different strains and plasmid vectors. These can be used directly to inoculate small starter cultures for liquid culture experiments. They can also be used to make glycerol stocks of the recombinant strain. However, it is recommended that the colonies on these plates are not used for long periods of time, as they will deteriorate by various mechanisms. For the best results, freshly transformed cells should be used for inoculation.

4.6.4 Transformation efficiency

The transformation efficiency measures the amount of colonies of cells that will be generated by a unit amount of the plasmid used in transformation. It will vary according to the type of plasmid and strain used, but also with various other parameters that can be improved by optimising the experimental technique. Parameters that will affect the transformation efficiency will include the length, and temperature of the heat shock and the purity of the plasmid used in transformation. The transformation efficiency can be calculated (in terms of transformants μg^{-1}) using the simple formula:

$$\frac{\text{Number of colonies on plate}}{\text{ng of DNA plated out}} \times 1000$$

For previously untested plasmids, it is difficult to approximate what the efficiency would be, but for commonly used pET vectors in commercially available cloning strains of *E. coli*, efficiencies of 1×10^6 colony forming units (cfus) per microgram of plasmid DNA are obtained.

4.6.5 Inoculation and growth of recombinant *E. coli*

There will of course be a number of variations on this theme (see below for assessing the optimum conditions for the heterologous expression of a particular gene in *E. coil*), but we provide some details here of a possible procedure for biotransformation using recombinant *E. coli* as a biocatalyst, based on some work using genes encoding Baeyer–Villiger monooxygenase enzymes expressed in *E. coli* described in reference [12]. In this example, a gene that encodes a Baeyer–Villiger monooxygenase Rv3049c from *Mycobacterium tuberculosis* has been introduced into *E. coli* B834 (DE3) and used to catalyse the resolution of the bicycle ketone in Figure 4.25.

Apparatus and reagents

- Established culture of recombinant *E. coli* with plasmid bearing gene of interest
- Sterile 15 mL or 50 mL Falcon tubes
- 2 L Erlenmeyer flasks
- Foam bungs and aluminium foil
- Sterile syringe barrels, needles and bacterial filters
- Bunsen burner or laboratory gas
- Orbital shaker at 37 °C; to be reduced to 20 °C after induction
- Automated pipettes
- Ingredients for LB growth medium (see Appendix 4)
- Aliquots of ampicillin at a 1000x concentration of 30 mg mL^{-1} in water
- 0.5 mL aliquots of 1 M isopropyl-β-D-thiogalactopyranoside (IPTG)

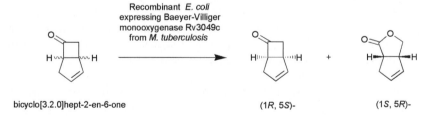

bicyclo[3.2.0]hept-2-en-6-one   (1*R*, 5*S*)-   (1*S*, 5*R*)-

Figure 4.25 Resolution of a bicyclic ketone by a recombinant strain of *E. coli* expressing a Baeyer–Villiger monooxygenase

Use *aseptic technique* throughout.

1. Using a sterile pipette tip, stab one of the colonies on the agar plate bearing the recombinant strain of *E. coli* and transfer this to 5 mL of sterile LB medium containing 5 μL of ampicillin (from the stock of 30 mg mL^{-1}) in a sterile Falcon tube (15 mL, but mixing, and hence oxygenation, will be better in a 50 mL tube). Grow the culture overnight at 37 °C. After 18–24 h growth, the culture should be appreciably turbid.
2. Prepare one or a multiple of 2 L Erlenmeyer flasks containing 500 mL of LB medium, bung and cover with foil and then sterilise by autoclaving. When cool, add 750 μL of a 30 mg mL^{-1} solution of kanamycin, using a sterile plastic pipette tip. Next, add the grown starter culture, flame the neck of the flask and leave to grow in an orbital incubator at 150 r.p.m. at 37 °C.
3. At 30 min intervals, using aseptic technique, remove 1 mL of the growing culture and measure the optical density at 600 nm in a spectrophotometer. Although the culture grows slowly at first, during the lag phase (see Chapter 3), the culture will grow quickly once the exponential phase begins and should be monitored carefully.
4. When the culture has achieved an optical density (A_{600}) of approximately 0.5, the culture has reached what is termed the 'mid-exponential phase' (see Chapter 3). Remove the culture(s) from the incubator and reduce the temperature of the incubator to 20 °C. At this time, the expression of the gene can be induced by the addition of 1 mM IPTG – the reasons for using this inducer are covered in Chapter 7. Transfer a 500 μL aliquot of 1 M IPTG of this solution to the flask using aseptic technique, and continue to incubate the flask at 20 °C overnight.

At this stage, there exist two options for the whole-cell biotransformation reaction. In the first option it is possible to perform a growing cell biotransformation, in which substrate is added directly to the growth flask about 30–60 min after induction. The reaction is then monitored using TLC or GC, then the cells are removed by centrifugation and the aqueous supernatant worked up as detailed below. In the second option one can leave the cells for a defined period – say 3 h–harvest the cells by centrifugation, wash them, and resuspend them for use in a resting cell biotransformation reaction as described in Section 4.4. In the first case, the greater volume allows the addition of substrate at perhaps a concentration of 1 g L^{-1}. With the resting cells, the reaction volume is reduced when the cells are resuspended in buffer and the overall scale of the reaction will be smaller consequently, although the reaction is cleaner and the work-up may be appreciably easier.

In either of these cases, when the biotransformation is considered to be complete, the cells can be removed by centrifugation at 5000 r.p.m, and the supernatant saturated with sodium chloride – this helps to eliminate emulsions during extraction. The supernatant can be extracted using ethyl acetate, the combined organic fractions dried over magnesium sulfate and evaporated, and, in the example used, the ketone and lactone can be easily separated using flash chromatography.

4.6.6 Assessing the best conditions for expression in recombinant strains of *E. coli*

The use of recombinant or designer strains of *E. coli* for biotransformations has become commonplace, as not only does the technique allow for the expression of a wide variety of biocatalysts that were previously not accessible by means of wild-type microbiology, but the methods used are also cheap and easy to carry out. There are some additional considerations of which it is useful to be aware at this stage, notably the factors that may affect the optimum expression of a foreign gene in *E. coli*. These will include the choice of plasmid vector that has been chosen for expression, and other more complex aspects of the molecular structure of the gene itself, including whether or not the gene has been 'tagged' (see Chapter 7) to assist in the purification of the encoded protein, or to increase its solubility. However, there are some simple, routine considerations for improved heterologous expression that may see greatly increased performance in a biocatalyst that are dependent on easy-to-adjust parameters that will involve no genetic engineering aspects whatsoever. Many of these have to do with improvements in the levels of soluble gene expression i.e. there are physical and chemical factors that will affect the extent to which the gene is expressed, and the extent to which the protein produced by the bacterium will be soluble within the cell, rather than trapped in insoluble, membrane enclosed 'inclusion bodies', and thus inactive for purposes of catalysis. In order to actually establish the extent to which a gene is expressed in the soluble fraction of the bacterial preparation, it is of course necessary to analyse the protein complement of the cells using standard biochemical techniques such as SDS-PAGE (see Chapter 6), but in appreciation of the limited biochemistry facilities that may exist within an organic chemistry laboratory, we may nevertheless suggest certain experiments that may lead to an improvement in biocatalyst performance, that may be judged empirically.

4.6.6.1 Choosing an expression strain of *E. coli*

Once in possession of a suitable expression plasmid, which may have been obtained either from collaborators or produced in-house (see Chapter 7), you will need to decide which strain or variant of *E. coli* to use for the expression experiment. There are a number of different so-called 'expression strains' of *E. coli* available from commercial molecular biology suppliers such as Stratagene,

Novagen, Promega and Qiagen. These strains are not only suitable for relatively simple heterologous expression of bacterial targets, but can also provide a means of improving the expression of more challenging targets, such as those from eukaryotic organisms. These different strains of *E. coli* allow for the possibility of improved gene expression, perhaps giving expression where there was previously none, or by greatly increasing the chances of solubility of the enzyme biocatalyst. Some commonly used strains, and their particular characteristics, are detailed below. This is not intended as a comprehensive list, rather as an initial guide to buying some selected strains that might be screened for each biocatalyst expression study. In practise, it is usual to screen a handful of different strains to find the one that gives the best expression, or in the context of this chapter, the best performance as a whole cell biocatalyst.

Most commercial expression strains of *E. coli* are derived from '*E. coli* B' strain, a model organism that has been used in laboratories for many years. The 'B' is retained in the name of many of the most commonly used strains including the most popular, BL21.

BL21

E. coli **BL21** is a common expression strain of *E. coli* that is commonly used for expression experiments in molecular biology labs, but also occurs commonly as a strain in recombinant biocatalysis studies. The BL21 strain is a derivative of another strain, B834 (see below) that has mutations that mean that two native proteases *lon3* and *ompT*, enzymes that might destroy the enzymes of interest in the cells once expressed, have been deactivated, leading to a higher yield of protein within the cell.

E. coli **BL21(DE3)** strains are variants of BL21 which possesses within its genome the gene encoding the T7 RNA polymerase that will act on the T7 promoter in pET vectors commonly used in cloning and expression protocols. (Some relevant details on promoters and vectors can be found in Chapter 7.) When using a pET vector construct, or other vectors that are dependent on the T7 promoter system for gene expression, it is best to use a strain of *E. coli* with the suffix (DE3) therefore.

EndA- strains of *E. coli* BL21 are also advantageous in that these lack a restriction endonuclease that can digest the plasmid in the expression system. Hence, as will be seen in Chapter 7, it has been common to clone a gene and produce plasmid in an EndA- cloning strain and then to perform a second transformation into the expression strain BL21 once the quality and sequence of the recombinant plasmid has been verified. There are now some BL21 strains, such as Stratagene **BL21 Gold**, that can be transformed directly with ligation mixes.

pLysS strains of *E.coli* BL21(DE3) have a gene which encodes the enzyme lysozyme, which bind to the T7 promoter and stops transcription until the addition of an inducer such as IPTG. Hence, it provides a strain which has transcription very tightly controlled, and may be of use where there is too much gene expression, or where the protein produced is toxic to the *E. coli* cells.

BL21 star (DE3 or not) cell lines may also offer improvements in expression owing to reduced degradation of mRNA after transcription, having a mutation in an RNAase enzyme that would degrade these transcripts.

Origami™

E. coli Origami™ strains are derivatives of *E.coli* K-12, another commonly used laboratory strain that was the first bacterium to have its genome sequenced (in 1997). Expression strains of Origami™, such as Origami™ (DE3) have adaptations which can assist in the formation of disulfide bridges in proteins and as such may prove useful in the expression of genes from eukaryotic sources. Owing to the specific adaptations however, plasmids that are dependent on kanamycin as resistance marker cannot be used. There are also Origami™ strains that are based on the BL21 strain of *E. coli*.

Rosetta™

We will see in Chapter 7 that amino acids are encoded within gene sequences by 'codons' – three-letter base-pair codes, all of which together form the 'genetic code'. There are 64 possible three-letter codons using the fours bases A, T, G and C, so many amino acids are coded for by more than one codon – the genetic code is thus said to be *degenerate*. *E. coli*, like other organisms, has a 'preference' for some codons that encode certain amino acids, and thus may not be a suitable host for expressing a gene wherein a codon not regularly used by *E. coli* encodes a certain amino acid. *Rosetta* strains of *E. coli*, derived again from BL21, provide additional cellular machinery, in the form of the required transfer RNA molecules, for the recognition of these unusual codons. If analysis of the gene sequence encoding the biocatalyst of interest reveals unusual codons, the use of Rosetta strains may result in improved expression and biocatalyst activity. Similar principles of rare codon usage are to be exploited in strains such as BL21-Codonplus from Stratagene. Strains of Rosetta (which may include DE3 derivatives) include Rosetta 2 and Rosetta-Gami (derived from *E. coli* K12), a strain which combines the advantages of the Origami strains described above, with the rare codon usage found in Rosetta.

B834

E. coli strain B834, again to be found in both DE3 and non-DE3 forms, is a methionine auxotroph. This means that it is unable to synthesise the amino acid methionine and is thus used in, for example, the preparation of selenium-labelled methionine-containing proteins that are used in X-ray crystallography. It can provide a useful alternative to BL21 strains for recombinant biocatalysis however, when transformed and grown normally on bacterial broth such as LB and in which the supply of methionine is not an issue.

In practice, in the event of a problematic expression, it is advisable to try a range of hosts that each have a distinctive property such as rare codon usage or less endogenous protease activity, and to use the most successful strain as the basis for further optimisations detailed below.

4.6.6.2 Temperature of incubation after induction

The temperature at which the recombinant *E. coli* is grown after induction with IPTG can have a great effect on, particularly soluble gene expression. It may be found that, instead of keeping the temperature at 37 °C for 3 h after addition of IPTG, the flask should be moved to an incubator at 30 °C (for 5 h), 20 °C, or even 16 °C – in the latter two cases overnight. The lower temperature is thought to engender slower protein folding and thus not result in improperly folded protein being sequestered in insoluble 'inclusion bodies'.

4.6.6.3 Concentration of the inducer

Instead of using 1 mM IPTG in induction, it may be worth trying a range of concentrations, from 0.05 mM through to 1 mM, to see if activity improves as the concentration of IPTG can often have an effect on the amount of gene expression, and thus the performance of the biocatalyst.

4.6.6.4 Time of induction

Soluble expression, and thus biocatalyst activity might be altered by the time during the exponential phase of growth at which the inducer is added. It may therefore be worth, once the optimum concentration of IPTG has been established, adding this at different stages of growth as determined by the A_{600} measurements, perhaps at 0.3, 0.7 or 1.0.

4.6.6.5 Use of auto-induction medium

Instead of using the IPTG-based system of induction to induce gene expression in T7 promoter-controlled vectors, a system of so-called 'auto-induction', developed by F. William Studier [13] can be used, in which, *E. coli* strains bearing pET-type plasmids with genes under T7 promoter control are grown in defined medium in which, as carbon source, have been added limiting concentrations of glucose and lactose. After the glucose has been exhausted as a growth substrate, the lactose is able to activate the T7 polymerase, leading to induction. The advantages of this method are very high cell densities and of course the non-dependence on the expensive and toxic IPTG as inducer. The recipe for an auto-induction medium is given in Appendix 4 and more details can be obtained from reference [13].

The guidelines provided above should provide the beginner with a starting point for the easy use of a recombinant biocatalyst for which the worker has been provided with a ready-made expression plasmid. It may also be possible to obtain the strain of recombinant *E. coli* from a collaborator – this would save having to transform the plasmid, and the strain can be treated in much the same way as a wild-type bacterium described earlier in the chapter (with the important exception of the GM containment issue that applies to such recombinant catalysts). There are of course a whole range of possible expression techniques using different plasmids and strains of *E. coli*, including temperature-dependent expression systems, and some of which are dependent on arabinose as an inducer. A recent review summarises some of the approaches that have been taken in high-throughput gene expression studies for structural proteomics [14]. There are of course a host of expression organism systems that are commonly used within academia and industry for recombinant biocatalysis, including other bacteria such as *Pseudomonas* and *Streptomyces*, fungi such as *Aspergillus* and yeasts, notably *Saccharomyces cerivisiae* and *Pichia pastoris*. Each of these systems will require alternative plasmid systems and culture and expression techniques and will not be dealt with further in this book.

4.7 Conclusion

At the end of this chapter, we have dealt briefly with the use of established recombinant strains for biocatalysis, but in Chapter 7, we will address how simple cloning experiments might be applied to generate recombinant biocatalysts of the workers' own choosing. This will inevitably involve the introduction of some simple molecular biology techniques, and also the techniques and facilities to study the subcellular component of cells including isolated enzymes, which are the subject of the following chapters.

References

1. A. M. Warhurst and C. A. Fewson (1994) Biotransformations catalyzed by the genus *Rhodococcus*. *Crit. Rev. Biotechnol.*, **14**, 29–73.
2. I. Osprian, A. Steinreiber, M. Mischitz and K. Faber (1996) Novel bacterial isolates for the resolution of esters of tertiary alcohols. *Biotechnol. Lett.*, **18**, 1331–1334.
3. A. J. Hoyle A. W. Bunch and C. J. Knowles (1998) The nitrilases of *Rhodococcus rhodochrous*. NCIMB 11216. *Enzym. Microb. Technol.*, **23**, 475–482.

4. G. Grogan, S. M. Roberts and A. J. Willetts (1993) Some Baeyer-Villiger oxidations using a monooxygenase enzyme from *Pseudomonas putida* NCIMB 10007. *J. Chem. Soc. Chem. Commun.*, 699–701.

5. R. E. Betts, D. E. Walters and J. P. N. Rosazza (1974) Microbial transformations of antitumour compounds, 1. Conversion of acronycine to 9-hydroxyacronycine by *Cunninghamellla echinulata. J. Med. Chem.*, **17**, 599–602.

6. W.-G. Liu, A. Goswami, R. P. Steffek, R. L. Chapman, F. S. Sariaslani, J. J. Steffens and J. P. N. Rosazza (1988) Stereochemistry of microbiological hydroxylations of 1,4-cineole. *J. Org. Chem.*, **53**, 5700–5704.

7. R. A. Johnson, M. E. Herr, H. C. Murray and G. S. Fonken (1968) Microbiological oxygenation of alicyclic amides. *J. Org. Chem.*, **33**, 3182–3187. [This was only one of many studies of bio-oxidations using *Sporotrichum sulphurescens* (now *Beauveria bassiana*) by the Upjohn group.]

8. S. J. Aitken, G. Grogan, C. S.-Y. Chow, N. J. Turner and S. L. Flitsch (1998) Biohydroxylations of CBz-protected alkyl substituted piperidines by *Beauveria bassiana* ATCC 7159. *J. Chem. Soc. Perkin. Trans.*, 3365–3371.

9. S. Pedragosa-Moreau, A. Archelas and R. Furstoss (1993) Microbiological transformations 28. Enantiocomplementary epoxide hydrolyses as a preparative access to both enantiomers of styrene oxide. *J. Org. Chem.*, **58**, 5533–5536.

10. I. A. Kaluzna, T. Matsuda, A. K. Sewall and J. D. Stewart (2004) A systematic investigation of *Saccharomyces cerevisiae* enzymes catalyzing carbonyl reductions. *J. Am. Chem. Soc.*, **126**, 12827–12832.

11. P. Ferraboschi, P. Grisenti, A. Manzocchi and E. Santaniello (1990) Baker's yeast-mediated preparation of optically active aryl alcohols and diols for the synthesis of chiral hydroxy acids. *J. Chem. Soc., Perkin Trans. I*, 2469–2474.

12. D. Bonsor, S. F. Butz, J. Solomons, S. C. Grant, I. J. S. Fairlamb, M. J. Fogg and G. Grogan (2006) Ligation-Independent-Cloning (LIC) as a rapid route to families of recombinant biocatalysts from sequenced prokaryotic genomes. *Org. Biomol. Chem.*, **4**, 1252–1260.

13. F. W. Studier (2005) Protein production by auto-induction in high-density shaking cultures. *Prot. Expr. Purif.*, **41**, 207–234.

14. N. S. Berrow, K. Büssow, B. Coutard, J. Diprose, M. Ekberg, G. E. Folkers, N. Levy, V. Lieu, R. J. Owens, Y. Peleg, C. Pinaglia, S. Quevillon-Cheruel, L. Salim, C. Scheich, R. Vincentelli and D. Busso (2006) Recombinant protein expression and solubility screening in *Escherichia coli*. A comparative study. *Acta Crystallogr., Sect. D*, **62**, 1218–1226.

Chapter 5
A Beginner's Guide to Biotransformations Catalysed by Commercially Available Isolated Enzymes

5.1 Introduction

For the synthetic organic chemist with little or no experience of biochemistry or molecular biology, the use of commercial preparations of isolated enzymes presents much less of a technical or conceptual problem than the use of micro-organisms, and indeed, there is almost no requirement for facilities, equipment (apart from, perhaps, a small temperature controlled orbital shaker) or techniques such as aseptic technique, that exceeds that routinely found in a synthetic chemistry laboratory.

In recent years, the number of transformations that might be performed under these low-tech conditions has been limited, largely by the limited range of commercially available enzymes. In this respect, much of the published work was restricted to the use of hydrolase enzymes (see Chapter 1), which are non-cofactor dependent, stable and are produced in vast quantities by the world's major enzyme suppliers, mostly for the bulk processing of materials in the food and materials industries. However, in recent years, companies such as Sigma-Aldrich, Boehringer and Biocatalytics (now owned by Codexis) have been responding to the demand for simple powdered biocatalysts for laboratory use, and have introduced new ranges of enzymes that can be tried out in a relatively simple fashion by the organic chemist.

A review of the biotransformation literature in the major organic chemistry journals will reveal that many of the groups that work in this area, either from industry or academia, are well-equipped to make their own cell-free or purified enzymes from either wild-type or recombinant strains. Approaches to making your

Practical Biotransformations: A Beginner's Guide   © 2009   Gideon Grogan

own cell free extracts and one or two simple methods for preparing enzymes from cells are given in Chapter 6, but as facilities and expertise for these experiments are often not within the compass of the synthetic organic chemistry laboratory, we provide a chapter on the use of enzymes that may simply be bought as either powders or solutions.

In this chapter, therefore, we will consider the application of commonly used hydrolytic commercial enzymes for synthetic reactions of interest, and will include the application of commercial enzymes for cofactor-dependent reactions, for which cofactor recycling will be required, and carbon-carbon bond formation. The methods in this chapter are of course extremely simple and do not involve much more that the sequential addition of enzymes and substrates to reaction vessels, followed by incubation and work-up, but it is useful to draw the reader's attention to the *amounts* and *concentrations* of substrate and biocatalyst, to the choice of reaction media, and to the incubation conditions in each case. It has been ensured that, in the interests of the reader being able to reproduce the methods, each example uses enzymes that should be obtainable from commercial suppliers.

5.2 Lipases

Lipases (E.C. 3.1.1.3) are a class of hydrolytic enzymes present in most organisms that in physiological systems are responsible for the hydrolysis of carboxyester bonds in triacylglycerols to fatty acids and glycerol. It has of course been found that these lipases will hydrolyse a wide range of carboxyesters with excellent catalytic rates, and also, because of the chiral nature of their active sites, will do this with exquisite enantioselectivity. Lipases are also robust, and are now used routinely in industry for bulk chemicals processing, as well as in the production of fine chemicals. Their stability, activity and ease of use have resulted in their being easily the best recognised and most widely used enzyme catalysts. The description of their use in organic solvents for the catalysis of *reverse* hydrolysis reactions, namely the acylation of alcohols and amines using the enzymes in low-water media, was revolutionary, and synthetic organic chemists have embraced lipases as organic-chemistry-friendly catalysts. There is an extensive discussion and review of suitable lipases for synthetic application (and other hydrolases) in Bornscheuer and Kaslauskas's textbook on hydrolases [1]. This textbook gives detailed information on the literature, methods and techniques and theory of the applications of hydrolases in synthetic organic chemistry. In the present book, we will restrict ourselves to a few initial examples to illustrate primary considerations and also include some additional examples of different cases that have been gleaned from the literature.

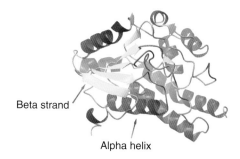

Beta strand

Alpha helix

Figure 1.9 The tertiary structure of an enzyme (pdb code 1ein) illustrating both of the secondary elements. The alpha-helices are shown in red and the beta-strands in yellow. The strands can be seen to lie together to form a beta-pleated sheet. Together with a variety of loops and turns, these elements make up the *tertiary* structure of the protein monomer as shown

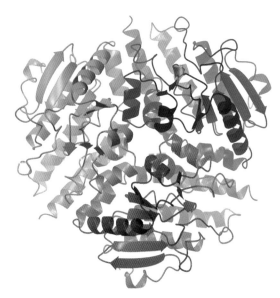

Figure 1.10 The quaternary structure of a protein as represented by the trimeric structure of a lyase enzyme (pdb code 1o8u). The constituent monomers are shown in red, blue and green

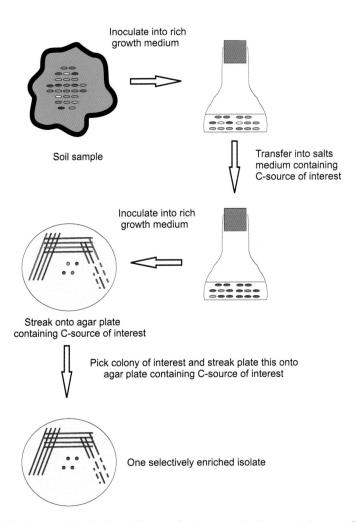

Inoculate into rich
growth medium

Soil sample

Transfer into salts
medium containing
C-source of interest

Inoculate into rich
growth medium

Streak onto agar plate
containing C-source of interest

Pick colony of interest and streak plate this onto
agar plate containing C-source of interest

One selectively enriched isolate

Figure 2.2 A strategy for selective enrichment of a micro-organism from an environmental sample

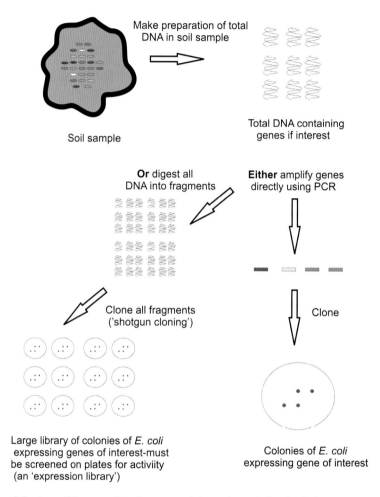

Figure 2.3 A possible strategy for a 'metagenomics' experiment designed to isolate a gene encoding a useful biocatalyst

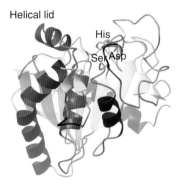

Figure 5.2 Structure of the lipase from *Thermomyces lanuginose* (pdb code 1ein) showing the helical lid near the active site that opens and closes over the catalytic triad (Ser, His, Asp) shown

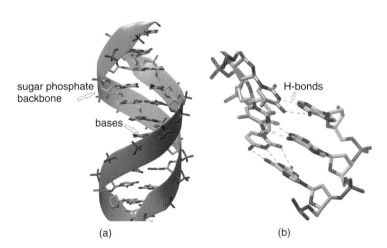

Figure 7.2 Structure of DNA. (a) Two complementary strands of DNA, one coral and one grey, make up the double helix. The backbone of each strand is made up of a chain of sugar (deoxyribose) phosphates. Attached to each sugar phosphate is an aromatic base – either a pyrimidine or a purine – the bases from complementary strands interact in the interior of the double helix through hydrogen bonding (b, dashed black lines)

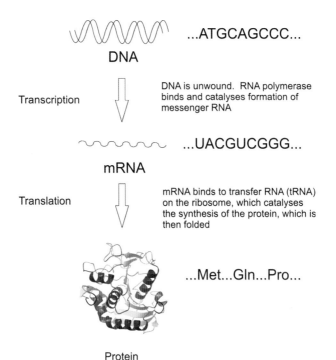

DNA

...ATGCAGCCC...

Transcription

DNA is unwound. RNA polymerase binds and catalyses formation of messenger RNA

mRNA

...UACGUCGGG...

Translation

mRNA binds to transfer RNA (tRNA) on the ribosome, which catalyses the synthesis of the protein, which is then folded

...Met...Gln...Pro...

Protein

Figure 7.4 *Transcription* of the DNA sequence into the mRNA message and *translation* into the amino acid sequence

```
CTTTGCGGTGATTAACATCCTTTGTTTCCTGTTTGTCGTTACGATCTGTCCAGAAACGAAGAACAAATCGC
TCGAGGAAATTGAAAAGCTTTGGATAAAATGAAAACGCTTTAATGAAACAGCCCTTTCTACGGGAAGGGCT
GTTTATATTGGGATGCACCATTTGGCGCTTTCTGTATAAGATAAAGATATATAGGATAAAATATTGCTGGA
TAAAACGACGCGGCATGAAAACTCTGCGAATATTGTCGATGAATTGGCTCTTAACAGTTGAATAAACAATT
CCACCCTGTTAAAATAATTAAAGAAAGCAGAAATGATTTTTTTTGGCTATGACGGGACGTTTT**TTGTCA**TA
GCGGGACA**TATAAT**GTCCAGCAAAAAAGGAAGGAACGTTTGAGTC**ATG**
```

Figure 7.5 Nucleotide sequence upstream of the gene *CggR* in *Bacillus* sp

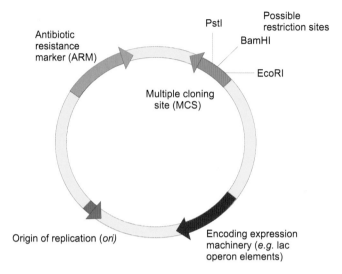

Figure 7.6 Diagram of a plasmid showing multiple cloning sites for digestion by restriction enzymes and antibiotic resistance marker (ARM)

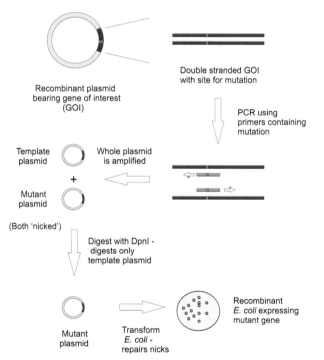

Figure 8.1 A contemporary method for site-directed mutagenesis using PCR

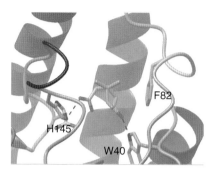

Figure 8.2 The structure of the active site of OCH containing the bound product of reaction and revealing a possible catalytic role for histidine residue in position 145, which is hydrogen-bonded to the pendant carboxylic acid moiety of the product

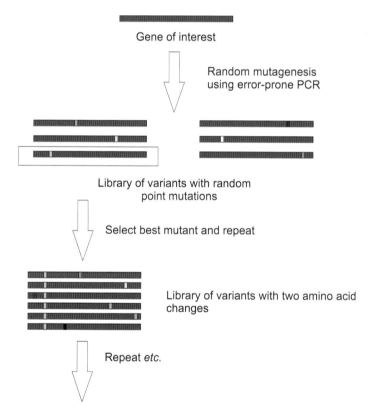

Figure 8.7 Generation of mutant enzyme libraries using random mutagenesis achieved through error-prone PCR

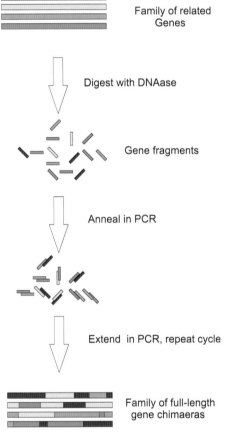

Family of related
Genes

Digest with DNAase

Gene fragments

Anneal in PCR

Extend in PCR, repeat cycle

Family of full-length
gene chimaeras

Figure 8.11   DNA shuffling

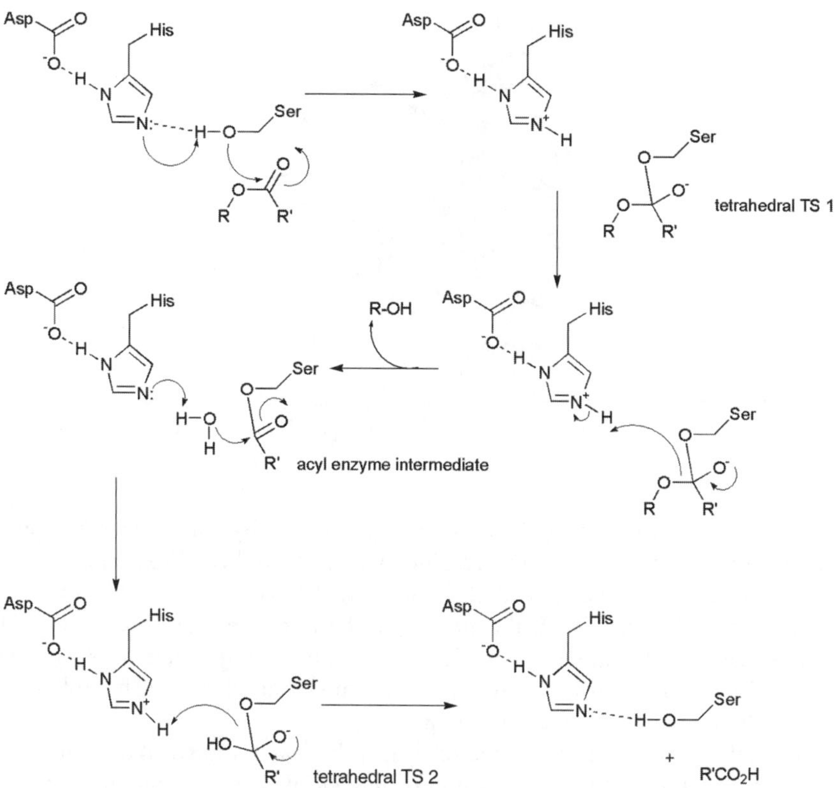

Figure 5.1 Mechanism of action of serine-triad dependent hydrolases

It is useful at this stage to consider some basic facts about the *mechanism* and *mode of action* of lipases, as these can help to inform synthetic strategies, and also explains the basis for their reversibility in low water systems.

The catalytic centre in the active site of the lipase is composed of a 'catalytic triad' – three amino acid residues (aspartate, histidine and serine) that act in concert to effect the nucleophilic attack of the hydroxyl side chain of a serine residue at the carbonyl group of the ester to be cleaved, essentially operating as a biological equivalent of hydroxide. The overall effect of this 'charge relay system' is to effect deprotonation of the serine such that its nucleophilicity is increased (Figure 5.1).

The serine forms a covalent bond yielding a tetrahedral oxyanion transition state, that loses the substrate alcohol residue on rearrangement to form a covalent

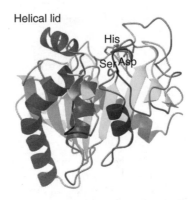

Figure 5.2   Structure of the lipase from *Thermomyces lanuginosa* (pdb code 1ein) showing the helical lid near the active site that opens and closes over the catalytic triad (Ser, His, Asp) shown

'acyl enzyme intermediate' (ACE). The ACE is attacked by water, which has been activated for attack by a histidine residue. A second tetrahedral oxyanion results that collapses to yield the free carboxylic acid and the free serine, liberated to effect another catalytic cycle. This is the mechanistic basis for the hydrolytic reaction of lipases that is exploited in the kinetic resolution of racemic esters. It is also the mechanism of simple protease and acylase enzymes that catalyse the hydrolysis of peptide bonds using serine as nucleophile.

The other famous characteristic of lipases that carries implications for their use as biocatalysts is the phenomenon of *interfacial activation*. Simply put, the three-dimensional structure of lipase enzymes has revealed that they possess a 'lid' (Figure 5.2) usually composed of an alpha helix, that sits over the active site in a closed conformation that does not allow access of the substrate to the catalytic residues.

Lipases in Nature act on hydrophobic substrates and the lid only opens when the enzyme comes into contact with that substrate at a critical concentration (the so-called 'critical micellar concentration' or CMC), and so the activity of lipases will change dramatically with the composition of the solution (see acylation reactions below). Hence, in the event of not identifying a lipase capable of hydrolysing the ester of interest, the addition of small amounts of organic co-solvents (below) might be worthwhile. Owing to the phenomenon of interfacial activation, it is not necessary for substrates to be soluble for lipase-catalysed hydrolytic reactions however, so one should proceed even if the substrate does not appear to be readily miscible with the buffer.

5.3 Hydrolytic Reactions using Lipases

5.3.1 General considerations

The general considerations for hydrolytic reactions are outlined below.

5.3.2 Enzymes

When considering a lipase for a hydrolytic reaction in buffered solution, the search will usually have been inspired by a relevant example from the literature. However, when considering a new substrate, it may be advisable to screen a small selection of commercial lipases for the reaction. Suppliers such as Fluka-SAFC and Immozyme will provide kits of ten or so lipases sold in powdered form and in sufficient quantity for assessing their suitability for the reaction of interest. Many of these enzymes are the most commonly used lipases, and indeed it is no coincidence that the same lipases recur in the organic synthesis literature for a wide range of substrate types. These include lipases from *Candida antarctica, Candida rugosa, Burkholderia* (formerly *Pseudomonas) cepacia* and *Rhizomucor miehei* from a range of suppliers listed in Chapter 2. Some of these will be freeze-dried powders of simple protein extracts from the bacterium or fungus and some will be immobilised *i.e.* attached covalently to a solid support such as Eupergit or other resin. Immobilisation can improve the activity of lipases and also raises the possibility of catalyst recovery and reuse through simple filtration.

5.3.3 Buffers, co-solvents and pH

As detailed in Chapter 3, there are a number of buffers that are suitable for use in lipase reactions, of which phosphate buffers are the most common. Water should work as a medium where control of pH is not such an issue, however, the generation of acidic products will of course depress the pH, leading in some cases to enzyme deactivation. Some of the 'biological' buffers (see Appendix 5), such as Tris-HCl, can be used for lipase reactions, but can in some cases inhibit enzymes. It is probably best to begin with a 0.05 M or 0.1 M phosphate buffer at neutral pH. Once the basics for a required transformation have been established, i.e. when the successful recognition of the substrate has been established and you are sure that the enzyme gives the desired enantiomer in sufficiently high optical purity, other parameters such as pH can then be investigated in a more detailed manner. Enzymes display pH optima (Chapter 1), usually (but not exclusively) with the best activities observed between physiological pHs of 5 and 9, so it may be that much-improved performance may be gained by adjusting this parameter. In the event of not identifying a lipase capable of hydrolysing the ester of interest, the addition of small amounts (1–10% v/v) of organic co-solvents, such as THF, acetonitrile or dioxane, acetone, isopropyl alcohol, or even nonpolar solvents, such as diethyl ether or TBME, might be worthwhile, in the light of the interfacial activation phenomenon described above. Most investigations of lipase-catalysed hydrolyses investigate the use of co-solvents as part of the study.

5.3.4 Temperature

Enzymes are said to work at ambient temperature, and indeed will often work at room temperature of 20–25 °C. However, both the rate of reaction, and indeed the enantioselectivity of a lipase (both magnitude and enantio-preference) can be affected by temperature, so once initial screens have been performed, it is wise to investigate the temperature on the resolution reaction.

Some specific examples of resolutions using commercial hydrolases follow. However, a good starting point for a trial for new reaction would be:

- Add 10 mg of ester substrate to 1 mL of 0.1 M phosphate buffer pH 7.0. Then add 10 mg of the commercial lipase powder. Incubate at room temperature or at 30 °C and sample by extracting 100 µL at 1, 2, 4, 8 h and overnight for analysis by TLC or GC.

5.3.5 Enantioselective hydrolysis of esters

The first example uses the lipase from *Candida rugosa* (CRL). CRL is known to exhibit good enantioselectivity towards a range of racemic esters, and serves as a good general example for resolution by hydrolysis. It can be purchased relatively inexpensively from SAFC or other chemical suppliers. As a simple example, we detail a method for the kinetic resolution by hydrolysis of the ethyl ester of (*R*, *S*)-2-phenylpropionic acid [2] (Figure 5.3) carried out in a buffered solution.

Apparatus and reagents

- 10 mL Round-bottom flask and magnetic stirrer
- *Candida rugosa* (formerly *Candida cylindracea*) lipase
- 0.1 M Tris-HCl buffer pH 7.1
- Phenylpropionic acid methyl ester
- Automated pipettes and tips

(NB *Candida rugosa* lipase can be purchased from Sigma-Aldrich, Immozyme, etc.)

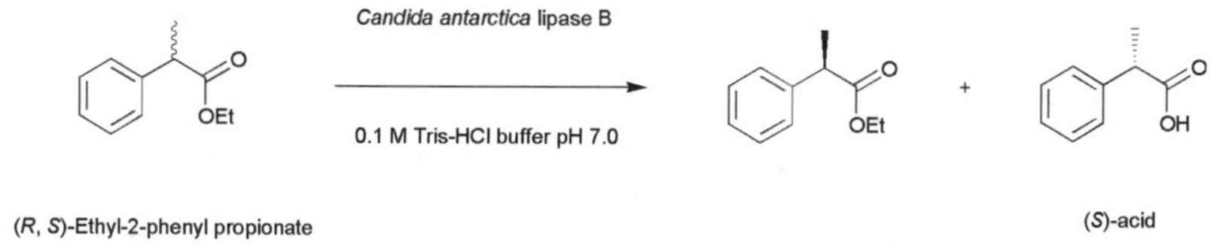

(*R, S*)-Ethyl-2-phenyl propionate   (*S*)-acid

Figure 5.3   Hydrolysis of the ethyl ester of (*R, S*)-2-phenylpropionic acid using lipase from *Candida rugosa*

1. Prepare a small amount of a 100 mM Tris buffer adjusted to pH 7.1 with HCl and transfer 4 mL to a 10 mL round-bottom flask.

2. Add 89 mg of racemic phenylpropionic acid methyl ester to the flask.

3. Weigh 17 mg of the powdered CRL into a glass vial, and add 1 mL of the Tris buffer and see the enzyme dissolve.

4. Transfer the enzyme solution to the reaction flask, stopper and incubate at 30 °C with stirring (shaking). Take 100 mL samples and extract them for analysis by GC or TLC and when the reaction has reached a plateau, the reaction can be stopped by the addition of a few drops of 0.1 M HCl, which will denature the enzyme. The reaction can then be worked up in the usual manner and the residual ethyl ester and product acid isolated.

The second example uses porcine pancreatic lipase (PPL) to catalyse the resolution of β-acetyloxymethyl-β-valerolactone [3] (Figure 5.4).

In this example, the extent of reaction is monitored by adding sodium hydroxide to maintain the pH of the reaction at 7.2. Thus, the reaction is quenched when 48% conversion (as estimated by this method) is achieved. This is of course most conveniently done in a pH stat, if such equipment is available.

Apparatus and reagents

- 25 mL Round-bottom flask and magnetic stirrer
- PPL
- 0.1 M Phosphate buffer pH 7.2
- β-Acetyloxymethyl-β-valerolactone
- 0.1 M Sodium hydroxide solution

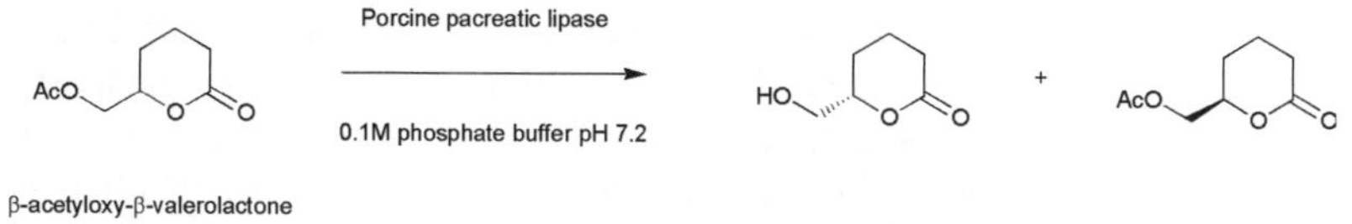

β-acetyloxy-β-valerolactone

Figure 5.4   Resolution of β-acetyloxymethyl-β-valerolactone by porcine pancreatic lipase

- Automated pipettes and tips
- Celite

(NB PPL can be purchased from Sigma-Aldrich, Biocatalytics, etc.)

1. Prepare a small amount of a 0.1 M phosphate buffer adjusted to pH 7.2. Transfer 12 mL to a 25 mL round-bottom flask.

2. Add 86 mg of racemic β-acetyloxymethyl-β-valerolactone to the flask.

3. Weigh 258 mg of the powdered PPL into a glass vial, and transfer this directly to the stirred substrate solution.

4. Incubate at 35 °C with stirring (shaking). Sample as for the first example.

5. When the reaction has reached approximately 50% conversion, quench the reaction with celite and ice and then filter the mixture through a celite pad, washing with ethyl acetate and water.

6. Wash the combined organic fractions with water, dry with magnesium sulfate, remove the solvent and subject the residue to flash column chromatography in the usual manner.

5.4 Using Lipases for Acylation Reactions

Although lipases have been shown to be valuable catalysts for hydrolytic reactions in aqueous media, they have become perhaps more noted for their ability to catalyse selective acylations or 'reverse hydrolytic reactions' in low water media, such as organic solvents, but also including both supercritical fluids such as carbon dioxide and ionic liquids. Such reactions most commonly take the form of transesterification reactions. In these processes, an *acyl donor*, that is, an ester or anhydride is included in the reaction medium, that is most often an organic solvent containing a small amount of water. The acyl donor is the auxiliary substrate, which is attacked by the lipase serine to give an acyl enzyme intermediate (as in Figure 5.1). In the absence of bulk water the other substrate, the chiral or prochiral alcohol (or amine) to be acylated acts as a nucleophile to break down the acyl enzyme intermediate, thus transferring the acyl group to this substrate to give a new ester product. Hence a whole range of lipase catalysed acylations have been described that use alcohol substrates and 'acyl donors' to perform the kinetic resolution of the alcohol substrate. The most widely used acyl donor is vinyl acetate. Vinyl acetate acts as the first substrate for the lipase – an acyl enzyme intermediate is formed and the vinyl alcohol portion tautomerises to acetaldehyde (Figure 5.5), which is volatile, and is removed from the reaction. Vinyl acetate may also serve as the solvent for the reaction in most cases, but is not suitable for all hydrolase enzymes, as aldehydes can react with enzyme lysine residues causing inactivation in some instances.

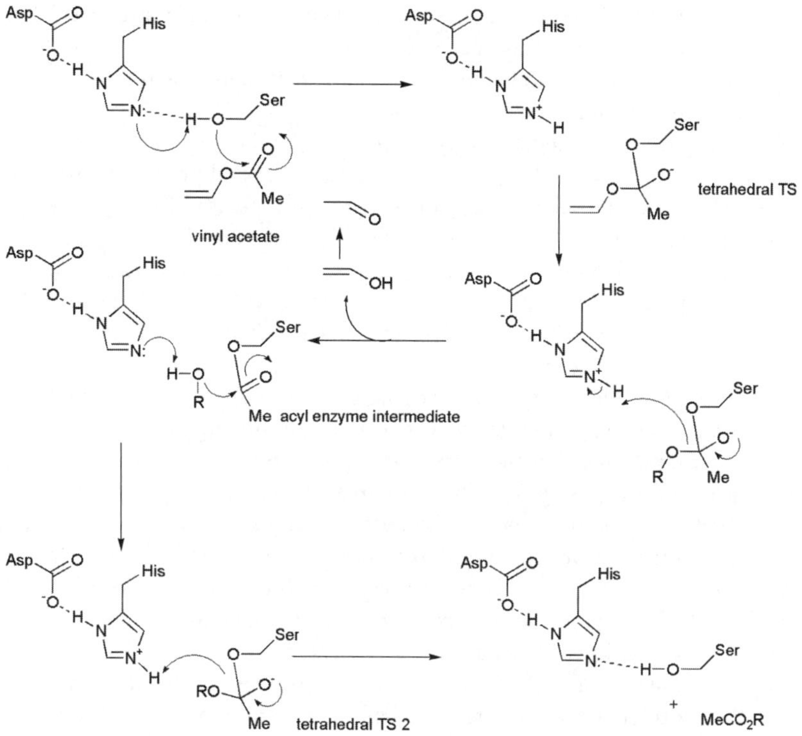

Figure 5.5 Mechanism of reverse action of serine-triad dependent hydrolases. Vinyl acetate is first attacked by serine, and the acyl enzyme intermediate is broken down by the attack of an alcohol R-OH

5.4.1 General considerations

In terms of enzymes, a similar range of commercial enzyme powders to that considered for hydrolytic applications can be screened in organic solvents for acylation reactions. A range of temperatures might also be considered for the reasons given earlier. However, for acylations and transesterifications, the choice of solvent, acyl donor and also the amount of water to be included in the reaction must also be considered.

5.4.2 Solvents and water content/water activity

A number of solvents can be considered for acylation reactions, and a number of physicochemical factors will need to be optimised for best performance [4]. These include primarily hydrophilic and hydrophobic solvents, but in many cases the acyl donor, particularly in the case of vinyl acetate, can be used as the solvent. Hydrophobic solvents such as toluene, diisopropyl ether and *tert*-butyldimethyl ether are commonly used and there is much evidence to suggest that the log P value (the octanol/water partition coefficient) of the solvent will have an influence on the reaction; solvents with higher log P values are usually best as they have been observed to give greater enantioselectivities and maintain the activity of enzymes for longer. Hence, n −heptane (log $P = 4.0$) and other medium-length chain alkanes, toluene (2.5), TBME (1.3) and DIPE (2.0) have all proved to be popular choices. However, the advantages gained through increased hydrophobicity of the solvent must be weighed against the solubility of substrate in any particular organic solvent, and in some cases a compromise must be found between substrate solubility and log P. The use of more polar solvents such as DMF, DMSO, THF or alcohols may result in inactivation of the enzyme, through stripping the water from the enzyme surface.

The role of water is of course crucial in enzyme structure, stability and mechanism. In any case, regardless of the hydrophobic organic solvent used, it has been demonstrated that the water content of the system is usually the more crucial factor in optimising the activity of an enzyme, but more particularly, the water activity a_w. The thermodynamic water activity of the system (where pure water has a value of 1.0 and a dry system, 0) has been shown to be the most reliable measure of the relevant water content in respect of the activity of biocatalysts [5]. It is certain that enzymes need a small amount of water to function if only to maintain structural integrity or as part of the catalytic mechanism – some water molecules will also be associated with the surface of the enzyme. It has also been shown that optimal activity of enzymes will be the same in different solvents at the same a_w. In a standard organic laboratory, and for the purposes of performing an initial screen of lipases for an acylation reaction, the measurement of water content, let alone water activity, is not straightforward in the absence of special apparatus., Consequently, when using commercial lyophilised enzymes directly from the bottle it may be useful, if positive results are not obtained, to repeat the reactions in the organic solvents of choice with the addition of small amounts of water (perhaps 0.5–5%) and incubation with vigorous mixing.

Finally, enzymes display pH optima in the aqueous phase as discussed above. The pH optimum will arise from all of the ionisable groups in the enzyme being in the right ionisation state to maintain optimum structural and functional integrity. If an enzyme is freeze-dried from a buffer that is at its pH optimum, it should retain the ionisation states of the relevant charged groups, and therefore its optimum activity when used in an organic solvent – a phenomenon termed 'pH memory'.

It can be seen that there is a large number of factors to be considered when undertaking an enzyme-catalysed reaction in an organic solvent therefore and indeed there has been a huge growth in research in this area in the last twenty-five years, resulting in a large amount of research literature. However, for the beginner, the use of lipase kits, in conjunction with trial and error tests and an informed use of the literature should yield at least an initial result from which optimisation through varying these factors can be attempted. If trying a reaction with a new substrate, the following simple protocol might be used:

- Add 10 mg of alcohol and 20 mg of vinyl acetate substrate to 1 mL of heptane (or other hydrophobic solvent). Then add 10 mg of the commercial lipase powder. Incubate with vigorous shaking/stirring at room temperature or at 30 °C and sample by extracting 100 μL at 1, 2, 4, 8 h and overnight for analysis by TLC or GC.

Some specific examples of acylation reactions in organic solvents follow.

5.4.3 Enantioselective acylation of 1-(2-pyridyl)-ethanol catalysed by lipase B from *Candida antarctica* (CAL-B)

This example uses the lipase B from *Candida antarctica*. CAL-B, as it is commonly known, is the most widely used lipase, having been shown to display excellent stability, a broad substrate tolerance and an excellent enantioselectivity in a wide range of reactions. CAL-B can be purchased from Novozymes (as Novozym 435) or Sigma-Aldrich and other suppliers and is most commonly used as a preparation of enzyme that has been immobilised on an inert solid matrix. The catalyst is extremely robust, and can be used straight from the bottle much as one would use a simple chemical reagent.

In this example, CAL-B is used as the catalyst with vinyl acetate as the acyl donor for the enantioselective acylation of a simple racemic secondary alcohol, β-1-(2-pyridyl)-ethanol (Figure 5.6) in dry diisopropyl ether as the solvent. This example is taken from reference [6].

Apparatus and reagents

- 50 mL Round-bottom flask and magnetic stirrer
- *Candida antarctica* lipase B
- β – 1(-2-Pyridyl)-ethanol)
- 4 Å Molecular sieves
- Vinyl acetate
- Automated pipettes and tips

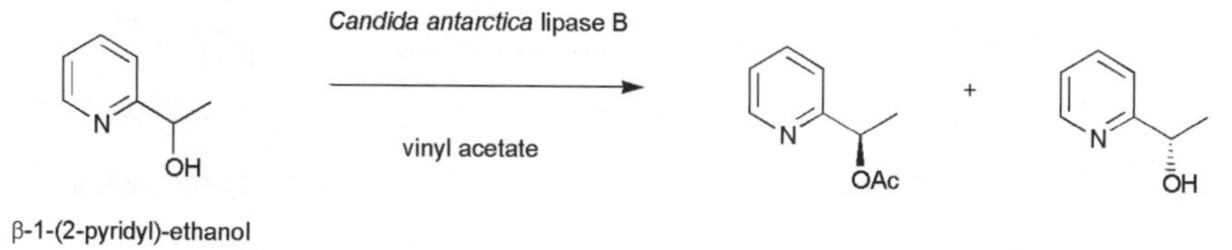

Figure 5.6 Enantioselective acylation of 1-(2-pyridyl)-ethanol

(NB *Candida antarctica* lipase B can be purchased from Novozymes, Sigma-Aldrich, etc.)

1. To 20 mL of dry diisopropyl ether in a 50 mL round-bottom flask, add 100 mg 4 Å molecular sieves.

2. To the reaction mixture, add 100 mg of the racemic 1-(2-pyridyl)-ethanol, 30 mg of the commercial lipase and 0.2 mL of vinyl acetate. Incubate at room temperature and sample at intervals for analysis by TLC or GC.

3. When the reaction has reached the desired level of conversion, filter through a plug of celite, evaporate the solvent, and purify the residual acetate and alcohol using flash chromatography.

5.4.4 Desymmetrisation of a meso-diol using an immobilised lipase from *Mucor miehei*

Kinetic resolutions are of course inherently disadvantaged by their limited maximum theoretical yield (50%). Hence, there is continuing interest in asymmetric quantitative transformations, many of which are carried out by enzymes. There are excellent examples of the use of dynamic kinetic resolution strategies [7], employing, for instance, a combination of chemo-catalysis (for racemisation of the starting material) and enzyme (for acylation). Another method is to employ a symmetrical starting material that, once modified by acylation, will yield 100% theoretical yield of an enantiomerically enriched product.

The following example uses Lipozyme®, an immobilised lipase from the fungus *Mucor miehei* as the catalyst, with vinyl acetate as the acyl donor and *tert*-butylmethyl ether as the solvent for the desymmetrisation of 1,2-dihydroxycyclohexane (Figure 5.7). This example is taken from reference [8].

Apparatus and reagents

- 100 mL Erlenmeyer flask and magnetic stirrer
- Lipozyme® (immobilised lipase from *Mucor miehei*)

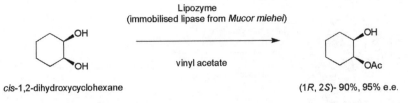

Figure 5.7   Desymmetrisation of 1,2-dihydroxycyclohexane by a lipase

- 1,2-Dihydroxycyclohexane
- Vinyl acetate
- Celite
- Automated pipettes and tips

(NB Lipozyme® can be purchased from Novozymes, Sigma-Aldrich, etc.)

1. Place 18 mL of *tert*-butylmethyl ether in a clean 100 mL Erlenmeyer flask. To this, add the diol (100 mg), vinyl acetate (0.36 mL) and Lipozyme® (760 mg).

2. Incubate with shaking at 300 r.p.m. at 45 °C and sample at intervals for analysis by GC or TLC.

3. When the reaction is complete, filter through a celite plug, evaporate the solvent and purify the product using flash chromatography.

5.4.5 Enantioselective lipase-catalysed acylation of amines in organic solvents

Given that the mechanism of lipases may be neatly intercepted in order to form esters via the attack of alcoholic substrates at the acyl enzyme intermediate (AEI), it is no surprise that the same AEI may be attached by other nucleophiles, notably chiral amines. This results in the resolution of a chiral amine by the formation of one enantiomer of product acyl amine and a residual amine enantiomer. Amines are of course much more nucleophilic than alcohols, and as such the choice of acyl donor is important as the amines will often react with the esters used as acyl donors spontaneously. In addition, the amides formed as products of enzymatic acylation reactions are not so readily cleaved to give the desired amines.

A number of strategies have been suggested to ameliorate the inherent problems associated with enzymatic amine acylation in organic solvents, including the use of specialised amine donors such as benzylisoprenyl carbonate or ethyl methoxyacetate and the ingenious use of penicillin acylase, coupled with (R)-phenylglycine amide as acyl donor in aqueous systems (so-called 'easy on, easy off' systems [9]).

5.4.6 Enantioselective acylation of 1-phenylpropan-2-amine using CAL-B

As a first example, we have chosen the acylation of amphetamine or 1-phenylpropan-2-amine using CAL-B and ethyl acetate in this case as the acyl donor (Figure 5.8). This example is adapted from reference [10].

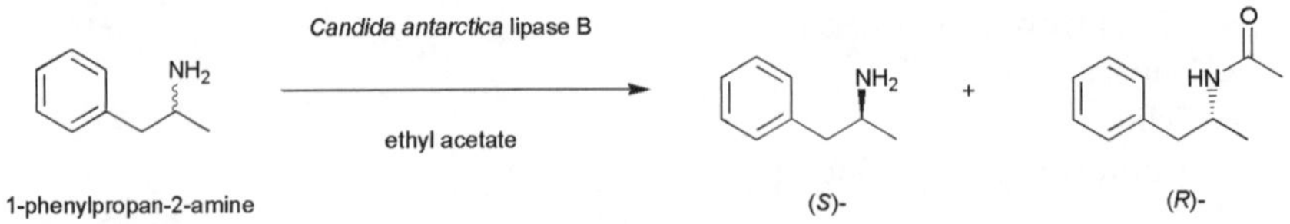

Figure 5.8   The acylation of amphetamine or 1-phenylpropan-2-amine using CAL-B

Apparatus and reagents

- 100 mL Round-bottom flask, nitrogen atmosphere and magnetic stirrer
- CAL-B (Novozym 435)
- Ethyl acetate
- Cotton wool plug
- Sodium hydroxide
- Automated pipettes and tips

1. Place 15 mL of ethyl acetate into a 100 mL round-bottom flask, under a nitrogen atmosphere.

2. Add 3 mmol of the racemic amine and 300 mg of CAL-B as provided by the supplier. Incubate at 28 °C and sample at intervals for analysis by TLC or GC.

3. When the reaction has reached a plateau, filter off the enzyme using a cotton wool plug and add 20 mL of 3 N HCl. Extract with dichloromethane. Treat the aqueous phase with solid sodium hydroxide until the pH has become basic, and extract with 4 × 20 mL dichloromethane. Evaporation of both organic phases should yield the amide, as a white solid, and the amine, respectively.

5.4.7 Resolution of phenylethylamine using CAL-B

This example, taken from reference [11], uses CAL-B again but this time in hexane with benzylisoprenylcarbonate as the acyl donor (Figure 5.9).

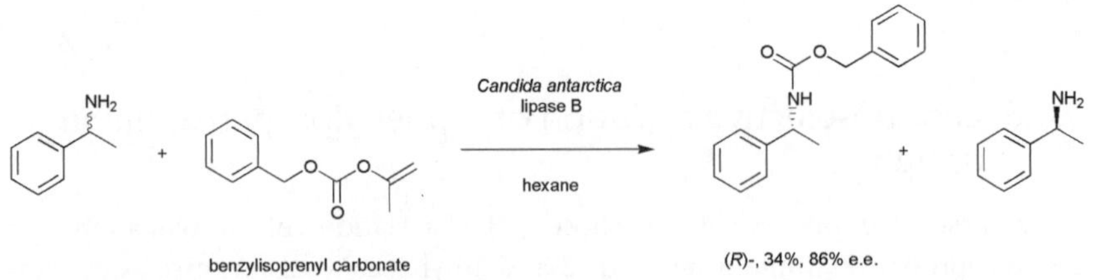

Figure 5.9   Resolution of phenylethylamine using CAL-B and benzylisoprenyl carbonate as the acyl donor

Apparatus and reagents

- 50 mL Round-bottom flask
- CAL-B (Novozym 435)
- Benzylisoprenylcarbonate
- 4 Å Molecular sieves
- Hexane
- Celite
- Automated pipettes and tips

1. Place 10 mL of ethyl acetate into a 50 mL round-bottom flask.

2. Add 168 mg of the racemic α-methylbenzylamine, 197 mg of benzylisoprenyl-carbonate, 920 mg of molecular sieves and 240 mg of CAL-B. Incubate at room temperature and sample at intervals for analysis by TLC or GC. In reference [11], the reaction reached 34% completion after 60 h incubation.

3. After 60 h, filter the reaction mixture through a celite plug, wash with diethyl ether, combine, dry and evaporate the organic residue and purify the acylated amine by flash chromatography. The acylated amine enantiomer in this example displayed an e.e. of 86%.

5.5 Other Hydrolases

The hydrolysis and formation of ester/amide bonds by lipases are just two examples of the types of hydrolase-based process available using commercially available enzymes. *Esterases*, whose mechanism is formally related to lipase, but which do not display interfacial activation, have also been used effectively for enzymatic resolutions of esters in aqueous systems [1]. *Peptidases* and *proteases*, notably subtilisin and thermolysin (a metalloprotease), have been used for the hydrolysis of esters and amides and are used industrially for the formation of peptides [1]. Most of these enzymes are serine-triad type hydrolases whose chemical mechanism is essentially similar to that of lipases. There are other E.C. class 3 enzymes of different mechanism and substrate specificity that are commercially available that are also very useful in biotransformations however.

5.5.1 Epoxide hydrolysis by epoxide hydrolase

One of the more recent interesting developments has been in enantioselective epoxide hydrolysis as illustrated by the hydrolysis of styrene oxide by whole cells of the filamentous fungus *Aspergillus niger* desribed in Chapter 4. The epoxide hydrolase from this organism (AnEH) is commercially available from Sigma-Aldrich, and it is therefore even easier to apply this catalyst to epoxide resolutions in the

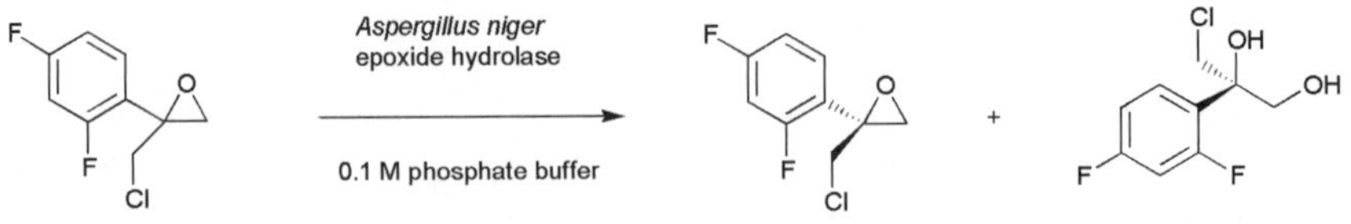

(S)-, 38%, 98% e.e.   (R)-, 41%, 98% e.e.

Figure 5.10   Enantioselective hydrolysis of an epoxide by an epoxide hydrolase

synthetic organic laboratory. The following example, adapted from reference [12], details the use of commercially available AnEH for the analytical scale resolution of a *bis*-fluorinated styrene oxide derivative (Figure 5.10), but could be applied to other systems as AnEH is known to display a fairly wide substrate tolerance.

5.5.2 Hydrolysis of 1-chloro-2-(2,4-difluorophenyl)-2,3-epoxypropane using the commercially available epoxide hydrolase from *Aspergillus niger*

Apparatus and reagents

- 50 mL Round-bottom flask and magnetic stirrer
- *Aspergillus niger* epoxide hydrolase (1.5 U mg^{-1})
- 1-Chloro-2-(2,4-difluorophenyl)-2,3-epoxypropane (or styrene oxide)
- Dimethylsulfoxide
- 0.1 M Phosphate buffer pH 7.0
- Acetonitrile
- Automated pipettes and tips

(NB *Aspergillus niger* epoxide hydrolase can be purchased from Sigma-Aldrich.)

1. Place 18 mL of 0.1 M phosphate buffer pH 7.0 into a 50 mL round-bottom flask.

2. Dissolve 8 mg of the epoxide substrate in 2 mL of DMSO and add this to the stirred buffer solution.

3. Dissolve 2.3 mg of the epoxide hydrolase powder in 400 μL of water and add this to the stirred reaction mixture. Incubate the reaction at room temperature with stirring and sample at intervals for analysis by TLC or GC.

4. When the conversion has reached approximately 50%, add 600 μL of acetonitrile to quench the reaction.

Whilst only at an analytical scale, this reaction can be scaled up to provide preparative amounts of enantio-enriched epoxide and diol. AnEH is known to tolerate substrate concentrations as high as 2.5 M when reactions have been optimised and performed in a suitable two-phase reaction, so even the identification of a useful hydrolytic process could lay the foundations for a scaleable resolution.

With analogy to lipases and other serine-nucleophile dependent hydrolases, it would of course be desirable to have an enzyme which was capable of employing nucleophiles other than water for the ring opening of epoxides, perhaps in the organic phase. Whilst there are enzymes which are capable of employing alternate nucleophiles for epoxide ring-opening, notably haloalkane dehalogenases, epoxide hydrolases cannot be exploited in this way as their mechanism is not dependent on a serine nucleophile attacking the electrophilc carbon of the epoxide, but rather on a carboxylate side chain of an aspartate to create an *alkyl enzyme intermediate*. The stereochemistry of epoxide-catalysed resolutions is complicated by both the *regioselectivity* and enantioselectivity of the enzyme, as some enzymes will attack, in the example of styrene oxide, the benzylic carbon or the terminal carbon, in the first instance giving rise to a retention of configuration in the product diol, and in the second, *inversion*. A mixture of enzymes displaying different regioselectivities may thus be exploited to catalyse enantiodivergent biotransformations, wherein each enantiomer of a racemic epoxide is converted to the same enantiomer of the diol product.

5.5.3 Hydrolysis of nitriles by commercially available nitrilases

We have already described the use of biocatalysts for nitrile hydrolysis in Chapter 4 and seen that there are two pathways for nitrile hydrolysis in biological systems (Figure 4.14). The use of isolated nitrilases should circumvent this problem by giving only hydrolysis of nitriles directly to their carboxylic acids. Nitrilases in cell-free form have been available for a number of years. We give an example of a study in which six different commercially available nitrilases were assayed for their ability to hydrolyse hydroxynitriles as precursor substrates for lactone products (Figure 5.11). This method is adapted from reference [13] but could be applied to the screening of other nitrile substrates.

Apparatus and reagents

- 25 mL Round-bottom flask and magnetic stirrer or vial with shaker
- Nitrilase NIT 1003

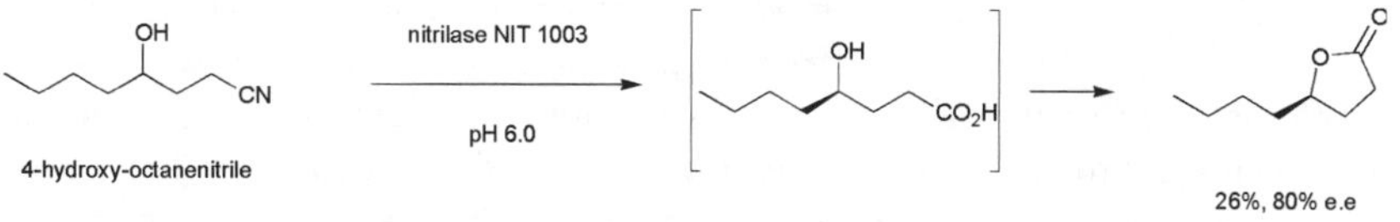

Figure 5.11   Hydrolysis of a hydroxynitrile substrate by a commercial nitrilase

- 4-Hydroxyoctanenitrile
- 0.1 M Phosphate buffer pH 6.0
- 1 M Hydrochloric acid
- Automated pipettes and tips

[NB Nitrilase NIT 1003 can be purchased from Biocatalytics (now Codexis).]

1. Place 5 mL of 0.1 M phosphate buffer pH 7.0 into a 25 mL round-bottom flask or capped vial.

2. To the reaction vessel, add 34 mg of 4-hydroxyoctanenitrile substrate and 7.3 mg of nitrilase NIT 1003.

3. Incubate the reaction at 30 °C with stirring/shaking for 1.5 h. Add a small amount of 1 M HCl to make the solution slightly acidic. Filter the reaction through a cotton wool plug or celite and wash with ethyl acetate. Work up the combined organic phases in the usual manner and purify the products using flash column chromatography.

5.5.4 Reactions catalysed by glycosidases or glycosyl hydrolases

The hydrolysis of glycosidic bonds, that is, ether linkages between monosaccharide residues of saccharides and aglycones is one of the most fundamentally important reactions in biological systems. The enzymes that catalyse the hydrolytic cleavage of these bonds are called glycosidases or glycosyl hydrolases (E.C. 3.2.X.X) and are an extremely widespread and diverse collection of enzymes. In terms of biocatalytic applications, there are two possible reactions of interest: the first is the selective cleavage of sugar residues from glycoconjugates; the second is the formation of glycosidic bonds in the reverse manner, either by straight formation of a glycosidic bond between two sugar residues, or by *transglycosylation*, in which a sugar residue is swapped from a donor oligosaccharide to an acceptor residue. In practice, the first, hydrolytic transformation is not used much in preparative biocatalysis, but the second has achieved a large amount of attention, because of the importance of synthesising the glycoconjugates of drug molecules and peptides/proteins. In the synthetic direction nature employs mostly the cofactor-dependent glycosyltransferase enzymes for the synthesis of glycosides, but it is possible to engineer glycosidase-catalysed reactions for synthesis, by engineering the medium conditions.

The ability to do this lies again with the mechanism of the enzymes. GHases are at one level divided into groups depending on their ability to catalyse the formation/cleavage of glycosidic bonds with either *retention* or *inversion* of configuration at the anomeric centre. If we consider the mechanism of a retaining glycosidase (Figure 5.12), we see that the retention of stereochemistry at the anomeric position results from a *double displacement* mechanism, in which the

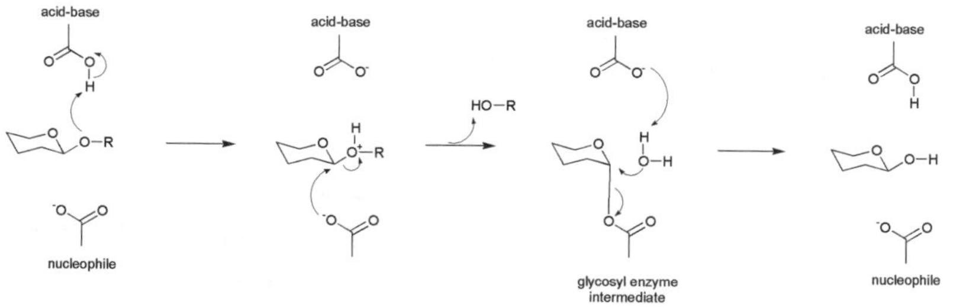

Figure 5.12 Mechanism of a retaining β-glycosidase, showing formation of glycosyl enzyme intermediate and double displacement mechanism

glycosidic oxygen is protonated, and a nucleophilic carboxylate residue attacks the anomeric position to form a glycosyl enzyme intermediate (GEI).

Base-catalysed activation of a water molecule for nucleophilic attack at the GEI then ensues, to give an anomeric – OH of the same configuration as the starting glycosidic link. Of course, it is theoretically possible to introduce a sugar residue, instead of water in the second step, giving a glycosidic link between the donor and a new sugar. However, there are obvious factors that militate against this reaction being feasible, the most obvious of which is the presence of water. However, by excluding water from the glycosidase system, perhaps through the use of organic solvents, or by increasing concentration of the donor sugar, it may be possible to favour the formation of the glycosidic link.

5.5.5 β-Galactosidase catalysed formation of galactosides of 2-hydroxybenzyl alcohol

In this example, the commercially available β-galactosidase from *Kluyveromyces lactis* is used to catalyse the formation of galactosides of 2-hydroxybenzyl alcohol, using high concentrations of the donor sugar lactose to optimise the chances of glycosidic bond formation. It can be seen that, in addition to the galactoside products required, both products of hydrolysis and unwanted regioisomers are also generated (Figure 5.13). This example is adapted from reference [14].

Apparatus and reagents

- 100 mL Round-bottom flask and magnetic stirrer
- β-Galactosidase from *Kluyveromyces lactis* (Lactozym 3000 L HP-G)
- 2-Hydroxybenzyl alcohol
- Lactose
- 0.05 M Phosphate buffer pH 6.5
- Magnesium chloride
- Automated pipettes and tips

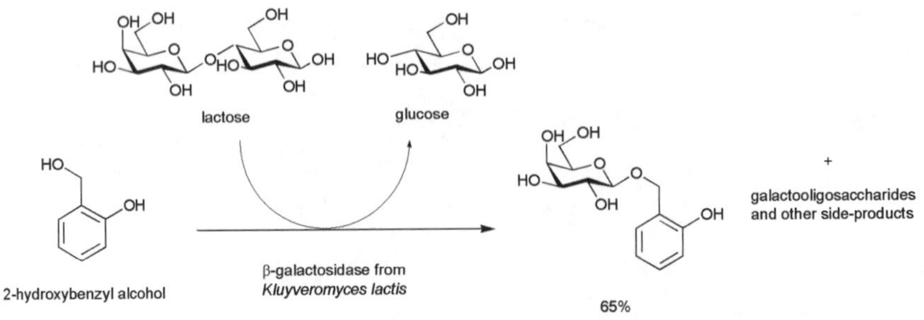

Figure 5.13 Formation of galactosides of 2-hydroxybenzyl alcohol using a bacterial β-galactosidase

[NB β-Galactosidase from *Kluyveromyces lactis* (Lactozym 3000 L HP-G) can be purchased from Novozymes.]

1. Place 50 mL of 0.05 M phosphate buffer pH 6.5 into a 100 mL round-bottom flask.

2. To the reaction vessel, add 1.65 g of 2-hydroxybenzyl alcohol (a final concentration of 250 mM), 17.1 g Lactose (1 M), 48 mg magnesium chloride (10 mM) and 500 U of β-galactosidase. In reference [14], the commercial enzyme has a specific activity (see Chapter 6) of 3530 U mL^{-1}, so this would correspond to 140 µL of the commercial enzyme solution.

3. Incubate the reaction at 40 °C with stirring. The reaction is continued until all the lactose is depleted (about 20 h) and monitored by HPLC.

4. After the reaction is complete, quench by heating to 100 °C for 5 min.

The mixture of products of the galactosylation reaction in this paper shows that the inherent lack of absolute selectivity, coupled with the necessary excess of lactose in the reaction medium, give rise both to di-galactosylated products (in 5% yield) and also the synthesis of large amounts of galacto-oligosaccharides (up to 192 g L^{-1} in the example). The search for a method of more selective glycosylation has therefore led to both the use of glycosyltransferase enzymes (below), which have their own disadvantages, and the engineering of the glycosidase enzymes themselves. 'Glycosynthase' enzymes [15] have been developed which lack the catalytic nucleophilic residue that forms the GEI in the retaining glycosidase mechanism (glutamate mutated to alanine). When used in conjunction with a donor sugar substrate with a good leaving group in the anomeric position (such as fluoride), it is possible to obtain *irreversible* glycosylations, still albeit with the possibility of unwanted extended oligomers. However, glycosynthases are not yet commercially available, but they can be created by quite simple site-directed mutagenesis experiments (see Chapter 8).

5.6 Commercially Available Coenzyme-Dependent Enzymes

As discussed in Chapter 1, many enzymes, particularly those that catalyse oxidation–reduction reactions, require the participation of non-proteinaceous coenzymes in order to effect catalysis. Many of these enzymes and coenzymes/cofactors are commercially available and some *in vitro* biotransformations will thus be dependent on the addition of cofactors/coenzymes in cases where these are not covalently bound to the enzyme. In this section, we examine nicotinamide-coenzyme based enzymes [NAD(P)/H], heme-dependent haloperoxidases and glycosyl transferases that are dependent on activated nucleotide sugars. It will be seen that whilst these enzymes have great potential, they are neither overall as robust as hydrolases nor as amenable to use in organic solvents, although some will be tolerant to the addition of co-solvents to aid substrate solubility.

5.6.1 Nicotinamide coenzyme-dependent biotransformations and coenzyme recycling

The coenzymes that require most consideration in terms of preparative biocatalysis are the nictonamide coenzymes NAD(H) and NADP(H). The structures of these cofactors were described in Chapter 1. They are usually either involved directly in catalysis, delivering hydride to a substrate (alcohol dehydrogenases) or may also be involved in transfer of hydride to a second coenzyme such as a flavin (some monooxygenases), which is then itself the active catalytic agent.

In either case, the coenzyme itself becomes oxidised, and requires that it be recycled in order that it may catalyse another reductive turnover of the ketone substrate, in the case of ketoreductases. When using whole cells to perform reduction reactions (see Chapter 4), there is no need to consider the necessary recycling of nicotinamide coenzymes, as this will be done in the cell – many enzymes require oxidised nicotinamide coenzymes, therefore there is a pool of such coenzymes that is continuously cycled through an oxidation–reduction sequence. However, in *in vitro* transformation systems, methods must be developed for recycling coenzymes that are usually dependent on both an auxiliary enzyme system, sometimes coupled with an auxiliary substrate. The major consideration here is cost: NADH is available from chemical suppliers at a price of approximately £70 g^{-1}; NADPH is approximately ten times more expensive, so the cost implications of using a stoichiometric equivalent of coenzyme are simply unworkable.

There are a few techniques for recycling oxidised nicotinamide cofactors [16]. Each is dependent on identifying an auxiliary enzyme-catalysed reaction that

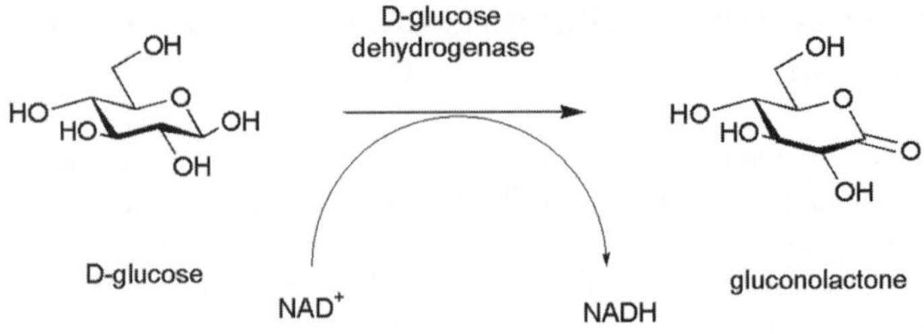

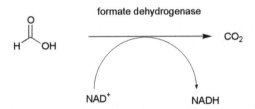

Figure 5.14   NAD-dependent transformation of glucose to gluconolactone catalysed by D-glucose dehydrogenase

Figure 5.15   Formate dehydrogenase (FDH) for the NAD-dependent conversion of sodium formate to carbon dioxide

uses oxidised nicotinamide coenzyme and a substrate–product system that will not interfere with the primary reaction of interest – for example, one that will result in an auxiliary product that will not unnecessarily complicate the process of product recovery. Unfortunately, many cofactor-dependent enzymes, including some that are used for cofactor recycling, cannot use both NAD(H) *and* NADP(H) as cofactor, hence there are separate systems for recycling each. For NAD(H) one of the most popular methods is the use of glucose dehydrogenase (GDH) for the NAD-dependent transformation of glucose to gluconolactone (Figure 5.14).

The co-substrate and auxiliary enzyme are relatively inexpensive, and the product gluconolactone, is water soluble and therefore not extracted in reaction work ups. The second example employs the enzyme formate dehydrogenase (FDH) for the NAD-dependent conversion of sodium formate to carbon dioxide, which diffuses freely from the reaction medium, therefore removing co-product considerations altogether (Figure 5.15).

This makes the FDH/formate system highly desirable, but whilst NAD(P)-dependent FDHs have been described [17], their use in industry is not widespread. A more usual example for cofactor recycling uses the alcohol dehydrogenase from *Thermoanaerobium brockii* for the NADP-dependent oxidation of isopropanol to acetone. Isopropanol is cheap of course, and the acetone evolved as product is volatile, and may well just evaporate from the reaction mixture (Figure 5.16).

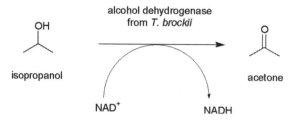

Figure 5.16 Alcohol dehydrogenase from *Thermoanaerobium brockii* for the NADP-dependent oxidation of isopropanol to acetone

The other common technique is the NADP-dependent transformation of glucose-6-phosphate to 6-phosphogluconolactone using glucose-6-phosphate dehydrogenase (G6PDH). Recently, the enzyme phosphite dehydrogenase has become an attractive alternative for NAD(P)H recycling also, and indeed, will accept both nicotinamide coenzymes.

Each system has its merits, but the formate and glucose (or G6P) dehydrogenase systems have become the most widespread in the literature and also popular in industry as activities are high and the co-substrates and auxiliary enzyme are relatively inexpensive. It is important that in an industrial setting, the recycling of cofactor is no longer seen as a barrier to an efficient and economical process. We will next provide some examples of the use of commercially available nicotinamide-coenzyme dependent enzymes, which include details on the formulation of reactions that include coenzyme recycling systems.

5.6.2 Alcohol dehydrogenases (ADHs) or Ketoreductases

There has been a resurgent interest in the use of ketoreductases for the reduction of carbonyl groups to produce chiral secondary alcohols in high enantiomeric excess. As described in Chapter 1, ADHs act essentially as biochemical equivalents of simple hydride reagents in organic chemistry, such as sodium borohydride. The hydride for reduction is delivered in a stereospecific manner from the dihydropyridine moiety of the nicotinamide coenzymes NAD(P)H. An isolated enzyme reaction using ADHs will therefore require: substrate; enzyme; co-substrate; coenzyme; and NAD(P)H. The following example uses a commercially available ADH from *Rhodococcus erythropolis* to reduce an acetophenone derivative (Figure 5.17), and incorporates the FDH/sodium formate coenzyme regeneration system. All of the enzymes used were commercial preparations from Biocatalytics (now Codexis). This example is adapted from reference [18].

Whilst in the example, the process was conducted in a somewhat sophisticated reactor, with automated control of pH, it does provide useful information on the amounts of production enzyme, auxiliary enzyme and substrate and

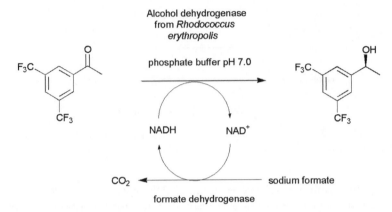

Figure 5.17   Alcohol dehydrogenase from *Rhodococcus erythropolis* catalysed reduction of an acetophenone derivative

co-substrate that might be used in a preparative scale coenzyme dependent reduction. The method uses a molar ratio of NAD (which is cheaper than the reduced coenzyme):substrate of approximately 1:70 for the reaction, illustrating the power of the recycling technique.

Apparatus and reagents

- 100 mL round-bottom flask
- 0.05 M Phosphate buffer pH 7.2
- NAD
- Sodium formate
- Alcohol dehydrogenase from *Rhodococcus erythropolis* (ADH RE)
- Formate dehydrogenase
- 3,5-Bistrifluoromethylacetophenone

[NB ADH RE can be purchased from Biocatalytics (now Codexis).]

1. To 40 mL of 0.05 M phosphate buffer pH 7.2 in a 100 mL round-bottom flask add: NAD (160 mg); sodium formate (0.6 g); ADH RE (120 units i.e. 3.4 mg at 35 U mg^{-1}); FDH (115 U i.e. 24.5 mg at 4.7 U mg^{-1}). Add the neat ketone (4.0 g) and stir the reaction at 30 °C, keeping the pH at a constant 7.0 by the careful addition of 2 M sulfuric acid.

2. Monitor the reaction by GC or TLC. When the reaction has reached completion, extract the mixture with 2 × 20 mL of hot heptane, followed by washing with water. After concentration, the product alcohol can be purified by crystallisation.

It was notable that the same reduction could be carried out using the GDH system of coenzyme recycling. In this case, the sodium formate and formate dehydrogenase were replaced with glucose and GDH, again available commercially.

5.6.3 Enzymatic reduction of a racemic aldehyde with dynamic kinetic resolution

This example employs the commercially available horse liver alcohol dehydrogenase (HLADH) as both the production enzyme and the coenzyme recycling catalyst. Racemic ibuprofenal is reduced selectively to the primary alcohol by the enzyme, but the residual (*R*)-enantiomer racemises in buffered solution, thus allowing more of the favoured (*S*)-enantiomer to be transformed (Figure 5.18), giving ultimately quantitative levels of conversion. This example is adapted from reference [19].

Apparatus and reagents

- 100 mL round-bottom flask
- 0.1 M Phosphate buffer pH 7.5
- NADH
- Ethanol
- Horse liver alcohol dehydrogenase (HLADH)
- Ibuprofenal
- Acetonitrile

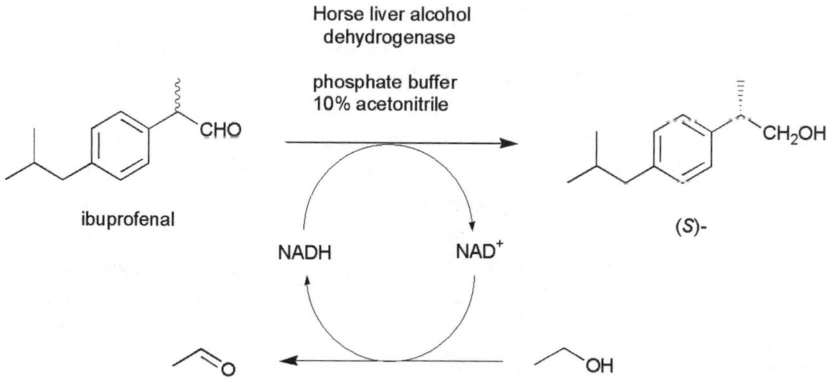

Figure 5.18 Dynamic kinetic resolution of ibuprofenal using ADHs

(NB HLADH can be purchased from Sigma-Aldrich.)

1. To 24 mL of a mixture of 0.1 M phosphate buffer pH 7.5 and acetonitrile (10% v/v) in a 100 mL round-bottom flask add: 1-(4-*iso*-butylphenyl)propanal (ibuprofenal) 2.8 mg (to a final concentration of 0.5 mM); NADH 0.17 mg (to a final concentration of 0.01 mM); ethanol 0.55 g (to a final concentration of 0.5 M); and 0.24 mg HLADH.

2. Monitor the reaction by GC or TLC. When the reaction has reached completion, extract the mixture with ethyl acetate, followed by washing with water. After concentration, the product alcohol can be purified by flash column chromatography.

Nicotinamide cofactor recycling systems are readily applicable to many types of oxidoreductases that have not been mentioned here, such as cytochromes P450 for hydroxylations, monooxygenases for the Baeyer–Villiger reaction, amino acid dehydrogenases and other enzymes, but there are few examples in the literature of the application of commercially sourced enzymes of these types in conjunction with recycling. Rather, they have mostly exploited home-grown biocatalysts of the type described in Chapter 6; we will return to the topic of coenzyme recycling for preparative-scale reactions there.

5.6.4 Haloperoxidases

Haloperoxidases, although not nicotinamide coenzyme dependent, are a type of oxidase enzyme that in Nature are usually involved in the halogenation (chlorination and bromination) of organic compounds, often in the marine environment [20]. Halogenation of the substrate is most commonly accomplished by first, the enzyme-catalysed formation of hypohalous acid from hydrogen peroxide and halide ion. Haloperoxidases are usually (but not exclusively) metal-dependent and if iron-dependent, that iron may be bound within a heme cofactor. The most frequently used haloperoxidase in the biotransformations literature is undoubtedly the chloroperoxidase (CPO) from the fungus *Caldariomyces fumago*, which is a heme-dependent haloperoxidase that is commercially available. In common with cytochromes P450, haloperoxidases are known to catalyse the oxidation of double bonds, to give epoxides and also heteroatoms such as sulfur, to give sulfoxides. The enzymes are sensitive to high concentrations of hydrogen peroxide and hence this must be added in aliquots during the course of the oxidation reaction. Alternatively, as seen below, *tert*-butyl hydroperoxide can be used for improved performance.

The first example uses CPO in conjunction with *tert*-butyl hydroperoxide for the asymmetric epoxidation of *p*-chlorostyrene (Figure 5.19). This example is derived from reference [21]. These reactions were performed on a small scale

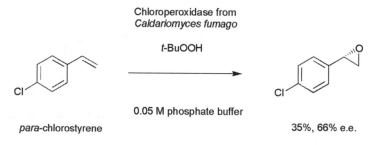

Chloroperoxidase from
Caldariomyces fumago

t-BuOOH

0.05 M phosphate buffer

para-chlorostyrene

35%, 66% e.e.

Figure 5.19   Asymmetric epoxidation of *para*-chlorostyrene using chloroperoxidase

in order to generate samples for chiral GC analysis in order to investigate the enantioselectivity of the enzyme.

Apparatus and reagents

- *p*-Chlorostyrene
- 0.05 M Potassium phosphate buffer pH 6
- Chloroperoxidase (CPO)
- *tert*-Butyl hydroperoxide
- Acetone

(NB CPO can be purchased from Sigma-Aldrich.)

1. Add 1 mg of *p* – chlorostyrene to 1 mL 0.05 M potassium phosphate buffer pH 6 containing *tert*-butyl hydroperoxide (15 µmol) and CPO from *C. fumago* (29 units).

2. Shake the mixture at room temperature and take samples for analysis by GC.

In addition to the formation of the (*R*)-epoxide (in 66% e.e.), some arylacetoaldehyde and diol products are also produced.

The second example describes the use of CPO in the asymmetric oxidation of 1,3-cyclohexadiene (Figure 5.20). This example is derived from reference [22].

Apparatus and reagents

- 1,3-Cyclohexadiene
- 0.1 M Sodium citrate buffer pH 5
- Chloroperoxidase (CPO)
- *tert*-Butyl hydroperoxide
- Diethyl ether

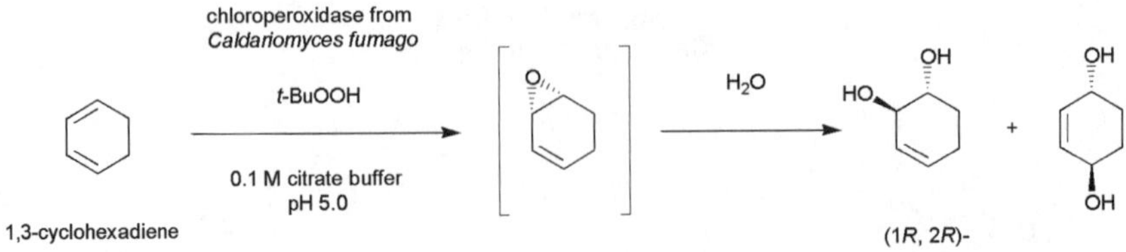

Figure 5.20   Asymmetric oxidation of 1,3-cyclohexadiene catalysed by chloroperoxidase

1. Add 300 mg of 1, 3-cyclohexadiene to 18 mL 0.1 M sodium citrate buffer pH 5.

2. Stir vigorously at room temperature for 5 min.

3. Add 1850 units of CPO to the mixture and 4.2 mmol of *tert*-butyl hydroperoxide. Stir for 24 h, then add an additional 4.2 mmol of the peroxide.

4. Extract the mixture with diethyl ether and purify the diol products by flash column chromatography.

5.6.5 Glycosyltransferases (GTases) and the use of activated sugar donors

In Nature, the formation of glycosidic bonds is carried out by GTase enzymes (E.C. 2.4.X.X). These enzymes transfer a sugar residue from an *activated sugar donor*, usually a sugar nucleotide or sugar phosphate, to an acceptor sugar (Figure 1.14). These enzymes have been used widely in the literature for the synthesis of oligosaccharides, but despite advantages of both *regio-* and *stereo*-selectivity in glycosidic bond formation, are problematic owing to the expense of the nucleotide sugars used as donors. They also often prove to be insoluble and consequently difficult to isolate (see Chapter 6). However, some GTases, notable those from bovine sources, are commercially available and have been used for a great many applications in the biocatalysis literature [23]. When it comes to substrates, GTases may not only exhibit preference for both the sugar moiety of the donor (e.g. glucose, galactose, mannose), but also the nucleotide (dUTP, dTTP, dGTP) and of course the *aglycone* to which the donated sugar residue is being transferred. If cofactor recycling is not being used (see below), in practise the glycosylation reactions are carried out quite simply on a milligram scale.

In a first example, the commercially available bovine β-1,4-galactosyltransferase is used for the galactosylation of an *N*-acetylglucosamine (GlcNAc) derivative (Figure 5.21). Manganese (in the form of the chloride) is added as a cofactor for the enzyme. This example is derived from reference [24].

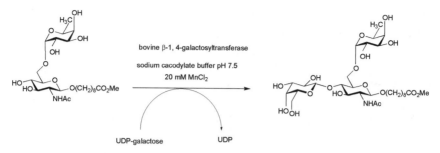

Figure 5.21   Galactosylation of *N*-acetylglucosamine (GlcNAc) derivatives using a galactosyltransferase

Apparatus and reagents

- 1.5 mL capped plastic tube
- 0.1 M Sodium cacodylate buffer pH 7.5
- UDP-galactose (30 µmol)
- Manganese (II) chloride (20 mM)
- Bovine β-1,4-galactosyltransferase
- 8-Methoxycarbonyloctyl 2-acetamido-2-deoxy$-6-O-\alpha$-L-fucopyranosyl-β-D-glucopyranoside.

(Bovine β-1,4-galactosyltransferase can be purchased from Sigma-Aldrich.)

1. To 300 µL of 0.1 M sodium cacodylate buffer, pH 7.5 in a 1.5 mL plastic-capped tube add: substrate (3 mg, 5.6 µmol); 0.1 units β-1,4-galactosyltransferase; 30 µmol UDP-galactose; and manganese chloride to a final concentration of 20 mM.

2. Incubate the reaction with gentle shaking at room temperature for 12 h and monitor by TLC. In the example, quantitative conversion to the trisaccharide product was observed after 12 h. When complete, the reaction is diluted with 10 mL water and purified using preparative HPLC.

In a second example, transfer of a sugar to the substrate, in this case a flavonoid, on a larger scale (44 mg) it was necessary to add not the activated sugar UDP-galactose but UDP-glucose plus an additional enzyme, an epimerase, which converts UDP-Glc to UDP-Gal (Figure 5.22). Alkaline phosphatase was also added to hydrolyse the UDP products to drive the equilibrium of the reaction towards products. α-Lactalbumin, an activator of Gal-T, is also added to optimise the enzyme activity. This example is derived from reference [25].

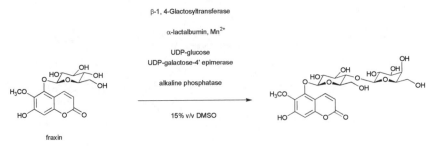

Figure 5.22 Enzymatic transfer of a sugar to a flavonoid substrate

The method is applicable to the transfer of galactose to other aglycones, and is useful in illustrating further considerations of the complexities inherent in GTase-catalysed biotransformations *in vitro*. It also demonstrates the use of small amounts of co-solvent with GTase-catalysed reactions as an aid to substrate solubility.

Apparatus and reagents

- 0.05 M Tris-HCl buffer pH 7.4
- UDP-glucose
- Manganese (II) chloride
- Flavonoid substrate (fraxin)
- Bovine β-1,4-galactosyltransferase
- Alkaline phosphatase
- UDP-galactose-4'-epimerase
- α-Lactalbumin
- Dimethylsulfoxide

1. To 4 mL 0.05 M Tris-HCl buffer, pH 7.4 in a 10 mL round-bottom flask add: flavonoid substrate fraxin (50 mg, 0.135 mmol); UDP-glucose (360 mg, 158 mM); 25 mM manganese chloride; 1 mg mL^{-1} α-lactalbumin; 2 units β-1,4-galactosyltransferase; 8 units UDP-galactose-4'-epimerase; 14 units alkaline phosphatase; 0.01% sodium azide; and 1 mM dithiothreitol.

2. Incubate the reaction with gentle shaking at 30 °C for 48 h and monitor by HPLC. When complete, the water was evaporated and the residue subjected to flash column chromatography to give an 8 5% yield of the product.

Whilst illustrating many of the complexities associated with the GTase-dependent strategy, the question of recycling of the UDP-activated sugar remains, and would be greatly beneficial to any preparative process. This can be solved to a degree by an even more complex system, first described by Whitesides and co-workers over 25 years ago [26]. The strategy is illustrated in Figure 5.23. In the synthesis of N − acetyllactosamine, bovine 1,4-Gal-T is used to transfer UDP-Gal to GlcNAc. The UDP product is phosphorylated by pyruvate kinase to give UTP at the expense of phosphoenolpyruvate. Glucose-1-phosphate is generated from glucose-6-phosphate by the action of phosphoglucomutase. UDP-glucose pyrophosphorylase catalyses the synthesis of UDP-glucose, which is then epimerised, as in the second example, by the action of UDP-galactose-4'-epimerase. These examples were notable for the scale (40 mmol) and make use of enzymes immobilised in-house.

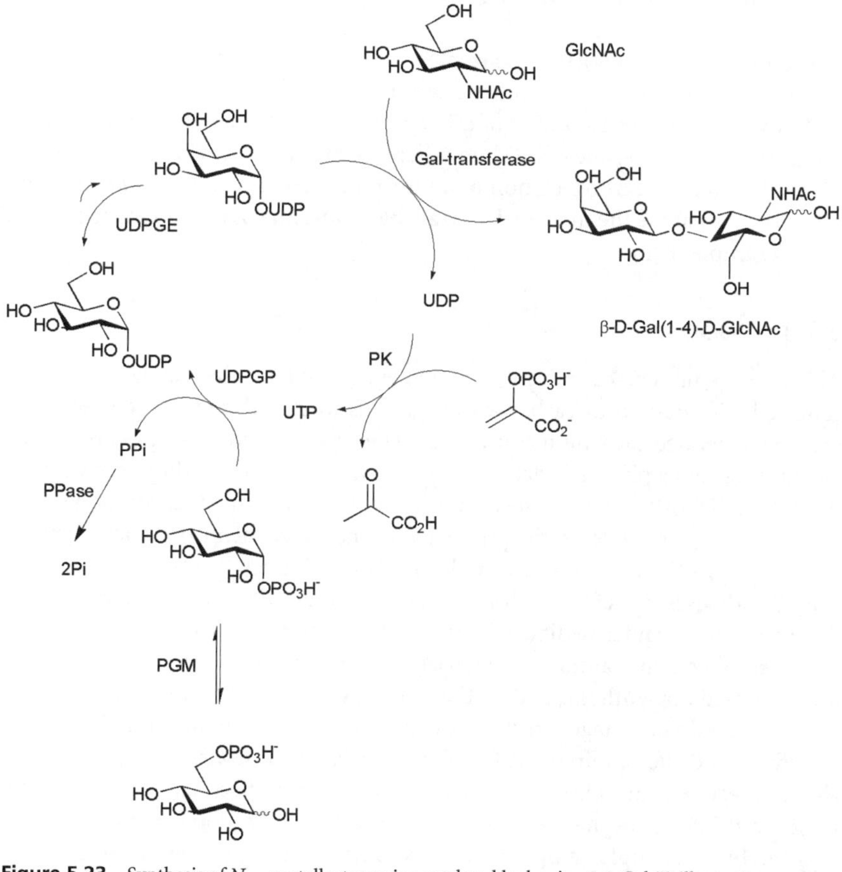

Figure 5.23 Synthesis of N − acetyllactosamine catalysed by bovine 1,4-Gal-T, illustrating strategy for recycling the UDP-activated sugar donor (derived from reference [25]). UDPGE, UDP galactose 4'-epimerase; Gal transferase, galactosyltransferase; UDPGP, UDP-glucose pyrophosphorylase; PGM, phosphoglucomutase; PK, pyruvate kinase; β-D-Gal-(1 → 4)-D-GlcNAc, N − acetyllactosamine

Such ingenious methods of nucleotide sugar regeneration mean that millimolar scale synthesis of oligosaccharides are possible, and are nowadays being engineered into complex recombinant strains of *E. coli* that contain all of the relevant activities. The impact of genomics has of course been as keenly felt in the world of GTases as with other enzyme types, and many new enzymes of different and interesting specificity are being described all the time. A website established in the University of Marseille (http://www.cazy.org/) is dedicated to documenting enzyme acting on carbohydrates including glycoside hydrolases and GTases and many new targets for cloning and expression (see Chapter 7) can be found there.

5.7 Carbon-Carbon Bond Forming Reactions

Carbon-carbon bond forming enzymes [27] have great potential in organic synthesis as they are able to synthesis carbon-carbon bonds with excellent stereoselectivity. There are a number of different types of enzymes, employing a wide range of mechanisms, and displaying different cofactor specificities. Of the commercially available carbon-carbon bond forming enzymes, it is the *aldolases* and *oxynitrilases* that are the most widely available, and some of their uses are detailed in the examples below.

5.7.1 Aldolases

Aldolases are ubiquitous enzymes in biochemical systems that catalyse the enantioselective formation of carbon-carbon bonds to give β-hydroxyketones using enzyme-generated carbanion equivalents. The substrates for aldolases are usually an aldehyde and a phosphorylated (or activated) alcohol, often dihydroxyacetone phosphate (DHAP). As with the chemical aldol reaction, the two components of reaction are ligated to form the product with the stereochemistry strictly defined at the beta position. There are, as traditionally considered, two types of aldolases: Class 1 aldolases, which use a lysine residue in the active site of the enzyme to form a Schiff base intermediate with the aldehyde substrate; and Class 2 enzymes, which employ a zinc atom for stabilisation of the reaction intermediates. They are both used for synthesis, and neither requires additional small molecule coenzymes. Their disadvantages are usually considered to be both the limited substrate specificity and the requirement for phosphorylated substrates. The products of aldolase reactions are phosphorylated sugars, but the phosphate can usually be removed using a phosphatase enzyme, included in the reaction mixture.

The first example employs the commercially available L-rhamnulose 1-phosphate aldolase from *E. coli* aldolase to catalyse the aldol reaction between DHAP and a derivatised proline aldehyde, *N*-CBz-prolinal (Figure 5.24). This example is derived from reference [28].

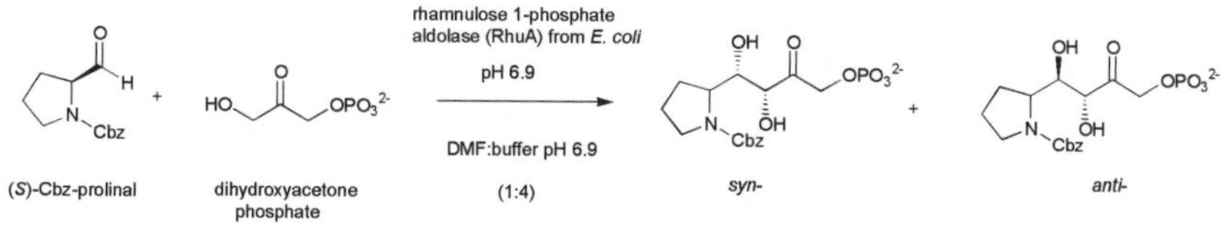

Figure 5.24 Use of *E. coli* aldolase to catalyse the aldol reaction between DHAP and *N*-CBz-(*S*)-prolinal

Apparatus and reagents

- Distilled water
- Dihydroxyacetone phosphate (DHAP)
- (*S*)-CBz-prolinal
- Rhamnulose 1-phosphate aldolase (RhuA)
- Methanol
- DMF
- Acid phosphatase

(NB DHAP can be purchased from Sigma-Aldrich and RhuA can be purchased from Boehringer-Mannheim.)

1. To 22.5 mL DMF add 3.75 g (*S*)-prolinal.

2. Make up 90 mL of a 105.2 mmol solution of DHAP at pH 6.9 and add this slowly to the stirred DMF solution of (*S*)-CBz-prolinal.

3. Shake the reaction at 4 °C until it reaches completion as judged by HPLC. Add 115 mL of methanol and the excess aldehyde is removed by extraction with ethyl acetate.

4. The resultant product solution is treated with acid phosphatase (50 mg at 7 U mg^{-1}) at pH 5.5. Hydrogenation of the product residue yielded alkaloid products of the hyacinthacine family.

The second example employs the D-fructose 1,6-bisphosphate aldolase (*FruA*) for the conversion of DHAP haloaldehydes into halo-xylulose products (Figure 5.25). The publication from which the method is derived was also interesting in having developed a convenient route for DHAP production *in situ* [29]. Phosphorylated substrates such as DHAP are expensive and their use in aldolase reactions is often thought to be one of their major disadvantages.

Apparatus and reagents

- Distilled water
- Disodium hydrogen phosphate

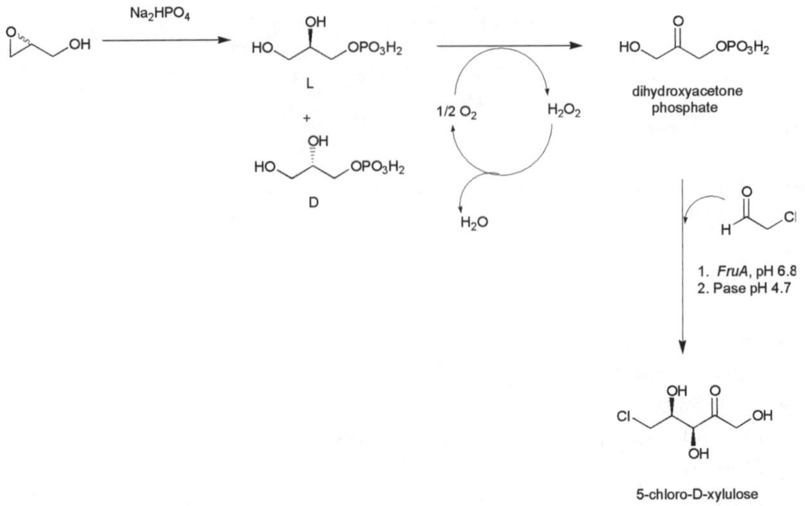

Figure 5.25 Aldol condensation of chloroacetaldehyde with DHAP catalysed by the fructose-1,6-bisphosphate aldolase *Fru*A and illustrating *in situ* generation of DHAP from epoxypropanol and removal of phosphte from the product by phosphatase (Pase)

- 3 N Hydrochloric acid
- GPO/catalyse mixture
- FruA from rabbit muscle
- Acid phosphatase
- 2-Chloroacetaldehyde
- Racemic 2,3-epoxypropanol

(FruA from rabbit muscle can be purchased from Sigma-Aldrich.)

1. Add racemic 2,3-epoxypropanol (0.38 g, 5 mmol) to 10 mL distilled water. To this add 0.74 g (5 mmol) disodium hydrogen phosphate.

2. Heat the mixture at 100 °C for 3 h.

3. Cool the reaction mixture and adjust the pH to 6.8 using 3 N hydrochloric acid. Add 0.1 mL of a mixture of L-glycerophosphate oxidase (45 units) and catalse (1800 units). Then add 70 μL of rabbit muscle adolase *Fru*A and 2-chloroacetaldehyde (2 mmol). Stir the reaction at room temperature overnight.

4. Adjust the pH to 4.7 with 1 N hydrochloric acid and add 50 units of acid phosphatase. Stir the reaction for a further 24 h.

5. Adjust the pH to 7.0 using 1 N sodium hydroxide and add 30 mL of methanol. Filter the mixture through celite. The products are recovered using flash column chromatography.

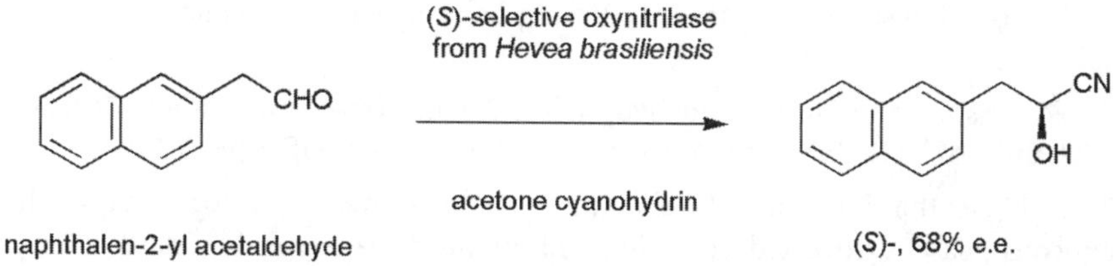

Figure 5.26 Enantioselective addition of HCN to napthalen-2-yl acetaldehyde catalysed by (S)-selective oxynitrilase from *Hevea brasiliensis*

5.7.2 Oxynitrilases

Oxynitrilases or hydroxynitrile lyases (HNLs) catalyse the enantioselective addition of cyanide to aldehydes [27]. They have proved popular owing both to their excellent selectivities in synthetic applications and also to their ready availability from natural plant sources. Hence, a crude preparation of crushed almonds (*Prunus amygdalus*) contains an (*R*)-selective oxynitrilase; rubber tree leaves (*Hevea brasiliensis*) contain an (*S*)-selective enzyme. Each is available commercially for applications in synthesis, although the genes encoding both have now been heterologously expressed. Oxynitrilases, like lipases, are robust enzymes and can be used with some success in two-phase systems and even organic solvent systems where substrate solubility may be limiting. Papers using HNLs routinely either employ solutions of HCN or acetone cyanohydrin as a source of cyanide. It is in fact the release of cyanide that is the clue to the natural function of HNLs in plants – the production of cyanide as a deterrent to consumption by animals. The protocols for using HNLs are therefore inherently hazardous, and the handling of cyanide or acetone cyanohydrin should only be undertaken by personnel with the requisite experience and with adequate protection e.g. in a fume hood and with an HCN detector. The following example uses the HNL from *Hevea brasiliensis* for the synthesis of a naphthyl derived cyanohydrin (Figure 5.26). This example is derived from reference [30].

Apparatus and reagents

- 0.1 M Sodium citrate buffer pH 5.5
- Acetone cyanohydrin
- Diisopropyl ether
- Hydroxynitrile lyase from *Hevea brasiliensis*

(Hydroxynitrile lyase from *Hevea brasiliensis* can be purchased from Novozymes.)

1. To 10 mL diisopropyl ether add 250 mg naphthalen-2-yl-acetaldehyde as the substrate.

2. To the stirred solution of aldehyde, add 1.3 equiv. of acetone cyanohydrin and 1500 units of HNL dissolved in 0.5 mL of sodium citrate buffer pH 5.5.

3. Incubate the reaction with shaking at room temperature for 5 days. The aqueous phase is extracted into ether and the products purified by flash column chromatography after work-up.

5.8 Conclusion

The use of commercially available enzymes in synthesis has certainly made it easier for those with no experience of biocatalysis to begin their first studies in their application for organic reactions. The increasing availability of these catalysts, which is taking the range of available powdered enzymes from simple hydrolases to more sophisticated enzymes such as those for redox processes and carbon-carbon bond formation should help to improve the uptake of this methodology even further. However, more work needs to be done on formulating stable and robust enzyme catalysts that can be delivered and used in powdered form. When it comes to even more sophisticated enzymes, such as cytochromes P450 and flavoprotein oxidases, their applications are still largely restricted to those laboratories that are able to express genes and make (partially) pure proteins in-house. In Chapter 6, we will consider simple methods for the preparation of such 'home-grown' biocatalysts from preparations of micro-organisms.

References

1. U. T. Bornscheuer and R. J. Kalauskas (2005) Hydrolases in Organic Synthesis. Regio and Stereoselective Biotransformations, 2nd Edn, Wiley-VCH, New York.
2. M. J. Hernáiz, J. M. Sanchez-Montero and J. V. Sinisterra (1994) Comparison of the enzymatic activity of commercial and semipurified lipase of *Candida cylindracea* in the hydrolysis of the esters of (*R,S*)-2-aryl propionic acids. *Tetrahedron*, **50**, 10749–10760.
3. H.-J. Ha, Y.-S. Park and G.-S. Park (2001) Lipase-catalysed enantioselective hydrolysis of β-acetyloxymethyl-β-valerolactone *ARKIVOC*, 55–61.
4. C. Laane, S. Boeren, K. Vos and C. Veeger (1987) Rules for optimization of biocatalysis in organic solvents. *Biotechnol. Bioeng.*, **30**, 81–87.
5. G. Bell, P. J. Halling, B. D. Moore, J. Partridge and D. G. Rees (1995) Biocatalyst behaviour in low-water systems. *Trends. Biotechnol.*, **13**, 468–473.
6. J. Uenishi, T. Hiraoka, S. Hata, K. Nishiwaki, O. Yonemitsu, K. Nakamura, and H. Tsukube (1998) Chiral pyridines: optical resolution of 1-(2-pyridyl)- and 1-[6-(2,2'-bipyridyl)]ethanols by lipase-catalyzed enantioselective acetylation. *J. Org. Chem.*, **63**, 2481–2487.

7. N. J. Turner (2004) Enzyme catalysed deracemisation and dynamic kinetic resolution reactions. *Curr. Opin. Chem. Biol.*, **8**, 114–119.
8. G. Nicolosi, A. Patti, M. Piattelli and C. Sanfilippo (1995) Desymmetrization of *cis*-1,2-dihydroxycycloalkanes by stereoselective lipase mediated esterification. *Tetrahedron: Asymmetry*, **6**, 519–524.
9. D. T. Guranda, A. I. Khimiuk, L. M. van Langen, F. van Rantwijk, R. A. Sheldon and V. K. Švedas (2004) An 'easy-on, easy-off' protecting group for the enzymatic resolution of (±)-1-phenylethylamine in an aqueous medium. *Tetrahedron: Asymmetry*, **15**, 2901–2906.
10. J. González-Sabín, V. Gotor and F. Rebolledo (2002) CAL-B-catalyzed resolution of some pharmacologically interesting β–substituted isopropylamines *Tetrahedron: Asymmetry*, **13**, 1315–1320.
11. S. Takayama, S. T. Lee, S.-C. Hung and C.-H. Wong (1999) Designing enzymatic resolution of amines. *Chem. Commun.*, 127–128.
12. N. Monfort, A. Archelas and R. Furstoss (2004) Enzymatic transformations. Part 55: Highly productive epoxide hydrolase catalysed resolution of an azole antifungal key synthon. *Tetrahedron*, **60**, 601–605.
13. J. A. Pollock, K. M. Clark, B. J. Martynowicz, M. G. Pridgeon, M. J. Rycenga, K. E. Stolle and S. K. Taylor (2007) A mild biosynthesis of lactones via enantioselective hydrolysis of hydroxynitriles. *Tetrhedron: Asymmetry*, **18**, 1888–1892.
14. N. Bridiau, S. Taboubi, N. Marzouki, M. D. Legoy and T. Margaud (2006) β-Galactosidase catalysed selective galactosylation of aromatic compounds. *Biotechnol. Prog.*, **22**, 326–330.
15. S. J. Williams and S. G. Withers (2002) Glycosynthases: mutant glycosidases for glycoside synthesis. *Aust. J. Chem.*, **55**, 3–12.
16. W. A. van der Donk and H. Zhao (2003) Recent developments in pyridine nucleotide regeneration. *Curr. Opin. Biotechnol.*, **14**, 421–426.
17. K. Seelbach, B. Riebel, W. Hummel, M.-R. Kula, V. I. Tishkov, A. M. Egorov, C. Wandrey and U. Kragl (1996) A novel, efficient regenerating method of NADPH using a new formate dehydrogenase. *Tetrahedron Lett.*, **37**, 1377–1380.
18. D. Pollard, M. Truppo, J. Pollard, C. Chen and J. Moore (2006) Effective synthesis of (*S*)-3,5-bistrifluoromethylphenyl ethanol by asymmetric enzymatic reduction. *Tetrahedron: Asymmetry*, **17**, 554–559.
19. D. Giacomini, P. Galletti, A. Quintavalla, G. Gucciardo and F. Paradisi (2007) Highly efficient asymmetric reduction of arylpropionic aldehydes by Horse Liver Alcohol Dehydrogenase through dynamic kinetic resolution. *Chem. Commun.*, 4038–4040.
20. V. M. Dembitsky (2003) Oxidation, epoxidation and sulfoxidation reactions catalysed by haloperoxidases. *Tetrahedron*, **59**, 4701–4720.
21. S. Colonna, N. Gaggero, L. Casella, G. Carrea and P. Pasta (1993) Enantioselective epoxidation of styrene derivatives by chloroperoxidase catalysis. *Tetrahedron: Asymmetry*, **4**, 1325–1330.
22. C. Sanfilippo, A. Patti and G. Nicolosi (2000) Asymmetric oxidation of 1, 3-cyclohexadiene catalysed by chloroperoxidase from *Caldariomyces* fumago. *Tetrahedron: Asymmetry*, **11**, 3269–3272.
23. M. M. Palcic (1999) Biocatalytic synthesis of oligosaccharides. *Curr. Opin. Biotechnol.*, **10**, 616–624.

24. M. M. Palcic, O. M. Srivastava and O. Hindsgaul (1987) Transfer of D-galactosyl groups to 6-*O*-substituted 2-acetamido-2-deoxy-D-glucose residues by use of bovine D-galactosyltransferase. *Carbohydr. Res.*, **159**, 315–324.

25. S. Riva, B. Sennino, F. Zambianchi, B. Danieli and L. Panza (1998) Effects of organic co-solvents on the stability and activity of the β-1,4-galactosyltransferase from bovine colostrum. *Carbohydr. Res.*, **305**, 525–531.

26. C.-H. Wong, S. L. Haynie and G. M. Whitesides (1982) Enzyme-catalysed synthesis of N − acetyllactosamine with *in situ* regeneration of uridine 5'-diphosphate glucose and uridine 5'-diphosphate galactose. *J. Org. Chem.*, **47**, 5416–5418.

27. J. Sukumaran and U. Hanefeld (2005) Enantioselective C-C bond synthesis catalysed by enzymes. *Chem. Soc. Rev.*, **34**, 530–542.

28. J. Calveras, J. Casas, T. Parella, J. Joglar and P. Clapés (2007) Chemoenzymatic synthesis and inhibitory activities of hyacinthacines A_1 and A_2 stereoisomers. *Adv. Synth. Catal.*, **349**, 1661–1666.

29. F. Charmantray, P. Dellis, S. Samreth and L Hecquet (2006) An efficient chemoenzymatic route to dihydroxyacetone phosphate from glycidol for the *in situ* alsolase-mediated synthesis of monsaccharides. *Tetrahedron Lett.*, **47**, 3261–3263.

30. G. Roda, S. Riva, B. Danieli, H. Griengl, U. Rinner, M. Schmidt and A. Mackova Zabelinskaja (2002) Selectivity of the (*S*)-oxynitrilase from *Hevea brasiliensis* towards α- and β-substituted aldehydes. *Tetrahedron*, **58**, 2979–2983.

Chapter 6

A Beginner's Guide to the Isolation, Analysis and Use of Home-Grown Enzyme Biocatalysts

6.1 Introduction

In Chapter 5 we outlined how simple cell-free biotransformations can be undertaken using commercially available isolated enzymes. It is certainly the case that one of the barriers to the uptake of biotransformations in organic chemistry has been the perceived difficulty or unfamiliarity associated with using microbes and the techniques used for isolating enzymes from them. Apart from the hydrolases and the odd oxidoreductase, until recently very few enzymes of other classes have been available in user-friendly form from enzyme suppliers. One of the consequences is that, if we review the biotransformations literature for an enzyme such as a Baeyer–Villiger monooxygenase, there are few examples of researchers in organic chemistry laboratories using the commercially available catalyst. The great majority of such papers report work from laboratories that are already equipped to perform, in many cases, recombinant DNA work, the growth of recombinant strains of E. coli and some form of enzyme isolation, if only as far as disrupting cells to obtain a cell-free protein preparation.

Whilst access to user-friendly preparations of enzymes of other classes is improving, we think that it would be useful to provide some short guidance on the isolation and analysis of protein biocatalysts using simple techniques, so that groups might attempt perhaps only the preparation of cell-free enzyme extracts from wild-type or recombinant microbial cultures. It is not intended to provide a detailed guide to protein purification, as the techniques required may be outside the technical facility offered by the standard organic laboratory, and indeed, the expense involved in purchasing the required machines for protein isolation is considerable. However, we feel it may be worthwhile for the worker to have an appreciation of, for example, whether the strain of organism with which they have

Practical Biotransformations: A Beginner's Guide   © 2009   Gideon Grogan

been supplied, is actually expressing the gene of interest. We will therefore describe general considerations for cell harvesting, disruption, crude protein fractionation and in some cases, purification of enzymes that have been engineered to possess purification-friendly tags. We also deal briefly with the major method of the analysis of protein purity (SDS-PAGE) and give some examples of enzyme assays and cell-free biotransformations in the literature.

6.2 Cell Growth and Harvesting

In order to make cell-free extracts for preparative biotransformations, the growth of the organism will need to be scaled up. We have seen that it is possible to obtain enough whole-cell biomass for a laboratory scale biotransformation by growing organisms in culture volumes of 50 or 100 mL, but the generation of sufficient amounts of biocatalyst for isolated enzyme experiments will require much more biomass. The volume of culture that needs to be grown will vary on a case dependent basis, but it will be necessary to resort either to one or a number of 2 L Erlenmeyer flask volumes, or, if available, that the organism is cultivated in a fermentor. Biomass can be harvested in a number of ways, but these may include simple filtration (either through muslin or a Büchner funnel) for fungi or centrifugation for bacteria. After obtaining a cell pellet, the organism will usually be resuspended in a volume of buffer for purposes of cell disruption.

6.3 Cell Disruption

A variety of suitable biological buffers exists that may be used for cell disruption purposes. Some of these are listed previously, in the discussion on choices of buffer for whole-cell biotransformation reactions. The choice of pH may well depend on the nature of the enzyme, which may for instance exhibit better stability and/or activity at a slightly alkaline pH, but it is probably best to use an approximately neutral pH as a starting point in the absence of other information. Buffers commonly used are phosphate, Tris-HCl, MOPS, HEPES and MES (see Appendix 5), usually at a concentration of 50 or 100 mM. Sodium chloride (100–500 mM) is often also added, if this is found to stabilise the protein.

The release of the cell components, and the consequent release of endogenous proteases, may demand that protease inhibitors are added to the buffer. Some consideration must be given here to the enzyme activity desired – for example, many protease inhibitors exert their action by covalent interaction with the catalytic serine of the catalytic triad in the protease active site that is very similar to that encountered in lipases and esterases. However, routinely, phenylmethylsulfonyl fluoride (PMSF) or 4-(2-aminoethyl) benzenesulfonyl fluoride (AEBSF) can be added to a final concentration of 0.02–1 mM, added in a small volume from a stock ethanolic solution. (NBSuch protease inhibitors are highly toxic.) Other stabilizing agents may also be added. These include thiol reagents such as β-mercaptoethanol, or more commonly dithiotreitol (DTT) at approximately 1 mM concentration, which prevent the oxidation of sensitive sulfur moieties in the proteins. Enzyme stability can in some cases be improved by the inclusion of essential cofactors – whilst this is done commonly when the cofactors are relatively cheap (thiamine diphosphate and magnesium ions for some classes of carbon-carbon bond forming enzymes, for example), it does of course become prohibitively expensive when they are not.

Additional reagents can also be added in an attempt to improve the overall solubility of the enzyme of interest, by removing this from the membrane fraction of the cell debris. To this end, 1–10% glycerol may be added to the buffer, or a small amount (0.5 or 1% v/v) of a detergent such as Triton X-100 or Tween 20.

The last consideration is that of temperature. Both the tendency of some proteins to denature at higher temperatures when removed from their natural cell environment, and the activity of protease enzymes (which will be higher at 20 °C than at 10 °C), may be addressed by carrying out all procedures on subcellular cell fractions at 4 °C, usually by storing proteins on ice when on the bench, and in the fridge or cold room when storing them.

There are a few commonly used methods of cell disruption that result in protein extracts suitable for biocatalysis studies. The method of choice usually depends on the organism used.

6.3.1 Ultrasonication

Many microbiology laboratories have an ultrasound generator that is used for the disruption of bacteria. In contrast to the bath type sonicators used routinely in organic chemistry laboratories, the sonicators used in microbiology laboratories consist of a small cabinet in which an ultrasound generator is connected to a vertically housed metal probe (Figure 6.1). The end of the probe is inserted into the cell suspension, usually just below the meniscus, and bursts of ultrasound are fired directly into the suspension, causing turbulence and also the generation of heat. In order to minimise thermal damage to the enzyme extract, such sonications are usually performed with the cell suspension contained in a glass beaker in an ice bath. Ultrasonication is suitable for the disruption of cells of *E. coli, Pseudomonas* and other bacteria such as *Acinetobacter*, but is sometimes not effective for the breaking of cells of strains of *Rhodococcus, Streptomyces* or filamentous fungi, for each of which mechanical methods of cell disruption must sometimes be used.

Figure 6.1   An ultrasonicator used for the disruption of bacterial cells

6.3.2 Pressure cells

An effective method of disrupting cells is provided by pressure cell equipment such as a 'French press'. In this type of apparatus, a cell suspension, some-times pre-frozen, is squeezed through a very small hole at very high pressures, resulting in disruption of the cells (Figure 6.2). Although quite common in biology–biochemistry departments, such presses are expensive, and may be beyond the scope of laboratories wishing to obtain extracts from recombinant strains of *E. coli*. As stated above, a mechanical method of disruption is necessary for organisms for which sonication is not successful.

6.3.3 Glass bead mills

An alternative piece of equipment for cell disruption in difficult cases is a glass bead mill. These take different forms, but can consist of a cooled chamber into which is poured the cell suspension plus a volume of small glass beads. When agitated together, sometimes using plastic propeller blades within the chamber, the cells are disrupted. Again, whilst hardly an everyday piece of equipment in a

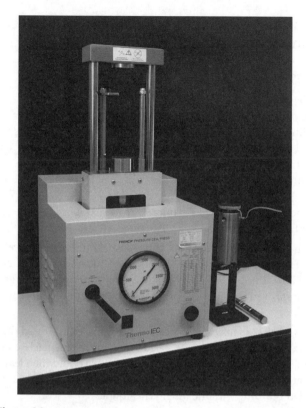

Figure 6.2   Pressure cell apparatus for the disruption of micro-organisms

synthetic laboratory, such a method should only prove necessary for organisms whose cell walls are difficult to break. It may also be possible to obtain some level of disruption by mixing glass beads with cell suspension in a capped plastic tube, followed by agitation on a whirlimixer.

6.3.4 Mortar and pestle

For filamentous fungi, and indeed, for some other organisms such as macroscopic algae (seaweed contains some very interesting biocatalysts), it may be possible merely to grind up the cell mass in buffer in a large mortar and pestle, containing some sand. It is usually possible to obtain fungal cell extracts in this way from species such as *Penicillium, Beauveria* or *Aspergillus* in the absence of a more effective method such as a pressure cell. The use of food blenders or other devices using sharp blades should be avoided, as the proteins within the extracts can be subject to shear stresses.

6.4 A Typical Procedure for Making a Cell Extract from a Recombinant Strain of *E. Coli*

Apparatus and reagents

- 2 L Erlenmeyer flasks
- Starter cultures of example strain of recombinant *E. coli*
- Centrifuge; large (250 mL) and small (50 mL) centrifuge tubes
- 0.1 M Tris-HCl buffer pH 7.1 containing 100 mM sodium chloride and 100 μM PMSF
- Ultrasonicator and ice bath

1. Using procedures detailed in Chapter 3, grow 4 × 500 mL of the recombinant organism in Erlenmeyer flasks.

2. Harvest the cells by centrifugation, and resuspend them collectively in 100 mL of a 0.1 M Tris-HCl buffer pH 7.1 containing 100 mM sodium chloride and 100 μM PMSF. (DTT, 1–10% glycerol and a small amount of detergent may also be added to improved the recovery of soluble enzyme if necessary, but begin with a simple buffer.)

3. Sonicate the cell suspension in 2 × 50 mL portions, each treated thus: Place the cells in a glass beaker, and place the beaker in a larger receptacle containing ice. Subject one 50 mL aliquot to a 45 s burst at high power, and then leave to rest for 1 min. Repeat twice, and then transfer the sonicated suspension to a fresh beaker, also on ice. Do the same with the other cell aliquot.

(NB The sonication of cells creates 'aerosols' or dispersions of micro-organisms in the air. It is important therefore, that sonication be performed with the cabinet door closed, and that the sonicator and probe are cleaned before and after use with 70% ethanol or disinfectant. Also, it is recommended that, when using cabinet sonicators of this type, suitable headphones are worn for ear protection.)

4. You should now have 100 mL of a sonicated cell suspension. This should be centrifuged at relatively high speed (12 000 r.p.m) in suitable tubes, to yield a cell extract supernatant and a pellet.

5. The cell supernatant should contain the enzyme activity, if contained in the soluble fraction. The best way to determine this is of course using gel electrophoresis (SDS-PAGE; see below), but certainly the supernatant, in the absence of disinfectant should be coloured slightly, if from *E. coli* and more so if from an organism such as *Pseudomonas*. The extract should also be somewhat frothy when shaken gently. A suitable enzyme activity assay of the supernatant, or indeed a small biotransformation of a compound previously established as a substrate for the enzyme, would also help to confirm or otherwise the success of the procedure.

Once the enzyme extract has been obtained, it may be used as a biocatalyst in a number of ways. It is of course possible, and usually best to use the enzyme fresh, but if it is necessary to store the enzyme for a short period of time, store at 4 °C. If storing for more than 24 h, it is recommended that the extract is either 'snap frozen', by rapidly freezing suitable aliquots in liquid nitrogen, or lyophilised (freeze-dried). The latter is an attractive option, as it reduces the enzyme material to a powder that may be easily applied in biotransformations.

Regardless of the organism from which the extract was obtained, or the method used, the cell disruption procedure has resulted in a 'crude extract' containing the desired enzyme activity. This enzyme is however, far from pure, even in the case of over-expressed genes in recombinant *E. coli* and as such, if greater levels of purity are required, the enzyme must be purified to obtain the 'homogeneous' catalyst.

6.5 Purification of Enzymes – A Brief Guide

We have mentioned on various occasions above that the purification of enzymes can be a laborious business, and also requires expensive equipment and trained personnel, thus being beyond the scope of the standard synthetic laboratory. This is still of course true – however, in some cases, protein purification can be inexpensive and facile, particularly where the enzyme to be purified has been engineered to facilitate purification.

The isolation of enzymes from *wild-type* strains of micro-organisms would of course be a challenge without the appropriate facilities and will therefore not be addressed in great detail in this section. However, it is felt that it would be of use to provide some information on the general techniques of protein chromatography in the interests of helping to engage with the relevant literature, and also to provide a bench-friendly protocol for the purification of recombinant enzymes that have been 'histidine tagged'.

6.5.1 Preparation of crude extracts – ammonium sulfate cuts and dialysis

6.5.1.1 Ammonium sulfate fractionation

As a crude technique for fractionating protein mixtures, ammonium sulfate precipitation can be used. In addition to being a cheap and convenient technique for concentrating protein, it can, if performed carefully, have some merit as a method of partial purification. The principle is that proteins form aggregates with each other and precipitate when ammonium sulfate is added to protein solutions, the salt effectively removing the water that hydrates the protein surface. Different proteins will aggregate at different concentrations, so it may be possible to 'salt' out the desired enzyme activity to make a concentrated preparation. Many biochemistry textbooks and practical guides will have an ammonium sulfate table that states how much ammonium sulfate to add to a fixed volume of solution to achieve a required level of saturation. Such a table can be found at the following link: http://www.science.smith.edu/departments/Biochem/Biochem_353/Amsulfate.htm; this was used to derive the table in Appendix 6. We will use these values as part of the following method.

Making an ammonium sulfate 'cut' of a cell extract

Apparatus and reagents

- 100 mL of a crude cell extract
- Solid ammonium sulfate
- 250 mL Erlenmeyer flask, magnetic stirrer and flea
- Ice bath
- Centrifuge; small (50 mL) centrifuge tubes

We will assume that 100 mL of a crude protein solution has been obtained from an amount of microbial culture. We will salt out various proteins by the gradual addition of ammonium sulfate, making three fractions constituting ammonium sulfate 'cuts' of 30, 50 and 80% saturation. As we are working with isolated protein, the procedure should be conducted at 4 °C, either by standing the Erlenmeyer flask in an ice bath, or by doing the precipitation in a temperature-controlled cold room.

1. Transfer 100 mL of protein solution to a clean 250 mL Erlenmeyer flask, place in an ice bath on a magnetic stirrer and set stirring.

2. Weigh out 16.4 g of solid ammonium sulfate. When this complete amount is dissolved in water, it will constitute a saturation of 30%. Add the salt slowly, beginning with one spatula-full at a time, to the stirring protein, which will froth as a precipitate begins to form. Leave to stir for a minimum of 45 min.

3. Transfer the suspension to suitably sized centrifuge tubes and centrifuge at 12 000 r.p.m for 20 min. Transfer the supernatant to a clean Erlenmeyer flask in an ice bath and add a magnetic stirrer. Resuspend the pellet in a minimum of buffer and retain at 4 °C.

4. Weigh out 11.7 g of solid ammonium sulfate. When this complete amount is dissolved in water, it will constitute a saturation of 50%. Add the salt slowly to the solution again and stir at 4 °C for 45 min. Centrifuge the resulting suspension as before, transfer the new supernatant to the cleaned Erlenmeyer flask, and retain the 50% pellet as before.

5. Weigh out 19.4 g of solid ammonium sulfate. When this complete amount is dissolved in water, it will constitute a saturation of 80%. Add the salt slowly to

the solution again and stir at 4 °C for 45 min. Centrifuge the resulting suspension as before, after which there will be a supernatant which can be discarded, as few proteins require >80% ammonium sulfate to precipitate. The pellet is resuspended in a minimum of buffer.

Using this procedure has efficiently concentrated the proteins from a volume of 100 mL to a few millilitres. It may also have resulted in some crude purification, which may be checked by analysing the various fractions by SDS-PAGE (see below). It may be useful on occasion, in the interests perhaps of time or just to concentrate the protein sample, rather than to attempt to fractionate the protein extract using $(NH_4)_2SO_4$, merely to take the initial crude extract to 80% saturation, and then to precipitate the majority of the protein in the sample in one step.

6.5.1.2 Dialysis of protein mixtures

Proteins that are aggregated in the presence of ammonium sulfate may be inactive or poorly active, so the ammonium sulfate is usually removed prior to using the enzyme. This is done using dialysis, which sounds both complicated and expensive, but is in reality a very simple and inexpensive technique routinely employed in biochemistry laboratories. The principle of dialysis as applied to the preparation of enzymes is to enclose the solution within a 'dialysis bag' or tube, made of a semi-permeable membrane that permits the migration of small molecules (such as salt, ammonium sulfate), then to suspend this in a comparatively large volume of buffer (say, 20 x the volume of the protein precipitate suspension, Figure 6.3). The tube within the dialysis buffer is then stirred overnight and, as the solutions within and outside the bag equilibrate, the concentration of ammonium sulfate in the protein sample is reduced. One can then change the buffer, in order to improve the gradient across the membrane and increase the efficiency of dialysis. We include below a simple method for setting up a dialysis.

Removal of ammonium sulfate using dialysis

Apparatus and reagents

- Ammonium sulfate-protein pellet obtained as above
- 2 L beaker, magnetic stirrer and flea
- 2 × 1 L 0.1 M Tris-HCl buffer pH 7.1 containing 100 mM sodium chloride and 100 μM PMSF (chilled)
- Short section (15–20 cm) of dialysis tubing (of required molecular weight cut-off) and of a diameter suitable for the protein volume.
- Length of cotton or plastic dialysis tube clips (Figure 6.4).

Figure 6.3   Dialysis tube containing a protein extract

1. Prepare 1 L of a buffer that is the same as the cell resuspension buffer used in the cell disruption step. Cool this to 4 °C.

2. Take a protein sample that has been prepared from some cells, and prepare an 80% ammonium sulfate cut of the extract by precipitation and then centrifugation.

3. Resuspend the pellet obtained in a minimum of 50 mM Tris-HCl buffer pH 7.1 containing any of the additives deemed necessary previously in the disruption step (likely to be 5–50 mL for 1 L of cell mass, depending on the organism and method of growth).

4. Cut a length of dialysis tubing appropriate to the volume of the suspension. A variety of molecular weight cut-offs for dialysis tubing can be purchased – usually permitting the diffusion of molecules with MW<3000 or 5000 Da. Tie one end of the tubing either directly or by using a piece of cotton. Using a Pasteur or

Figure 6.4 Plastic dialysis clips

automatic pipette, transfer the suspension to the tube. Tie the other end of the tubing, attempting to exclude air from the top of the liquid column (Figure 6.5).

5. Place the dialysis tube inside the cooled buffer prepared in (1) above and place on a magnetic stirrer in the cold room overnight. If required, one can change the buffer in the dialysis flask for fresh buffer after an interval of several hours.

6. After dialysis, remove the dialysis tubing from the buffer and place in an empty beaker. Cut one end of the dialysis tubing, being careful that the contents do not spurt out of the bag.

This concentrated enzyme preparation is now ready to use or to store at −20 °C or −80 °C, preferably after snap freezing in liquid nitrogen. It may be that a greater degree of purification is required however, and that the extract needs to be subjected to one or more of the techniques described below.

6.6 Techniques for Protein Purification

What follows is a brief list of some of common techniques that may be encountered when reading about enzyme isolation in the biocatalysis literature. A more detailed description of these techniques may be obtained from reference [1]. Enzymes, being proteins, are polymers of different lengths and shapes and are made up of different of amino acids (see Appendix 1), each of which has a side chain that is chemically and structurally distinct from the other. Enzymes are distinguished from each other therefore by both chemical and physical differences and these

(a)   (b)

(c)   (d)

Figure 6.5 Preparing a dialysis of a protein extract. The dialysis tube is moistened in buffer then sealed at one end. The protein suspension is transferred by pipette into the dialysis tube and this is then stirred in a large volume of buffer

variations will become helpful when we try to separate them from one another using chromatographic techniques. Most methods of protein chromatography exploit one or other of these differences: in particular, *size, charge,* or *hydrophobic properties.* In each case, a resin or slurry made of polymer-based beads is used for purification, usually packed into a column that is directly analogous to a silica column used for the separation of organic molecules in synthesis. Nowadays, this column is usually fitted to an automated HPLC system that has been specifically designed for the higher pressures and volumes associated with preparative protein chromatography, sometimes called fast protein liquid chromatography (FPLC). These machines consist of a system of pumps that are connected to buffer solutions, which connect then to a mixer, which can control the composition of the eluant, and then to the column (Figure 6.6).

At the end of the column is a UV detector that is normally set to 280 nm, a useful wavelength for detecting proteins, as it detects the aromatic side chains of the amino acids tyrosine, tryptophan and phenylalanine. The column output is then directed to a fraction collector, where the successive eluted fractions are stored in tubes. Both the automated chromatography machines and their columns are expensive, but we list below the major techniques used for separation, as these will often be encountered in papers describing the isolation of enzymes.

Figure 6.6   A fast protein liquid chromatography machine (Akta^{TM} purifier from GE Healthcare)

6.6.1 Size exclusion chromatography

Enzymes widely range in size, from monomeric structures of perhaps 10 000–100 000 Da, to large complexes of more than one polypeptide chain that may be megadaltons in size. It is comparatively simple to exploit these differences in size for separation, by using a technique called size exclusion chromatography or gel filtration. In this technique, a column is packed with inert polymer beads that themselves contain tiny pores. The protein is loaded onto the column and eluted in isocratic fashion with a buffer solution that is unchanged in composition throughout the duration of elution. Large proteins cannot enter the small pores, and travel around them and are thus eluted first (Figure 6.7). Small proteins enter the pores and must travel a more complex route and are thus eluted last.

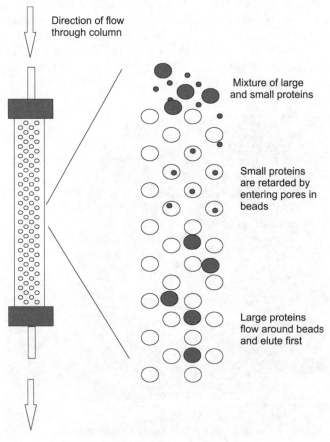

Figure 6.7   Gel filtration chromatography

6.6.2 Ion exchange chromatography

Enzymes have many different side chains, some of which are positively and negatively charged at different pHs, dependent on the pK_a of the relevant ionisable groups. The consequence of this is that each protein has an 'isolectric point' or a pH at which the overall charge of the protein is neutral. It also means that at any one pH, different proteins will have different signs and magnitudes of charge and can thus be separated on these bases.

In ion exchange chromatography, the column is packed with beads that are either *positively* (anion exchange chromatography) or *negatively* (cation exchange chromatography) charged. The protein is loaded in a buffer solution that contains no salt, binds to the column beads, and is then eluted with a gradient that increases the salt concentration (usually sodium or potassium chloride). In the case of anion exchange, positively charged proteins are repelled by the beads and elute first; negatively charged proteins interact with the positively charged beads and are then eluted as the negatively charged chloride ions compete with them for the beads as the salt concentration increases (Figure 6.8).

6.6.3 Hydrophobic interaction chromatography

The different composition of amino acid side chains in enzymes will also confer distinct hydrophobic properties on the enzymes, as many of the side chains, such as phenylalanine, leucine and isoleucine, are themselves hydrophobic (see Appendix 1). In hydrophobic interaction chromatography (HIC), the column is packed with beads to which are attached to hydrophobic groups (such as octyl or phenyl groups). The protein mixture in this case is loaded in a high salt (usually ammonium sulfate in the case of HIC) concentration, as hydrophobic interactions are greatest at higher salt concentrations. The most hydrophobic proteins thus bind to the column beads. The column is eluted with a *decreasing* gradient of ammonium sulfate, the least hydrophobic proteins eluting last and the most, first.

6.6.4 Affinity chromatography

Any ideal purification method would of course entail the isolation of an enzyme in a single step. The general approach to this is using an engineered protein, one that bears a histidine tag (see below) or other tag that has been introduced by genetic engineering. However, another approach, suitable for single step purifications of natural enzymes, may be to use an affinity resin. In affinity chromatography, the resin beads have had attached to them a ligand such as an enzyme substrate or analogue. When a crude mixture of proteins is loaded onto the column, the enzyme recognises the substrate ligand and binds to the beads. The remainder of the proteins do not bind to the column. The enzyme of interest can then be eluted

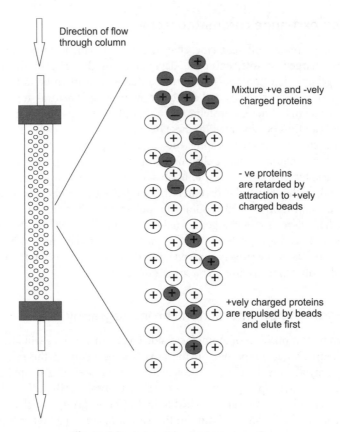

Direction of flow
through column

Mixture +ve and -vely
charged proteins

- ve proteins
are retarded by
attraction to +vely
charged beads

+vely charged proteins
are repulsed by beads
and elute first

Figure 6.8   Anion exchange chromatography

from the column by washing the beads with a solution of the substrate/analogue, which competes for the enzyme with the beads, thus effecting the release of the enzyme. Overall, this technique is simpler than it sounds, as it does of course rely on the successful design and implementation of an appropriate ligand for the beads. However, it has proved very successful in some cases.

The first three methods of protein purification highlighted above may be seen as 'conventional' protein chromatography and each is still useful in the purification of proteins from wild-type or recombinant sources. In any campaign of protein purification, particularly from wild-type sources, it will be necessary to combine these techniques in order to effect protein purification, possibly requiring three or four columns from crude protein to the 'homogeneous' or pure enzyme. As such, a whole sequence of protein purification may be time-consuming and expensive and, because losses of material are associated with each step, also result in meagre amounts of pure catalyst.

However, in an age when most biocatalysts derived from cells are now coming from recombinant systems, methods have been found to engineer the proteins produced by *E. coli* for ease of purification. These methods are collectively known as 'tagging' and the most widely employed is the use of histidine tags.

6.7 Isolation of Recombinant Enzymes Using Histidine Tags

The use of histidine tags constitutes a special case of affinity chromatography, although in this case, one which is not dependent on the specificity of the enzyme as for the affinity chromatography described above.

In Chapter 7, we supply a brief introduction to the molecular biology principles and techniques that have done much to revolutionise the science of biocatalysis within the last twenty years. In Chapter 8 we outline how, through facile modification of the gene sequence encoding an enzyme, it is possible to engineer changes in amino acids at specific sites in an enzyme sequence, and also to introduce new sequences of amino acids either at the end or in the middle of a protein sequence. Many commercial plasmids, the vectors that are used to carry foreign DNA into an *E. coli* cell in order to make it express new genes, carry with them the ability to fuse the protein sequence of interest to a number of histidine residues (or other tags) at either the N-terminus or the C-terminus of the protein as an aid to purification.

The consequence of engineering an expression vector to produce a histidine-tagged protein is that, when the gene is expressed, the protein that results will, in addition to its natural sequence have 4, 10 or more usually 6 histidine residues attached at either its N- or C-terminus. A hexahistidine structure at the end of a protein will have the ability to chelate to divalent metal ions, in this case, particularly nickel ions, to which the protein will bind as a result.

The strategy for single step purification is thus analogous to the affinity chromatography methods outlined above. The polymer beads used for separation are, in this case, charged with nickel ions. The protein is loaded in a standard buffer solution, and binds to the nickel resin. Importantly, none of the other proteins in the mixture bears a histidine tag, so all of these are eluted without binding. After washing with buffer, the enzyme of interest is eluted with an increasing gradient of imidazole. Imidazole of course has the same structure as the side chain of histidine (see Appendix 1), and it competes with the histidine-tagged protein for the binding sites. Thus, at a certain concentration of imidazole (usually between 0 and 500 mM), the enzyme of interest elutes as a clean peak from the column.

The best results from nickel affinity chromatography are often obtained using an FPLC machine to provide the imidazole gradient for purification. One of the advantages of this, and other affinity methods, is that the 'yes/no' mode of binding can be exploited in a simple batch preparation that needs no FPLC machine at all. Consider as an example the method given below for isolating a histidine-tagged enzyme from a recombinant strain of *E. coli*. We will assume that cell material from a 2 L culture of the bacterium has been prepared, containing a protein with an engineered histidine tag.

6.7.1 Benchtop protocol for isolating a histidine-tagged protein

Apparatus and reagents

- 200 mL 50 mM Tris-HCl buffer pH 7.1 containing 300 mM sodium choride and 100 µM PMSF ('the buffer')
- 200 mL volumes of a series of versions of 'the buffer' containing, respectively, 30, 100, 250, and 500 mM imidazole
- 200 mL of a solution of 0.5 M sodium chloride
- 200 mL of a solution of 0.5 M sodium chloride and 100 mM EDTA
- 50 mL of a crude cell extract of *E. coli* that contains the histidine-tagged protein
- Ni-NTA Sepharose beads
- Some deionised water
- Some 25% ethanol in water

[NB Ni-NTA Sepharose beads can be purchased from e.g. Qiagen (Ni-NTA CL-B6 Sepharose), Sigma-Aldrich and other (bio)chemical suppliers.]

1. Prepare 20 mL of a crude cell extract from the cells by ultrasonication, using 'the buffer' as described above. Filter through a 2 µm filter to obtain a clear extract.

2. The nickel resin will be prepared by a series of washing and recovery steps. Dispense 1 mL of the resin slurry (250 µL per 5 mL of extract) into a 50 mL Falcon tube and centrifuge at 3000 r.p.m. for 1 min.

3. Add 20 mL of 50 mM Tris-HCl buffer pH 7.1 containing 300 mM sodium choride and 100 µM PMSF. Mix and then centrifuge again at 3000 r.p.m. for 1 min. Discard the supernatant.

4. Add the clear cell extract to the resin, resuspend and incubate for 1 h at room temperature.

5. Centrifuge at 3000 r.p.m. for 1 min, then add 20 mL of 50 mM Tris-HCl buffer pH 7.1 containing 300 mM sodium choride and 100 µM PMSF. Resuspend, then centrifuge as before. Repeat this wash three times. Retain the pooled supernatants for gel analysis (below).

6. Wash the resin another three times, this time using the buffer containing 30 mM imidazole. Retain the pooled supernatants for gel analysis as before.

7. To the washed resin now add 1 mL of buffer containing 250 mM imidazole. Incubate with shaking at room temperature for 10 min, and then remove the resin by centrifugation. The supernatant should contain the histidine-tagged protein of interest, but to ensure maximum recovery, you might wash the resin again with buffer containing 500 mM imidazole.

8. The resin can be reused after cleaning. In order to clean the resin, it can be washed successively with (i) 100 mM EDTA plus 0.5 M NaCl, (ii) 0.5 M NaCl and (iii) deionised water and then store under 20% ethanol, which should stop the growth of organisms in the resin.

This is a simple procedure that may be used on the bench without the need for expensive chromatography apparatus. The recovery of protein may not be as good as that obtained with a column and pump apparatus, but should at least yield some protein for initial biotransformations tests. The result of the fractionation is that you have an enzyme that is dissolved in a solution containing both sodium chloride and imidazole, either or both of which may be detrimental to the stability or activity of the biocatalyst. You may observe some protein denaturation as precipitation in the tube containing enzyme. Both the salt and imidazole can be routinely removed using a dialysis procedure as described in the previous section.

The success, or otherwise of the protocol, can be established by analysis of the wash fractions obtained during the experiment by gel electrophoresis, but first we will consider a method for determining the quantity of enzyme obtained from such a procedure.

6.8 Estimation of Protein Concentration

Having obtained an extract it will then be useful to know the concentration of enzyme, so that activities (Chapter 1) can be calculated and so it can be estimated how much enzyme might be added to a biotransformation reaction. There are several methods of estimating the concentration of proteins in a given sample. These range from simple and convenient UV spectroscopy to more time-consuming, yet arguably more accurate, chemical methods. It is important to emphasise that these are methods of protein *estimation* and that they should not be considered accurate measures of protein concentration. They are however, accepted as the routine methods by which the calculation of protein concentration is performed.

6.8.1 UV spectrophotometry

The measurement of protein concentration by ultraviolet spectrophotometry is most useful in cases of enzymes that are pure or at least have been through some stages of purification. Simply reading the UV absorbance at 280 nm can give a rough estimation of protein content – an absorbance of 1.00 is often thought of as shorthand for a concentration of 1 mg mL^{-1} although for pure proteins, a better estimate may be gained by adjusting this figure using a correction factor. This factor can be simply calculated by accessing the EXPASY website (Chapter 2) and under 'Proteomics and Sequence Parameter Tools' click on 'ProtParam'. This will open a window into which can be inserted the amino acid sequence of the enzyme in FASTA format. After pasting the sequence, press 'Compute parameters' and a list of various parameters corresponding to the enzyme will be listed, including molecular weight, amino acid composition and extinction coefficients. Under this last heading will be found the statement:

$$\text{Abs } 0.1\%(= 1g/l) \quad 1.470$$

This is an indication that, in theory, a concentration of 1 mg mL^{-1} of that particular protein would give an absorbance at 280 nm of 1.470 and that the rule of thumb would be an underestimate. Multiplying the figure obtained by 1/1.47 will possibly give a more reliable figure for the protein concentration in this instance.

6.8.2 The Bradford assay

The Bradford assay is probably the most popular of the protein assay methods and derives its name from the original procedure described in 1976 in a famous and much-cited paper in *Analytical Biochemistry* by Marion M. Bradford [2]. It essentially exploits the ability of a blue dye, Coomassie Brilliant Blue, to form complexes with some, but not all, amino acids in proteins, and is hence an estimate rather than an absolute measurement of protein concentration. The principles of the Bradford assay have been adopted by commercial kits, including one that is available from Bio-Rad, with which one may simply add a preformed reagent to a protein sample.

Each chemical method of protein determination, such as the Bradford and Lowry assay is dependent on comparing the results obtained to a standard curve, which is usually calculated using bovine serum albumin, a common laboratory standard protein, as the model. We have supplied below a general method for the Bradford assay using the Bio-Rad reagent as this is more user-friendly, if more expensive, than assembling the reagents according to the original publication.

6.8.3 Protocol for estimation protein concentration using Bio-Rad reagent

Apparatus and reagents

- 50 mM Tris-HCl buffer pH 7.1 containing 300 mM sodium choride and 100 µM PMSF or the base buffer used to prepare the protein
- Bottle of Bio-Rad protein assay. This concentrated reagent should be diluted 4:1 with distilled water prior to use
- 1.5 mL capped plastic tubes
- Deionised water
- Bovine serum albumin (BSA; the stock solution of BSA at a concentration of 10 mg mL^{-1} available from molecular biology suppliers such as New England Biolabs or Promega provides a convenient source of BSA for standard curves)
- Protein sample (different dilutions will be used in order to determine a mean)

1. Prepare 15 mL of the diluted Bradford reagent by adding 12 mL of distilled water to 3 mL of the reagent in a clean 50 mL Falcon tube.

2. Prepare a series of aqueous dilutions of BSA in 1.5 mL capped plastic tubes representing 0, 20, 40, 60, 80, 100, 150, 200, 250, 300, 400, 500 µg. Each of these should be in a final volume of 200 µL of deionised water.

3. Prepare three dilutions of the protein sample of interest, by taking 5, 10 and 20 µL of the protein sample and diluting this to 200 µL with deionised water. It may be necessary to prepare additional dilutions, as the result obtained with the unknown protein sample must fall within the linear portion of the standard curve prepared using BSA.

4. To each of the fourteen samples, add 800 µL of the commercial reagent. Mix thoroughly on a whirlimixer and leave to stand for 5 min at room temperature.

5. Using a simple benchtop spectrophotometer, measure the absorbance of the samples at 595 nm. The colour in the samples of low protein concentration will be light brown turning pale blue; it will turn to a strong blue colour for concentrated protein samples.

6. Plot the absorbance of the samples against the protein concentration – a curve is obtained that is mostly linear throughout the first ten samples or so (Figure 6.9). Now read the absorbance of the samples of unknown protein concentration. If the readings exceed the linear portion of the graph, repeat using three dilutions of lower concentration. Ideally, a sample that contains 20 µL of the protein should have twice the absorbance of one that has 10 µL, etc., but in many cases, an average has to be calculated. The readings should give a value for your protein sample (in µg per mL of protein), which has to be correct for the dilutions that you have made.

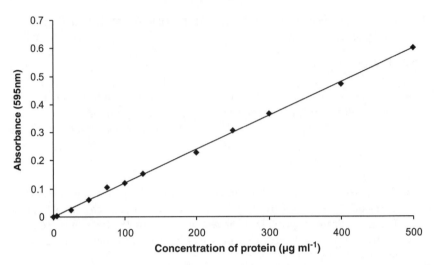

Figure 6.9 Calibration curve for a protein estimation experiment, in which the concentration of protein in a series of samples of known concentration is plotted against the absorbance of the sample at 595 nm

6.9 Concentrating Protein Samples by Centrifugation

Pure or partially pure enzyme that has been recovered from dialysis following ammonium sulfate fractionation or nickel affinity chromatography will be of a certain protein concentration that can be estimated using one of the methods described above. Should the concentration of the enzyme be below that required for the biotransformation reaction, the enzyme solution can be concentrated in a number of ways. You could of course either precipitate the solution with ammonium sulfate as described above, and re-suspend the protein in much less buffer, then dialyse to remove the salt. Also, where appropriate, one may lyophilise or freeze-dry the solution to a powder that can be added in milligram quantities to the reaction buffer or solvent. However, enzyme solutions can also be concentrated in centrifuges using membrane concentrators – so-called 'ultrafiltration'. These commercially available devices come in many different volumes and designs (Figure 6.10), but the principle of action is roughly the same.

The enzyme solution is placed in a chamber adjacent to a membrane filter of a specified molecular weight cut-off (say, 10 000 Da). The device is then placed in a centrifuge and spun at low speed (perhaps 3000 r.p.m. so as not to break the membrane). The progress of centrifugation is checked at intervals. A filtrate will accumulate on the other side of the membrane, whilst the enzyme solution is concentrated in the chamber. This is a quick and easy method for enzyme concentration, but care should be taken with unstable enzymes and enzymes may

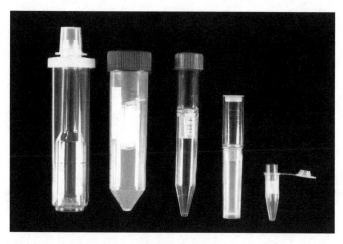

Figure 6.10 A range of membrane concentrator units for concentration of protein solutions in centrifuges

also exhibit a tendency to associate with the membranes, and can be washed with a minimum of buffer to recover any residue.

6.10 Analysis of Protein Samples by Sodium Dodecylsulfate Polyacylamide Gel Electrophoresis

We have presented a number of techniques for the isolation and fractionation of protein extracts, analogous to the reaction work up and chromatography used routinely in synthetic organic chemistry. In order to assess the purity of small organic molecules, or to evaluate the progress and/or success of purification columns, TLC is used. In enzyme biochemistry, the standard equivalent of TLC for the analysis of protein purity is SDS-PAGE.

In SDS-PAGE, in its most accessible form, proteins are loaded into lanes in a vertically positioned slab gel that is connected to electrodes in a tank full of buffer, and a voltage is then applied across the gel. The proteins migrate down the gel according to their size, for reasons explained below, which is usually estimated by including a range of molecular weight markers on the same gel. The gel is removed from the tank and stained with Coomassie Brilliant Blue, which interacts with proteins in much the same way as is observed for the Bradford protein estimation assay described above. The excess stain is then removed to reveal bands that correspond to proteins within the sample. A series of many bands of different sizes is observed for a series of crude samples (Figure 6.11a) – the objective of any enzyme purification strategy is of course to reveal only one band on the gel, of the molecular weight corresponding to the expected enzyme (Figure 6.11b).

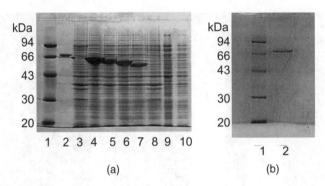

Figure 6.11 SDS-PAGE gels showing low molecular weight markers (lane 1, a), crude cell extracts (lanes 3–10, a) and low molecular weight markers (lane 1, b) against a pure protein (lane 2, b)

Before the protein sample is loaded onto the gel, it is boiled with an amount of 'loading buffer' (see Appendix 5) that contains a blue dye and a detergent compound, sodium dodecyl sulfate (SDS). The effect of the SDS is to bind to hydrophobic regions of the enzymes, in a ratio common to all proteins. This ensures that the proteins migrating down the gel all have the same *charge*, and thus migrate on the basis of size only, negating the possible effect of the different charges of the proteins when subjected to an electric field. As all the proteins in the samples are denatured, the consequence of this is that the bands that arise due to the presence of the expected enzyme are due to the monomeric single polypeptide of the enzyme in question, and not in quaternary complex either with itself, in oligomers, or with other proteins, for ease of interpretation of the gel.

The gel itself is made from polyacrylamide, made from a solution of acrylamide that contains the cross-linker *bis*-acrylamide. When treated with two additives, ammonium persulfate and TEMED (*N,N,N',N'*-tetramethylethylenediamine), the gel solidifies. The composition of the gel can be altered to either speed up the migration of large proteins (by including less acrylamide) or slow down the migration of smaller ones (by using a larger percentage of acrylamide in the gel). Given that most useful enzymes, when denatured, have monomer molecular weights within the range 10 000–100 000 Da, a gel containing 12% acrylamide is a useful all-purpose gel for getting started. The buffer in which the gel is run is a simple Tris-glycine buffer that also contains a small amount of SDS, the same denaturant added to the loading buffer. After a period of about 45–60 min, the gel is removed from the apparatus and stained with Coomassie Brilliant Blue dye, after which the stain is removed using a mixture of methanol, acetic acid and water to reveal the protein bands.

We provide below a simple method for the analysis of a protein sample using an SDS-PAGE gel. It should be noted that gels of the required percentage can be purchased and merely inserted into the electrode apparatus and used. However, these are expensive and so we provide a standard method and a list of the equipment and reagents that will be required.

6.10.1 Procedure for running a 12% SDS-PAGE gel

Apparatus and reagents

- Hoefer Mighty Small Multiple Gel Caster for 8 × 7 cm gels
- Hoefer Mighty Small II gel apparatus for 8 × 7 cm gels
- Small and large glass plates for 8 × 7 cm gels
- Plastic gel spacers
- Plastic gel comb
- Some filter paper
- 50 or 100 μL glass-barrelled syringe of type used for HPLC injection
- 30% Acrylamide solution
- 1 mL of a 10% w/v solution of ammonium persulfate (APS)
- A bottle of TEMED
- Wash bottle of deionised water
- Small bottle of butanol
- Loading buffer (see Appendix 5)
- Gel running buffer (see Appendix 5)
- Low molecular weight marker mixture from BioRad or other biochemicals supplier
- Coomassie stain solution (see Appendix 5)
- Destain solution (see Appendix 5)

1. Make a sandwich of one small and one large glass plate and separate them at the edge with the plastic gel spacers. Mount this vertically into the gel stand, making sure that the bottom of the glass sandwich makes a neat fit with the rubber strip at its base. Gently screw in the sides of the apparatus that hold the sandwich. Do not overtighten as this will crack the glass plates (Figure 6.12).

2. Check for leaks in the assembly with water. Squirt some water in between the glass plates and wait for a few minutes to check that the level of the meniscus does not drop. If it does, then unscrew the apparatus, reassemble and repeat the check.

3. When you are satisfied that the apparatus does not leak, mix 5 mL of the resolving gel mix for a 12% polyacrylamide gel (see Appendix 5) and to this add

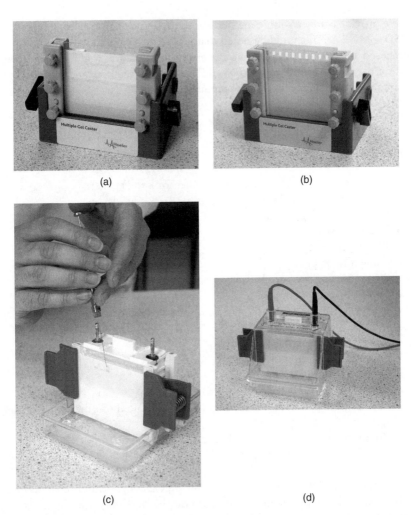

(a)

(b)

(c)

(d)

Figure 6.12 Preparation of an SDS-PAGE gel. The resolving gel is first poured into the gel caster and left to set (a). The stacking gel, mixed with a blue dye, is then poured on top and a plastic comb inserted to form the wells (b). The gel is placed in the electrode apparatus, buffer is added, and samples of protein, mixed with loading buffer are injected into the wells (c). The gel apparatus is connected to a power pack and run until the line of blue dye reaches the base of the gel (d)

$30\,\mu L$ of the APS solution, and $16\,\mu L$ of the neat TEMED. Mix gently and, using a pipette, squirt the mixture in between the glass plates until about two-thirds of the space is full. The top of the liquid may have bubbles and be uneven. This may be evened out by the addition of a small amount of butanol, which sits on top of the gel solution. Leave the gel to stand for 30 min.

4. Check that the gel has set by inclining the apparatus – the butanol should move but the top of the gel should be set. Remove the butanol by pouring off the excess and blotting the remainder away with the edge of some filter paper. Mix together 2.5 mL of the stacking gel solution (see Appendix 5) and add 15 μL of 10% APS and 8 μL of TEMED. Mix gently and squirt between the glass plates. Immediately take the plastic gel comb and insert between the gel plates – the excess stacking gel mix will run out and down the glass. Make sure that the teeth of the comb are straight and that there is a small distance between the base of the comb teeth and the start of the resolving gel. Leave to stand for 30 min.

5. Fill the gel tank with the gel running buffer. Remove the solidified gel in its supports from its stand and place it into a gel electrode. Submerge this in the buffer and only then remove the comb, very carefully from between the plates.

6. To 10 μL of each protein sample for analysis, add 10 μL of the gel loading buffer (see Appendix 5) in small plastic-capped tubes. Boil the samples for 5 min, as well as 10 μL of a solution of low molecular weight markers, diluted according to the manufacturer's instructions.

7. In the first lane on the left of the gel, carefully dispense 10 μL of the markers solution using the HPLC syringe. A blue solution should settle within the well – the carbohydrate in the loading buffer makes the samples dense for this purpose. Wash the syringe thoroughly with deionised water, then transfer 10 μL of the first protein mixture into the well adjacent to the ladder. Do this with the remaining samples, and then place the lid on top of the apparatus.

8. Set the power pack to run the gel at a voltage of 200 V. It will take approximately 45–60 min for the blue dye in the samples, which is small and will migrate more quickly than any of the proteins in the sample, to reach the bottom of the gel. When the blue dye in a line is observed to reach the base of the gel, switch off the power.

9. Remove the gel sandwich from the apparatus, and be careful to clean all the plastic apparatus in water, as lasting exposure to the SDS can damage the plastic over time. Ease the glass plates apart by slowly turning one of the plastic spacers and remove one glass plate from the gel. Carefully ease the gel off the other glass plate into a suitably sized beaker or plastic receptacle.

10. Add 100 mL of the Coomassie stain solution to the gel and leave to stain for 30–60 min. The staining process can be sped up if the gel in stain solution is heated carefully in a microwave oven; do *not* boil the stain solution.

11. Pour off the excess stain – this can be reused many times – and wash the gel with water. Then add 100 mL of destain solution and either shake the gel gently at intervals or place on a rocking platform if available. Discard the destain (this can be recyled if filtered through activated charcoal in a filter funnel and collected) at intervals, and replace with fresh destain solution. A series of changes of destain

will result in the finished gel. The gel can then be washed and kept in water and photographed as a record of the experiment. For longer-term storage, the gels can be suspended in 20% glycerol.

There are many other ways to analyse proteins by gel electrophoresis, including 'native gel' electrophoresis, which is done in the absence of SDS, and therefore shows proteins in their non-denatured form – this can give information on subunit association for example. It is also possible to exploit the differences in pI of each enzyme for separation in a technique known as isoelectric focusing (IEF). In 2-D electrophoresis, SDS-PAGE and IEF are combined – the latter is run first to effect separation of proteins by pI, then the gel is turned 90° and subjected to SDS-PAGE. These techniques are beyond the scope of this book and would not be used routinely by workers interested in merely assaying the purity of either home-grown or commercial biocatalysts.

6.11 Examples of Enzyme Assays

When an enzyme catalyst has become the focal point of an important industrial biotransformation, significant efforts will have been made to understand the kinetic parameters affecting catalysis, including a determination of turnover rates, the Michaelis constant (see Chapter 1) and also the possible effects of high substrate and product concentrations on the inhibition of enzyme activity. As this is just a beginner's guide, we will not provide extensive details on the quantitative characterisation of enzymatic activity, or the optimisation of enzymatic reactions. However, once an enzyme extract has been obtained, it may be useful at least in basic terms, to quantify the amount of activity present, as an accessory to demonstrating the activity by TLC, GC or other analytical technique, before committing the precious enzyme to a preparative biotransformation reaction.

The most convenient quantitation of enzyme activity is usually (but not exclusively) provided using UV spectrophotometry, wherein the transformation of substrate into product will result in a measurable change in the absorbance at a defined wavelength. Such assays are now routinely performed in high-throughput manner using 96 (or more) well-plate based assays and automatic plate readers. However, where a single enzyme is concerned, a few readings on a simple UV spectrophotometer can provide a lot of useful information. The monitoring of substrate disappearance or evolution of a product requires, of course, that the relevant entity has a suitable chromophore. Observation of either of these phenomena might be thought to constitute a primary observation of enzyme activity, however, it is important to remember that, particularly with crude extracts, the disappearance of substrate may be due to another enzyme altogether.

It is also true that not all enzyme substrates or products will have a suitable chromophore, but it may be possible to observe the enzyme activity more indirectly. A simple example of this is with nicotinamide-dependent enzymes: NAD(P)H absorbs strongly at 340 nm, but its oxidised counterpart does not. Hence the reductive activity of an alcohol dehydrogenase might be observed by monitoring the decrease in UV absorbance at 340 nm. Of course enzymes other than that expected may also use NAD(P)$^+$, so the assay, especially in crude extracts, needs some sort of validation, perhaps by HPLC or GC (even TLC) by which the evolution of the desired product can be verified.

More complex examples, still relatively simple to perform, involve coupled enzyme assays. In these examples, the activity of the enzyme of interest is not directly observable, but the transformation of the evolved product by a different enzyme may yield a suitable assay. For example, the activity of the carbon-carbon bond forming enzyme transketolase has not usually been followed directly, but rather by exploiting two additional enzyme reactions (Figure 6.13); the final observation is of the oxidation of NADH by the enzyme glycerophosphate dehydrogenase acting on dihydroxyacetone phosphate [3].

6.11.1 Enzyme activity and specific activity – the unit

In Chapter 5, when describing literature methods for the application of commercially available biocatalysts, we came across the measure of enzymatic activity known as the 'unit'. One unit of enzyme activity is described as the amount of enzyme that converts 1 μmol of substrate to product (or evolves 1 μmol of product) in 1 min per unit amount of enzyme. If quoted as an activity per milligram of protein, the activity is defined as *specific activity*, but it is fairly common to see enzyme activity quantified as U mL^{-1}. Recently, the international biochemistry community has moved to encourage the uniform use of the term *katal* (kat), which denotes 1 mol of substance converted per minute. Enzyme activities are commonly cited in terms of *nanokatals* (nkat).

6.11.2 Examples of spectrophotometric enzyme assays

Although we cannot provide an exhaustive list of spectrophotometric assays of enzyme activity, below is a short list with some valuable references, that should aid in starting enzyme assays using the enzymes most commonly used in the biotransformations literature. In each case, the activity is calculated using Beer's law as a basis:

$$\Delta A = \varepsilon \Delta c \; l$$

where ΔA is the change in absorbance per unit time,; ε the molar absorption coefficient of the chromophoric species (acquired either from the literature or from

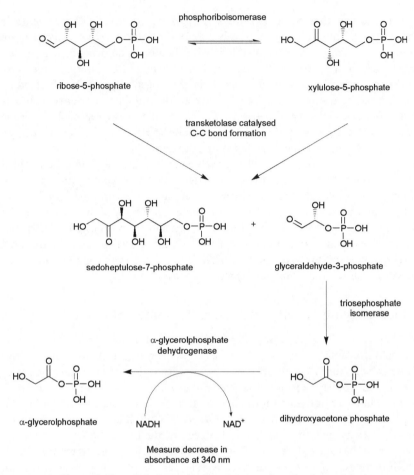

Figure 6.13   A linked assay for the estimation of transketolase activity

a plot of absorbance versus concentration, *c*, at the required wavelength) and *l* is the path length of the cell in centimetres (which in most UV methods is equal to 1). A simple rearrangement allows calculation of the change in concentration of the absorbing species per unit time, thus the change in amount based on the volume of the cuvette. When the value is corrected for the amount of protein, as determined by the protein estimation assay described in Section 6.10, then unit values for the activity of the enzyme are easily determined. Assays are routinely performed in duplicate or triplicate and an average taken to calculate the final value.

6.11.3 Hydrolases

Lipases, esterases and proteases are routinely assayed using hydrolytic reactions that depend upon the release of *para*-nitrophenol (p – NP) from a suitably designed substrate. Hence, the esterolytic activity of esterases and lipases is assayed using the hydrolysis of *para*-nitrophenyl acetate; proteases by the hydrolysis of p – NP from a suitable amide.

6.11.3.1 Lipase

Lipase activity is most usually assayed by measuring the evolution of acetic acid or butyric acid from triacetin or tributyrin, respectively, using either titration or a pH stat, but activity can also be assessed using the hydrolysis of *para*-nitrophenyl acetate or propionate according to the example given in reference [4]. Here, the activity of CAL-B was measured using the increase in absorbance at 348 nm in a 25 mM phosphate buffer at pH 7.5, with 0.4 mM of the substrate, *para*-nitrophenyl propionate.

6.11.3.2 Esterase

Spectrophotometric assays for esterases are often based on the detection of p-NP released from a p-NP ester of the carboxylate for which the esterase shows substrate specificity. For example, in the assay of feruloyl esterase, an enzyme that releases ferulic acid from plant polysaccharides, 4-nitrophenyl ferulate is used as the substrate (Figure 6.14).

There are some solubility issues that can be addressed using co-solvents and/or detergents such as Triton X-100 [5]. In the assay, the substrate solution was first prepared by mixing 9 vol of 0.1 M potassium phosphate buffer at pH 6.5 containing 2.5% Triton X-100 with 1 vol of 10.5 mM 4-nitrophenylferulate. The release of p-NP was monitored at 410 nm, by adding 5% final volume of enzyme solution to the substrate solution in a cuvette.

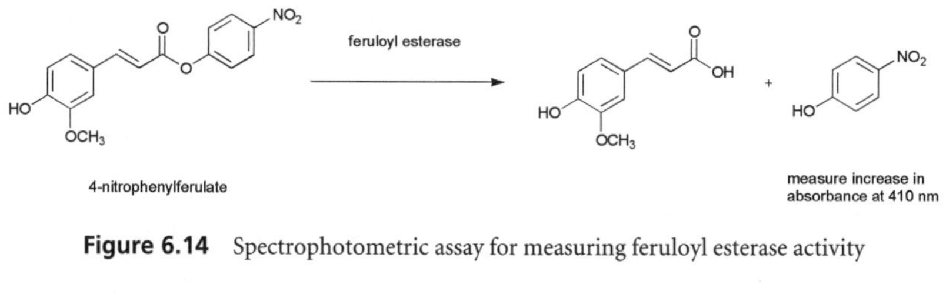

Figure 6.14 Spectrophotometric assay for measuring feruloyl esterase activity

6.11.3.3 Epoxide hydrolase

Epoxide hydrolases are difficult to assay using spectrophotometry, as the reactions rarely result in the generation or alteration of a suitable chromophore. They have thus routinely been assayed using either GC or HPLC methods. However, a recent report describes an assay that can be performed in a spectrophotometer that is suitable for any epoxide that has at least one hydrogen substituent [6]. The assay is dependent on the reaction of ketones or aldehydes that result from periodate cleavage of diols with Schiff's reagent (Figure 6.15).

The dye that forms can be assayed spectrophotometrically by reading at 560 nm. For the assay a 90 mM solution of sodium periodate in 0.1 M sodium acetate buffer pH 5.0 is prepared and a 0.8 M solution of sodium sulfite pH 5.0 and Schiff's reagent. Then 100 µL of the reaction mixture is mixed with 100 µL of sodium periodate solution and incubated at room temperature for 10 min. Then 80 µL of sodium sulfite was added and the solution centrifuged. To 200 µL of the clear supernatant was added 20 µL of the Schiff reagent and this was incubated overnight. An additional 50 µL of sodium sulfite solution was then added and the absorbance read at 560 nm. It was necessary to perform careful control experiments as the amount of background absorbance was related to the amount of cell extract in the assay. Another spectrophotometric assay of epoxide hydrolase activity was reported in reference [7].

6.11.3.4 Glycosidase

The release of *p*-NP is also very suitable as the basis of an assay for glycosidase or glycosyl hydrolase enzymes, where the *p* – NP ether is made in the correct

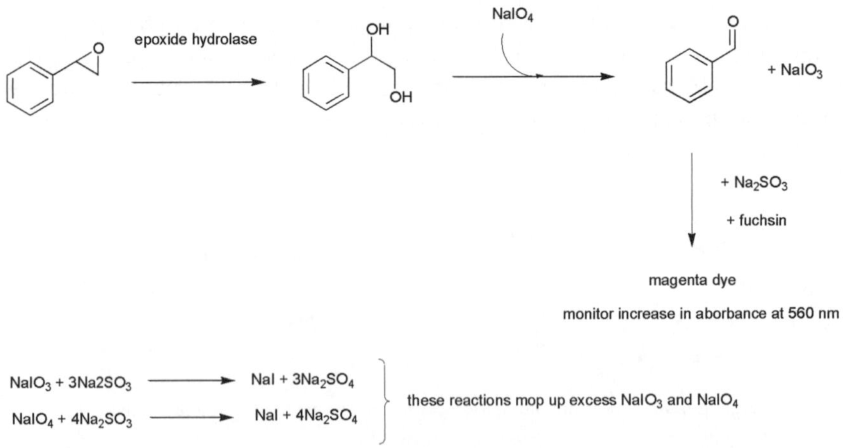
epoxide hydrolase
OH
OH
NaIO4
O
+ NaIO3
+ Na2SO3
+ fuchsin
magenta dye
monitor increase in aborbance at 560 nm
NaIO3 + 3Na2SO3
NaI + 3Na2SO4
NaIO4 + 4Na2SO3
NaI + 4Na2SO4
these reactions mop up excess NaIO3 and NaIO4

Figure 6.15 Spectrophotometric assay for measuring epoxide hydrolase activity

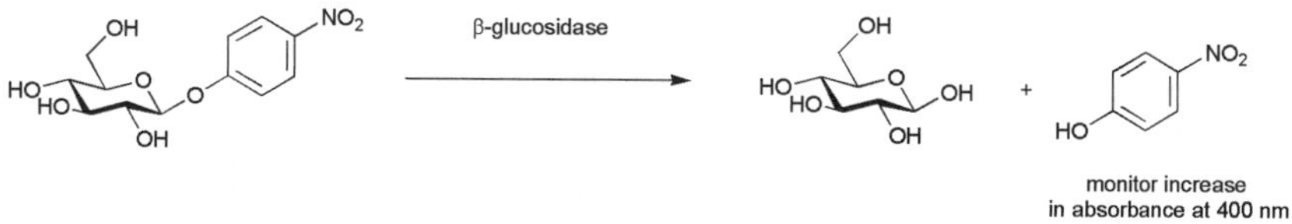

Figure 6.16 Spectrophotometric assay for measuring glycosidase activity

anomeric configuration to suit the stereoselectivity of the particular enzyme. As an example, the activity of a human cytosolic β-glucosidase (hCBG) was assayed using $p-$NP-β-D-glucopyranoside as substrate. Substrate concentrations of $10-500\,\mu M$ were used in $50\,mM$ phosphate buffer pH 7.0 (Figure 6.16) [8]. The addition of substrate was followed by monitoring the increase in absorbance at $400\,nm$.

6.11.3.5 Nitrilase

As with epoxide hydrolases, the lack of a suitable chromophore change means that direct monitoring of activity is difficult, but some spectrophotometric assays have been developed. One method is dependent on the addition of a buffered solution of $CoCl_2$ to the reaction mixture (Figure 6.17).

The $CoCl_2$ reacts with the ammonia produced as a by-product of nitrilase activity and the resultant colour change can be measured in a spectrophotometer at $375\,nm$ [9]. For the assay, a $1\,mL$ volume of substrate solution consisting of $20\,mg$ of the nitrile substrate in $10\,mM$ Tris-HCl buffer pH 7.0 was prepared. Enzyme was added and the reaction left for 2 h at room temperature. One volume of this mixture was then mixed with 1 vol of $CoCl_2$ solution, which consisted of $10\,mM$ cobalt in $10\,mM$ Tris-HCl buffer pH 7.0. The absorbance at $375\,nm$ was read after incubating the mixture for 5 min at room temperature.

6.11.4 Oxidoreductases

6.11.4.1 Ketoreductase or Alcohol Dehydrogenase (ADH)

Alcohol dehydrogenase assays are the most straightforward examples of enzyme assays that are dependent on monitoring the substrate-coupled oxidation or

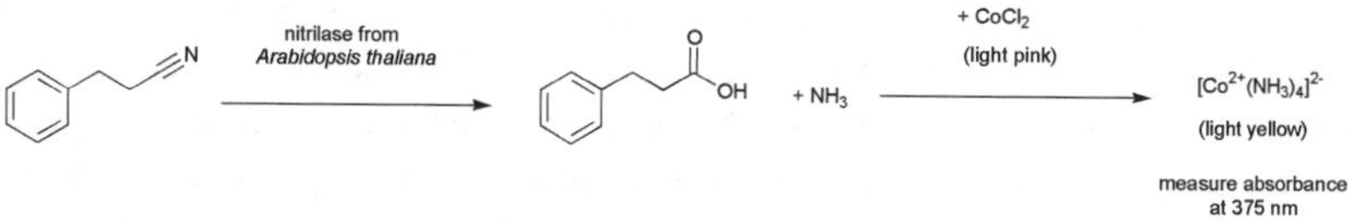

Figure 6.17 Spectrophotometric assay for measuring nitrilase activity

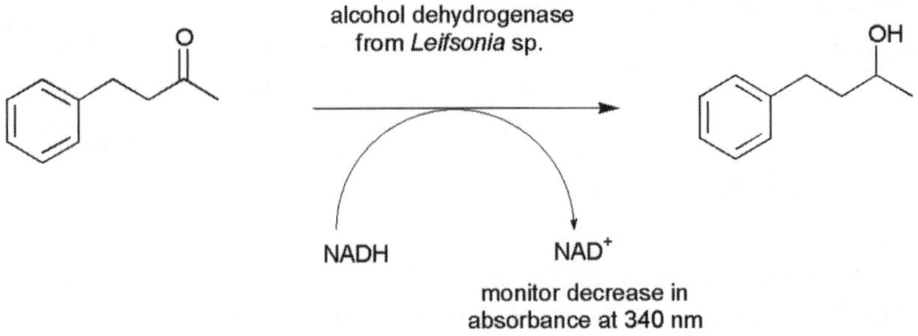

Figure 6.18 Spectrophotometric assay for measuring alcohol dehydrogenase (ketoreductase) activity

reduction of NAD(P)/H. ADHs function by delivering hydride from NAD(P)H to the substrate in a manner analogous to that achieved using e.g. sodium borohydride. The change from reduced to oxidised coenzyme can be monitored at 340 nm in the spectrophotometer (Figure 6.18).

The molar extinction coefficient of NAD(P)H is usually given as $4220 \, \mathrm{mol^{-1}}$ $\mathrm{dm^3 \, cm^{-1}}$. A simple example is given by the recent characterisation and assay of a wide-spectrum NADH-dependent ADH from a *Leifsonia* species [10] that reduces a number of aldehydes and secondary carbonyl compounds including benzaldehyde, 2-heptanone and 1-phenyl-3-butanone. The assay mixture, in a final volume of 1.5 mL consisted of 0.05 M potassium phosphate buffer, 0.4 µmol of NADH, 3 µmol of ketone substrate and 10 µL of the enzyme solution. In practice, particularly with crude cell extracts, it is sometimes advisable to measure the oxidation of the cofactor *before* the addition of substrate, as the extract may well contain other enzymes capable of oxidising NADH. After measuring this, the substrate is added, and it is the *substrate-stimulated* oxidation of the cofactor, derived by subtracting the control rate from the second rate, that will give the accurate measure of activity.

6.11.4.2 Baeyer–Villiger Monooxygenase (BVMO)

The assay of BVMOs constitutes a slightly more complex challenge for spectrophotometric assay than ADHs. The mechanism of BVMOs starts first with the NADPH-dependent reduction of a flavin coenzyme FAD, which is usally covalently bound to the enzyme. The reduced flavin, $FADH_2$, then reacts with molecular oxygen to form a flavin 4a-hydroperoxide which is the source of the biological peroxidate that catalyses the oxygen insertion reaction from ketone to lactone, analogous to that performed routinely wth peracid reagents in organic chemistry. It is thus possible to assay BVMOs by the substrate stimulated oxidation of NADPH in a similar experimental fashion to that used with ADHs.

4-Hydroxyacetophenone monooxygenase (HAPMO) is a BVMO that catalyses the oxygenation of 4-acetophenone to its corresponding acetate ester [11] but also accepts a range of aromatic ketones and sulfides as substrates. In the assay, a 1 mL volume of 50 mM potassium phosphate buffer pH 7.5 contained 250 μM NADPH and to this would be added 1 mM of the ketone substrate. In the particular case of HAPMO, a wavelength of 370 nm was used, in conjunction with an altered molar extinction coefficient, to monitor the oxidation of the cofactor, as hydroxy-acetophenone demonstrated a pronounced absorbance at 340 nm. In most cases, 340 nm would be appropriate.

6.11.4.3 Amino Acid Oxidase (AAO)

AAOs are flavin-containing oxidoreductases that catalyse the oxidation of amino acids to their corresponding imino acids, which are hydrolysed quickly in the aqueous medium to give keto acids. They do not require nicotinamide cofactors for activity and hence a direct spectrophotometric assay is unsuitable. One by-product of the enzyme activity is hydrogen peroxide and it is the evolution of this that can be measured by including in the assay a peroxidase enzyme that uses H_2O_2 to oxidise the substrate O-anisidine (Figure 6.19). The evolution of hydrogen peroxide is also used as the basis for measuring the activity of a number of other oxidase enzymes.

In one example, an L-AAO from snake venom was characterised using the coupled assay described [12]. The assay contained: 0.1 M Tris-HCl buffer pH 8.0, 20 μM L-leucine as AAO substrate, 10 μM anisidine, and horseradish peroxidase. Enzyme was added to initiate the reaction, which was monitored spectrophoto-metrically at a wavelength of 436 nm.

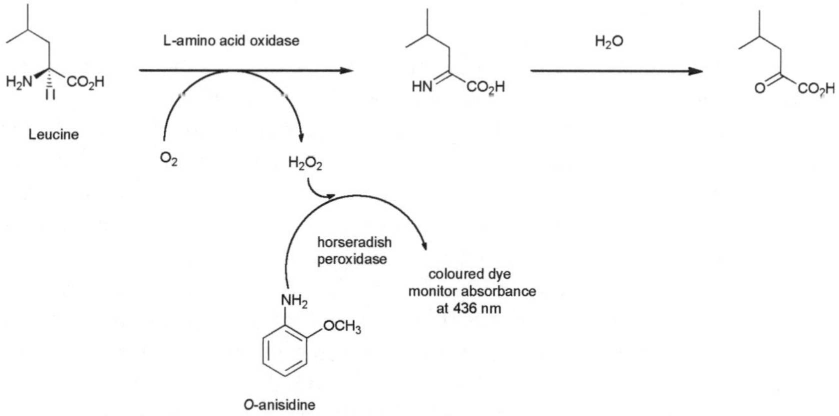

Figure 6.19 Spectrophotometric assay for measuring amino acid oxidase activity

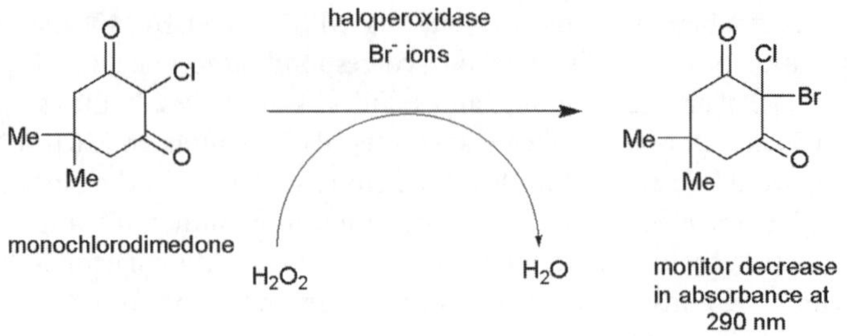

Figure 6.20 Spectrophotometric assay for measuring haloperoxidase activity using monochlorodimedone as substrate

6.11.4.4 Chloroperoxidase (CPO)

CPOs catalyse the halogenation of organic compounds by generating hypohalous acid from hydrogen peroxide and halide ions, but can also catalyse enantioselective epoxidation reactions as described in Chapter 5. The assay used routinely for CPOs is the monochlorodimedone (MCD) assay (Figure 6.20) which measures the rate of chlorination of the substrate 2-chloro-5,5-dimethyl-1,3-dimedone. In one example [13], mutants of a vanadium-containing CPO from the fungus *Curvularia inequalis* were assayed using, in a 1 mL volume: 100 mM Tris-sulfate buffer pH 8.0; 100 μM sodium vanadate (required only for vanadium-containing CPOs); 1 mM hydrogen peroxide; 1 mM of bromide ions (sodium or potassium bromide); and 50 μM of monochlorodimedone. The decrease in absorbance was monitored at 290 nm.

6.11.5 Transferases

6.11.5.1 Transaminase

Transaminase (or aminotransferase) enzymes are transferases that are dependent on the covalently bound cofactor pyridoxal phosphate (PLP; see Chapter 1) that catalyse the deamination of amino acids in the first half of their catalytic reaction cycle. They are then able to use the sequestered ammonia equivalent to transfer an amine group to a keto acid or a ketone in the second half of the cycle. As with some enzymes described above, a direct measure of the activity is often not possible and a coupled spectrophotometric assay is performed. In the example, an aspartate aminotransferase (AAT) from the archaebacterium *Haloferax mediterranei* converts aspartate into the keto acid product oxaloacetate [14]. Oxaloacetate is converted into malate using malate dehydrogenase, the activity of which is dependent on NADH. Hence, the activity of the AAT is measured indirectly by the oxidation of NADH at 340 nm in the spectrophotometer (Figure 6.21). The assay mixture in the cuvette consists of a 50 mM potassium

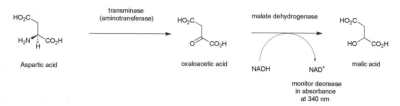

Figure 6.21 Spectrophotometric assay for measuring transaminase (aminotransferase) activity

phosphate buffer, at pH 7.8, 100 mM L-aspartate, 0.05 mM PLP, 0.3 mM NADH, 60 units of malate dehydrogenase from pig heart (in this case prepared in 20% v/v glycerol) and 10 mM 2-oxoglutarate.

6.11.5.2 Glycosyltransferase

Glycosyltransferases transfer sugar residues from an activated sugar bound to a nucleotide, to a sugar or aglycone acceptor as seen in Chapter 5. Once again, the direct observation of enzyme activity was by using a UV spectrophotometer so a form of coupled assay must be used. The assay used most routinely is based on that established in reference [15] and depends on a coupled method dependent on indirectly measuring the phosphate evolved when the sugar nucleotide is spent after the catalytic cycle (Figure 6.22).

Pyruvate kinase catalyses the transfer of phosphate from phosphoenolpyruvate to UDP, leaving pyruvate, the evolution of which can be monitored by the NADH-dependent activity of lactate dehydrogenase converting pyruvate to lactate and monitored at 340 nm. One recent example concerned the kinetic characterisation of two macrolide antibiotic modifying enzymes, OleI and OleD, which catalyse the transfer of glucose from UDP glucose to, in the first case, oleandomycin [16]. The assays in a volume of 500 μL contained therefore: 20 mM Tris-HCl buffer pH 8.0 containing 13 mM manganous chloride; 1 mg mL^{-1} bovine serum albumin; 0.7 mM potassium phosphoenolpyruvate; 0.15 mM NADH; 1.8 units of pyruvate kinase; 3.6 units of lactatae dehydrogenase; and 10 nM to 55 μM of the GTase enzyme. The concentration of the aglycone, oleandomycin was 0.5 mM and UDP glucose was added at concentrations of 0.2 to 10 times the K_M value determined for the enzyme.

6.11.6 Carbon-carbon bond forming enzymes

6.11.6.1 Aldolase

Aldolases, which catalyse the coupling of aldehyde and an alcohol to form a β-hydroxyketone are routinely divided into two classes as described in Chapter 5:

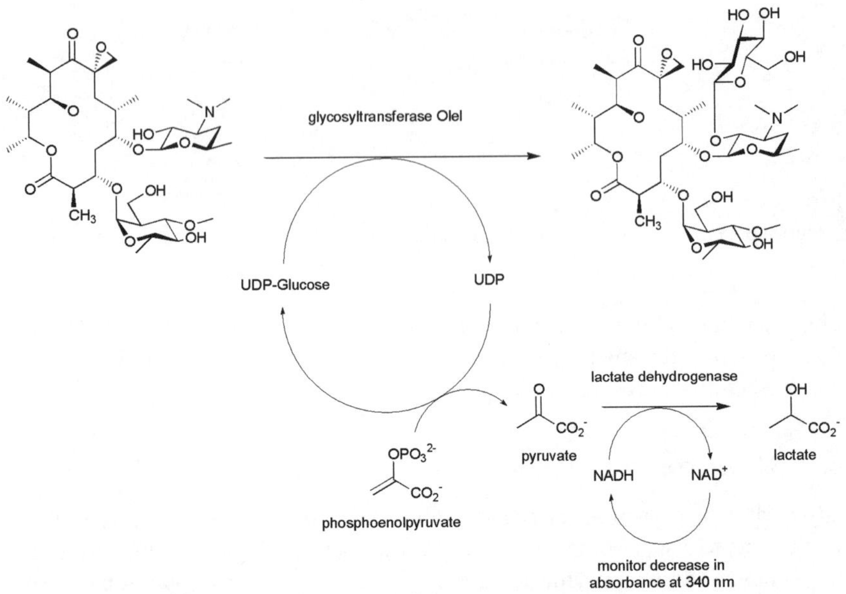

Figure 6.22   Spectrophotometric assay for measuring glycosyltransferase activity

those that employ a covalent Schiff base in catalysis and those that stabilise reaction intermediates using a zinc atom in the active site. Such enzymes can again be assayed using coupled methods. In the example [17], the enzyme 2-keto-4-hydroxyglutarate aldolase from bovine kidney catalysed the aldol coupling of pyruvate and glyoxylate to form 2-keto-4-hydroxyglutarate (KHG) (Figure 6.23).

The enzyme assay reported used, in a total volume of 1 mL: 100 μmol Tris-HCl buffer pH 8.4; 5 μmol of glutathione; 0.5 μmol of racemic 2-KHG; 0.165 μmol of NADH; 12.5 μg of lactate dehydrogenase; and a small sample of the aldolase enzyme. It is the activity of the lactate dehydrogenase acting on pyruvate evolved as a product of the reaction that is taken as a measure of the activity of the aldolase, by measuring the rate of NADH-dependent oxidation of pyruvate at 340 nm.

6.11.6.2 Hydroxynitrile Lyase (HNL)

HNLs or oxynitrilases catayse the reversible addition of cyanide to aldehydes, giving chiral cyanohydrins as seen in Chapter 5. A convenient spectrophotometric assay for HNLs was recently described [18]. In this method, the cyanide ions evolved from cyanohydrin cleavage are oxidised by N-chlorosuccinamide to cyanide cations. These react with isonicotinic acid forming the dialdehyde (Figure 6.24),

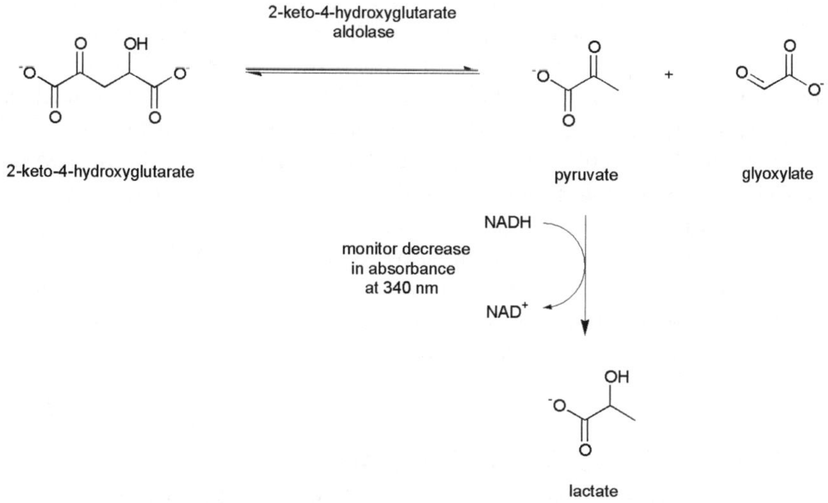

Figure 6.23   Spectrophotometric assay for measuring aldolase activity

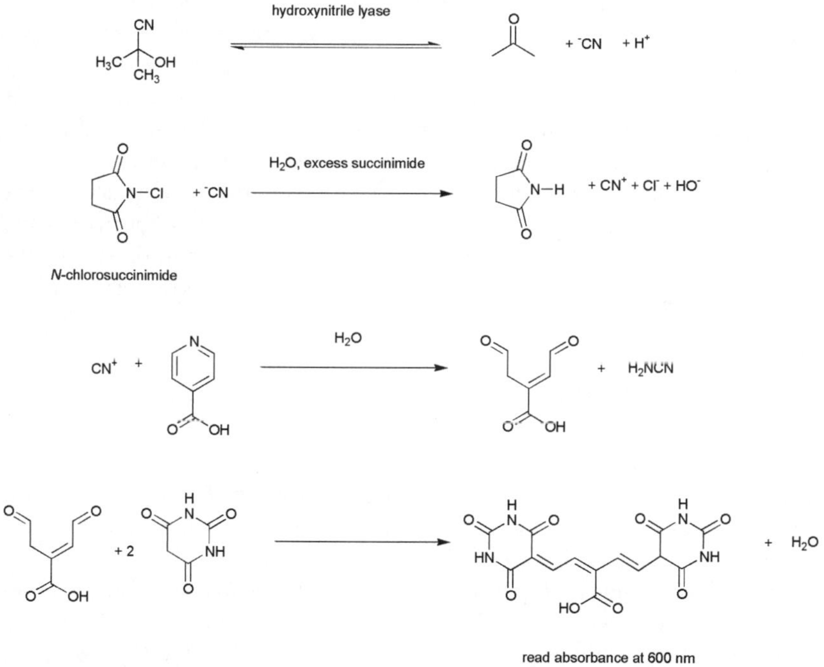

Figure 6.24   Spectrophotometric assay for measuring hydroxynitrile lyase activity

which reacts with barbituric acid to form a dye, the evolution of which can be followed spectrophotometrically at 600 nm. The assay is constituted thus: 140 μL 0. 1 M citrate-phosphate buffer pH 5.0 with 10 μL of cell extract containing the enzyme and 10 μL of a 15 mM solution of a suitable substrate such as benzaldehyde cyanohydrin are mixed and incubated for 5 min at room temperature. Then 10 μL of a mixture of N – chlorosuccinimide and succinimide (100 mM NCS with a 10-fold excess w/w of succinimide) is added to stop the HNL-catalysed reaction. After incubation for 2 min, 30 μL of a solution containing 65 mM isonicotinic acid and 125 mM barbituric acid in 0.2 M sodium hydroxide is added, and the evolution of the dye monitored at 600 nm over a period of 20 min. The pH of the assay was deemed to be crucial, as hydroxynitriles decay spontaneously at pH above 6.0, yet the activity of HNLs is poor below pH 5.0.

6.12 Using Home-Grown Enzymes for Biotransformations – Some Recent Examples

The procedures listed in this chapter should enable the preparation and analysis of crude enzymes that are suitable for application in preparative biotransformations. Again, it should be emphasised that, at this stage, most of the examples of biotransformations in the literature that do not involve commercial hydrolases are being performed by experienced biotransformation laboratories. However, it is hoped that this chapter has given workers some confidence in preparing their own enzymes for application. The following is a brief summary of some methods of application of home-grown biocatalysts in biotransformations from the literature, which detail the source of the enzyme, the degree of purity in which it was prepared and applied and the amounts and reaction conditions employed in the biotransformation. Many of the techniques and concepts described in detail in the chapter are illustrated.

6.12.1 Hydrolytic Reactions

6.12.1.1 Esterase

Zocher and co-workers described the use of the cell extract of a recombinant strain of *E. coli* expressing an esterase from *Pseudomonas fluorescens* for the hydrolysis of esters such as α-phenylethyl acetate [19] (Figure 6.25).

The esterase in this example is from a strain of *Pseudomonas fluorescens* and was cloned and expressed in *E. coli* containing a gene, the expression of which can be induced by the addition of the sugar rhamnose. The *E. coli* was grown on LB broth and the cells, after harvesting and washing, were disrupted by sonication in a 50 mM sodium phosphate buffer pH 5.0 to yield a crude extract that was used

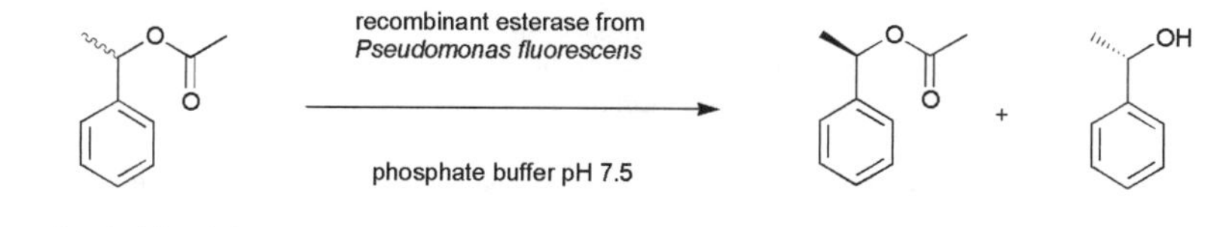

Figure 6.25 Hydrolysis of α-phenylethyl acetate by recombinant esterase from *Pseudomonas fluorescens*

in biotransformations. The activity of the enzyme was determined by measuring the rate of hydrolysis of *para*-nitrophenol acetate in a spectrophotometer. Small biotransformation reactions contained 3 mL of 50 mM sodium phosphate buffer pH 7.5 containing 200 U of enzyme, 2 mL of toluene (to aid substrate solubility) and 0.5 mmol of α-phenylethyl actetate and their progress was monitored by GC.

6.12.1.2 Epoxide Hydrolase

Monfort and co-workers described the use of an epoxide hydrolase for the enantioselective hydrolysis of a hindered aromatic epoxide [20] (Figure 6.26). The enzyme was sourced originally from the fungus *Aspergillus niger* and the gene encoding it had been cloned and expressed in *E. coli*. In a preparative scale experiment, 13 mg of the crude freeze-dried enzyme with an activity of 8 U mg^{-1} from the recombinant strain of *E. coli* was added to 1 L of 100 mM phosphate buffer pH 7.0 containing 10% DMSO. Then 400 mg of the substrate was added and the reaction incubated at room temperature with shaking. The reaction was monitored by extracting samples and analysing them by GC. The resolution reaction (i.e. to approximately 50% conversion) was complete after 4 h 45 min.

6.12.1.3 Nitrilase

Nitrilases, as seen in Chapter 4, are very often used in whole-cell form, however, they can also be employed as cell-free catalysts. Rustler and co-workers reported the

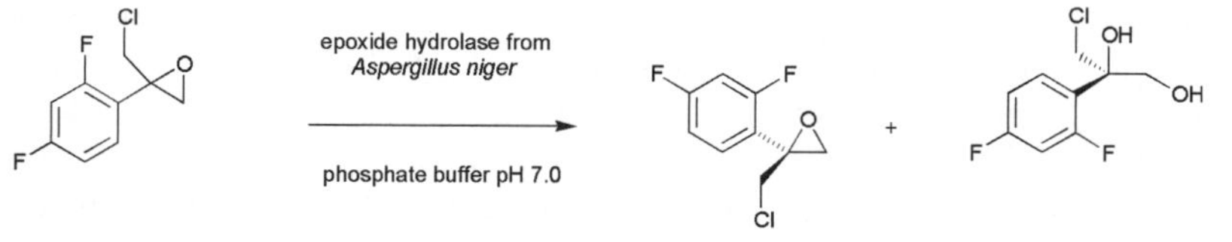

(S)-, 41%, 98% e.e.   38%

Figure 6.26 Enantioselective hydrolysis of a hindered aromatic epoxide by epoxide hydrolase from *Aspergillus niger*

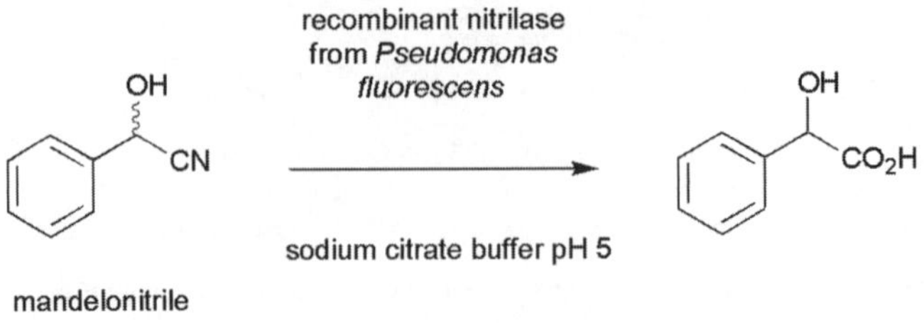

Figure 6.27   Hydrolysis of mandelonitrile by a recombinant nitrilase

use of a nitrilase from *Pseudomonas fluorescens* EBC191 that had been cloned and expressed in *E. coli* EBC191 [21]. This enzyme could be used for the hydrolysis of aromatic substrates such as mandelonitrile and phenylglycinenitrile (Figure 6.27). The cell extracts were derived from the recombinant strain of *E. coli* JM109 (pIK9) containing relevant plasmid and that had been induced by the addition of rhamnose. The cell extract was obtained using a French press on a cell sample that had been resuspended in 100 mM sodium phosphate buffer pH 7.0, and the cell debris removed by centrifugation. The reaction mixtures contained 1.5 mL of 10 mM mandelonitrile (from a methanolic stock solution) in 100 mM sodium citrate buffer at pH 5, with 0.1 mg mL^{-1} of the cell extract. The pH was significant as, at higher pH, the mandelonitrile was found to decompose spontaneously. The reaction was incubated at 34 °C until completion as analysed by HPLC.

6.12.2 Oxidoreductases

6.12.2.1 Ketoreductases

As the popularity of ketoreductase enzymes continues to grow, there are an increasing number of examples in the literature of the cloning, expression and use of cell-free enzymes. Yang and co-workers have described the cloning and expression in *E. coli* of an alcohol dehydrogenase YMR226c from *Saccharomyces cerivisiae* [22]. The enzyme had broad substrate specificity but was particularly selective for the reduction of aryl-substituted acetophenones (Figure 6.28).

The use of the NADPH-dependent enzyme will of course necessitate the use of a cofactor recycling system – the one chosen was the NADP-dependent transformation of D-glucose to D-gluconic acid using D-glucose dehydrogenase. The enzyme was obtained from a Rosetta2 (DE3) strain of *E. coli* that contained a plasmid into which the YMR226c gene had been inserted, and that had been induced using IPTG. The cell pellet resulting from growth was resuspended in a 100 mM potassium phosphate buffer at pH 7.4 containing 0.1% 2-mercaptoethanol and the cells were disrupted using an EmulsiFlex® -C5 high pressure homogeniser (essentially another form of high-pressure cell disruption apparatus). In order

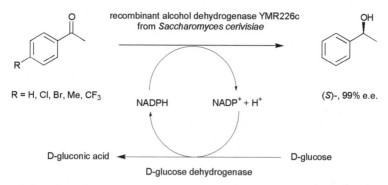

Figure 6.28 Reduction of aryl-substituted acetophenones by a recombinant alcohol dehydrogenase from *Saccharomyces cerivisiae*

to remove lipids from the cell extract, it was mixed with an equal volume of a polyethyleneimine solution, which was made up of 0.25% polyethyleneimine, 6% sodium chloride and 100 mM borax pH 7.4. The cell extract was then taken to 55% ammonium sulfate saturation and the resultant pellet was dissolved in 100 mM potassium phosphate buffer at pH 7.4 containing 0.1% 2-mercaptoethanol. The enzyme was then desalted by gel filtration – but this could be done by dialysis – and freeze-dried for use in biotransformation reactions. A 1 mL reaction contained a 100 mM phosphate buffer pH 6.5 with 4 mg of D-glucose, 0.5 mg of D-glucose dehydrogenase, 0.5 mg NADPH, 0.5 mg of the freeze-dried ADH and 50 μL of a 0.25 M solution of the ketone, made up in DMSO. The reactions were incubated at room temperature and extracted prior to analysis by GC.

6.12.2.2 Baeyer–Villiger Monooxygenases (BVMOs)

Owing to the requirement for coenzymes, BVMOs are also most often exploited as whole-cell preparations, but systems that use cell free enzymes plus coenzyme recycling have also been reported. Zambianchi and co-workers used a cell-free preparation of the cyclohexanone monooxygenase (CHMO) from *Acinetobacter calcoaceticus*, which had been cloned and expressed in *E. coli* [23]. After growth and cell disruption by sonication, a 40–85% ammonium sulfate cut was obtained from the crude cell extract. This pellet was dissolved in 0.02 M potassium phosphate buffer pH 7.0, dialysed to remove the salt and then freeze-dried for use in biotransformation reactions. The specific activity of this powder was determined using the spectrophotometric assay described earlier in this chapter. The enzyme was used to oxidise the aromatic sulfide thioanisole (Figure 6.29), in a mixture that also contained the cofactor recycling system composed of glucose-6-phosphate dehydrogenase and glucose-6-phosphate, although recycling systems based on

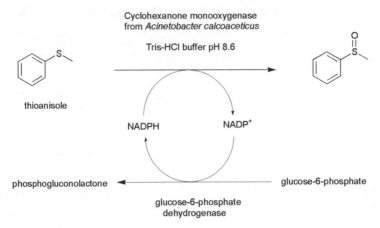

Figure 6.29 Oxidation of the aromatic sulfide thioanisole by cyclohexanone monooxygenase from *Acinetobacter calcoaceticus*

alcohol dehydrogenase from *T. brockii* and formate dehydrogenase systems were also tested. To 2 mL of 0.05 M Tris-HCl buffer pH 8.6 were added 20 mM thioanisole, 0.5 mM NADP, 50 mM glucose-6-phosphate, 4 units of CHMO and 4 units of glucose-6-phosphate dehydrogenase. Reactions were monitored first by extracting with ethyl acetate and analysis by GC.

6.12.2.3 Amine oxidase

Amine oxidases catalyse the oxidation of amines to imines, which are then hydrolysed very quickly to form ketones. A recombinant form of the amine oxidase from *Aspergillus niger* expressed in *E. coli* has been used in cell free form in a chemo-enzymatic oxidation–reduction cycle for the deracemisation of chiral amines such as α-methyl benzylamine and 1-methyltetrahydroisoquinoline (Figure 6.30).

The amine oxidase oxidises only one enantiomer of the starting material to yield an intermediate mixture of 50% imine, and 50% residual amine enantiomer. The inclusion of a chemical reductant, ammonia-borane, in the system results in a reduction of the imine to a racemic mixture of amine enantiomers – the residual enantiomer is thus enriched in one cycle to a proportion of 75% of the starting mixture. Successive cycles result in almost complete deracemisation of the amine. The amine oxidase was expressed in an *E. coli* BL21 star strain and grown on LB medium without specific induction [24]. The cells were resuspended in a 25 mM Tris-HCl buffer pH 7.8 containing 300 mM sodium chloride, 1 mM β-mercaptoethanol, 1 mM PMSF and 10 mM imidazole, as the protein was then to be subjected to nickel affinity chromatography on a small Ni-NTA column. The

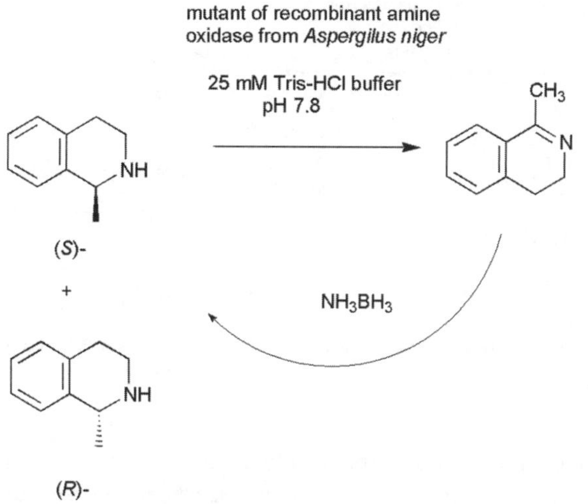

Figure 6.30 Deracemisation of 1-methyltetrahydroisoquinoline by recombinant amine oxidase from *Aspergillus niger*

protein was loaded onto the column and eluted with a buffer containing 200 mM imidazole. The enzyme was then desalted using a P10 desalting column to remove salts and imidazole (this, once again, could be performed using dialysis) and then frozen at −80 °C for use in biotransformation reactions. To a solution of 600 μL of a 25 mM Tris-HCl buffer pH 7.8 containing 1 mM DTT, 1 mM PMSF and 300 mL sodium chloride was added 20 mM 1-methyltetrahydroisoquinoline and 400 mM ammonia-borane ($NH_3.BH_3$) and was shaken at 30 °C for 2 h. To this was then added 0.215 mg of the amine oxidase and small samples of the reaction removed at intervals for analysis by HPLC, which confirmed complete deracemisation of the amine after 90 h.

6.12.3 Transferase reactions

6.12.3.1 Transaminase

In the section on enzyme assays we saw how transaminase or aminotransferase enzymes can be used to transfer ammonia to ketones, resulting in chiral amines. Shin and Kim have described the use of a wild-type transaminase from *Vibrio fluvialis* for the amination of acetophenone to α-methylbenzylamine using L-alanine as the amine donor [25]. The inclusion of lactate dehydrogenase helped to improve the reaction, as it removed the pyruvate from the reaction mixture and consequently drove the equilibrium of the reaction toward product formation (Figure 6.31).

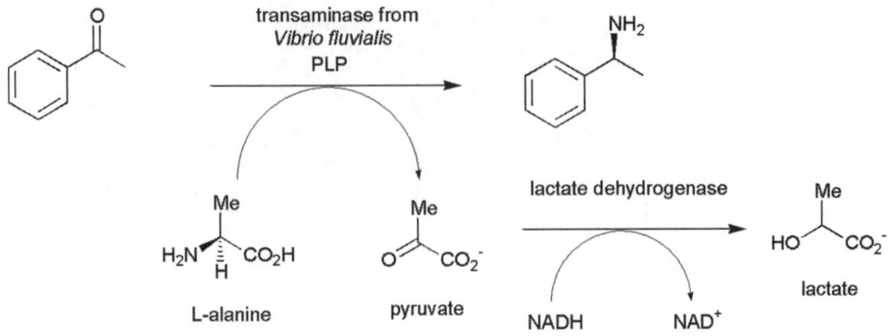

Figure 6.31 Amination of acetophenone to α-methylbenzylamine using a transaminase from *Vibrio fluvialis*

The cell-free enzyme was simply produced from cells of *V. fluvialis* that had been grown on LB medium at 37 °C and a cell extract derived using ultrasonication. A cell pellet was resuspended in a 10 mM potassium phosphate buffer pH 7.2 containing 20 μM pyridoxal-5-phosphate monohydrate (the essential cofactor for the transaminase), 2 mM EDTA, 1 mM PMSF and 0.01% v/v β-mercaptoethanol and the cell suspension was sonicated. After removal of the cell debris by centrifugation, the enzyme was dialysed against a larger volume of the cell resuspension buffer and then stored at −20 °C for use in biotransformation reactions. For the biotransformation of acetophenone the reaction contained 30 mM acetophenone, 300 mM L-alanine, 6.12 units mL^{-1} of the enzyme extract, 10 units mL^{-1} of lactate dehydrogenase and 10 mM NADH.

6.12.3.2 Glycosyltransferase

For glycosyltranferase (GTase)-catalysed reactions, it is necessary to add both the substrate and the nucleotide sugar donor as seen in Chapter 5. Ko and co-workers described the use of a recombinant GTase, *BcGT-1*, from *Bacillus cereus* that was cloned and expressed in *E. coli* BL21 (DE3) [26]. Cells of *E. coli* grown with IPTG-based induction were resuspended in a 20 mM sodium phosphate buffer containing 0.5 M sodium chloride pH 7.4 and the enzyme, produced with a histidine tag, was isolated using a nickel affinity column. The enzyme was used to catalyse the glycosylation of flavonoids such as kaempferol (Figure 6.32). The biotransformation mixtures contained, in a 100 mM potassium phosphate buffer at pH 7.4, 3 μg of the isolated GTase, 5 mM magnesium chloride, 500 μM UDP-glucose and 70 μM of the flavonoid substrate. The reactions were incubated at 37 °C and analysed using either TLC or HPLC.

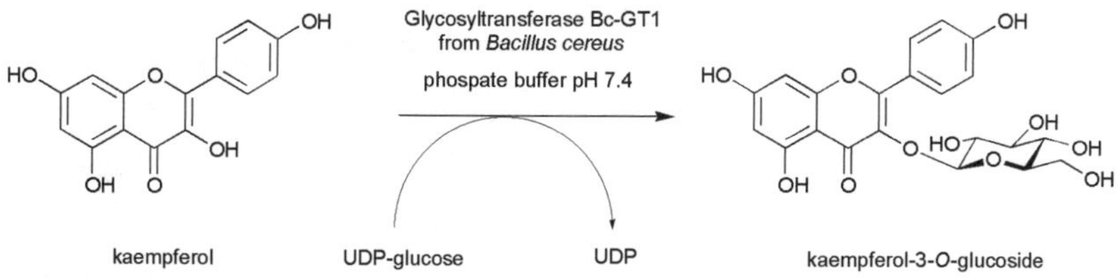

Glycosyltransferase Bc-GT1
from *Bacillus cereus*

phospate buffer pH 7.4

kaempferol   UDP-glucose   UDP   kaempferol-3-O-glucoside

Figure 6.32   Glycosylation of kaempferol by glycosyltransferase Bc-GT1 from *Bacillus cereus*

6.12.4 Carbon–carbon bond formation

6.12.4.1 Oxynitrilase

As the most common sources of oxynitrilase have been from plant sources such as almonds or the rubber tree, the enzymes, whether purchased commercially or derived in-house, have been used in cell-free form as a matter of course. The enzyme from almond, *Prunus amygdalus*, has also been produced from a recombinant source, with the relevant gene being cloned and expressed in the yeast *Pichia pastoris* [27]. Weis and co-workers describe the cloning and expression of the gene and also its application to the transformation of 2-chlorobenzaldehyde [27] (Figure 6.33).

Cloning and expression in yeast is not dealt with in this book, but the *Pichia* strain was grown in a fermentor with the HNLase expressed using a methanol-based system of induction. The system also has the advantage of secreting the HNLase into the culture supernatant, removing the need for cell disruption. The supernanant was merely concentrated using ultrafiltration in a centrifuge as described earlier in the chapter, applied to an anion exchange column and eluted with a simple salt gradient. The enzyme was then applied to gram-scale

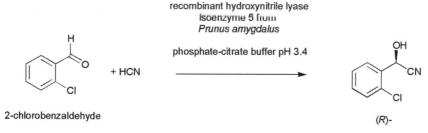

recombinant hydroxynitrile lyase
Isoenzyme 5 from
Prunus amygdalus

phosphate-citrate buffer pH 3.4

+ HCN

2-chlorobenzaldehyde

(R)-

Figure 6.33   Transformation of 2-chlorobenzaldehyde by a recombinant hydroxynitrile lyase from *Prunus amygdalus*

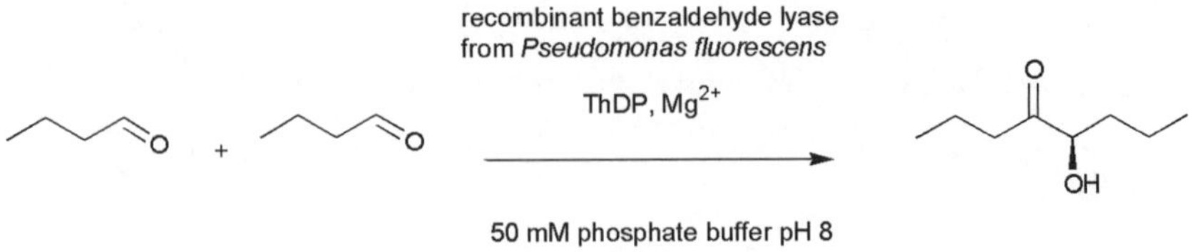

Figure 6.34   Coupling of two molecules of *n*-butanal to form an optically enriched acyloin, catalysed by recombinant benzaldehyde lyase from *Pseudomonas fluorescens*

syntheses of cyanohydrins such as that derived from 2-chlorobenzaldehyde. Then 150 units of the enzyme were diluted with 50 mM potassium phosphate-citrate buffer pH 3.4 to a final volume of 3.7 mL. To this was added 15 mmol of the substrate that had been dissolved in 2.1 mL of *tert*-butylmethyl ether. The reaction was cooled to 10 °C and 1.2 mL HCN added. Samples were removed at intervals and the mixtures acetylated before analysis by GC.

6.12.4.2 Benzaldehyde Lyase

There are many carbon-carbon bond forming enzymes that employ the organic cofactor thiamine diphosphate. Benzaldehyde lyase (BAL, E.C. 4.1.2.38) catalyses the coupling of two aldehyde molecules to form optically enriched acyloins (Figure 6.34).

The enzymes have been sourced mostly from bacteria and some have been cloned and expressed in *E. coli*. Benzaldehyde lyase from *Pseudomonas fluorescens* biovar I has been cloned and expressed in *E. coli* and applied to the enantioselective synthesis of aliphatic 2-hydroxyketones by Domínguez de María and co-workers [28]. Cells of a recombinant strain of *E. coli* transformed with a plasmid carrying the BAL gene were grown, harvested and resuspended in a 50 mM phosphate buffer containing 2.5 mM magnesium sulfate (magnesium ions are also essential for BAL activity), and 0.3 mM thiamine diphosphate. The crude cell extract was used directly for biotransformation reactions. To 50 mL of the same buffer at pH 8.0 were added 500 units of BAL in the presence of 20% DMSO and 50 mM substrate such as propanal or *n*-butanal. Samples were extracted at intervals into solvent and analysed by GC.

6.13 Conclusion

It can be seen from the contemporary literature that whilst the use of isolated enzymes in biocatalysis is widespread, most of the academic or industrial groups that employ non-hydrolytic enzymes are using home-grown biocatalysts, and in most cases, these are derived from recombinant organisms, usually *E. coli* but also yeasts such as *Pichia pastoris*. The techniques and facilities associated with producing recombinant organisms would not of course be within the scope of the usual organic synthesis laboratory, but in the interests of providing both a guide to engaging with the literature relevant to the area, and also a theoretical and practical introduction for groups interested in embarking on this type of work, we provide in Chapter 7 an introduction to the production and application of recombinant organisms as biocatalysts.

References

1. R. K. Scopes (1994) Protein Purification: Principles and Practice, 3rd Edn, Springer-Verlag, Berlin
2. M. M. Bradford (1976) A rapid and sensitive method for the quantitation of microgram quantities of protein utilizing the principle of protein-dye binding. *Anal. Biochem.*, **72**, 248–254.
3. J. J. Villafranca and B. Axelrod (1971) Heptulose synthesis from non-phosphorylated aldoses and ketoses by spinach transketolase. *J. Biol. Chem.*, **246**, 3126–3131.
4. R. Torres, C. Ortiz, B. C. C. Pessela, J. M. Palomo, C. Mateo, J. M. Guisán and R. Fernándes-Lafuente (2006) Improvement of the enantioselectivity of lipase (fraction B) from *Candida Antarctica* via adsorption on polyethylenimine-agarose under different experimental conditions. *Enzym. Microb. Technol.*, **39**, 167–171.
5. V. Mastihuba, L. Kremnicky, M. Mastihubová, J. L. Willett and G. L. Côté (2002) A spectrophotometric assay for feruloyl esterases. *Anal. Biochem.*, **309**, 96–101.
6. K. Doderer, S. Lutz-Wahl, B. Hauer and R. D. Schmid (2003) Spectrophotometric assay for epoxide hydrolase activity toward any epoxide. *Anal. Biochem.*, **321**, 131–134.
7. C. Mateo, A. Archelas and R. Furstoss (2003) A spectrophotometric assay for measuring and detecting an epoxide hydrolase activity. *Anal. Biochem.*, **314**, 135–141.
8. S. Tribolo, J.-G. Berrin, P. A. Kroon, M. Czjzek and N. Juge (2007) The crystal structure of human cytosolic β-glucosidase unravels the substrate aglycone specificity of a family 1 glycoside hydrolase. *J. Mol. Biol.*, **370**, 964–975.
9. D. R. Yazbeck, P. J. Durao, Z. Xie and J. Tao (2006) A metal-ion based method for the screening of nitrilases. *J. Mol. Catal.*, B-Enzym., **39**, 156–159.
10. K. Inoue, Y. Makino and N. Itoh (2005) Purification and characterisation of a novel alcohol dehydrogenase from *Leifsonia* sp. strain S749: a promising biocatalyst for an asymmetric hydrogen transfer bioreduction. *Appl. Environ. Microbiol.*, **71**, 3633–3641.
11. N. M. Kamerbeek, A. J. J. Olsthoorn, M. W. Fraaije and D. B. Janssen (2003) Substrate specificity and enantioselectivity of 4-hydroxyacetophenone monooxygenase. *Appl. Environ. Microbiol.*, **69**, 419–426.
12. S. A. Ali, S. Stoeva, A. Abbasi, J. A. Alam, R. Kayed, M. Faigle, B. Neumeister and W. Voelter (2000) Isolation, structural and functional characterisation of an apoptosis-inducing L-amino acid oxidase from leaf-nosed viper (*Eristocophis macmahoni)* snake venom. *Arch. Biochem. Biophys.*, **384**, 216–226.

13. Z. Hasan, R. Renirie, R. Kerkman, H. J. Ruijssenaars, A. F. Hartog and R. Wever (2006) Laboratory-evolved vanadium chloroperoxidase exhibits 100-fold higher halogenating activity at alkaline pH. *J. Biol. Chem.*, **281**, 9738–9744.

14. F. J. G. Muriana, M. C. Alvarez-Ossorio and A. M. Relimpio (1991) Purification and characterisation of aspartate aminotransferase from the halophilic archaebacterium *Haloferax mediterranei*. *Biochem. J.*, **278**, 149–154.

15. S. Gosselin, M. Alhussaini, M. B. Streiff, K. Takabayashi and M. M. Palcic (1994) A continuous spectrophotometric assay for glycosyltransferases. *Anal. Biochem.*, **220**, 92–97.

16. D. N. Bolam, S. Roberts, M. R. Proctor, J. P Turkenburg, E. J. Dodson, C. Martinez-Fleites, M. Yang, B. G. Davis, G. J. Davies and H. J. Gilbert (2007) The crystal structure of two macrolide glycosyltransferases provides a blueprint for host-cell antibiotic immunity. *Proc. Natl Acad. Sci. USA*, **104**, 5336–5341.

17. E. E. Dekker and R. P. Kitson (1992) 2-Keto-4-hydroxyglutarate aldolase: Purification and characterisation of the homogeneous enzyme from bovine kidney. *J. Biol. Chem.*, **267**, 10507–10514.

18. J. Andexer, J.-K. Guterl, M. Pohl and T. Eggert (2006) A high-throughput screening assay for hydroxynitrile lyase activity. *Chem. Commun.*, 4201–4203.

19. F. Zocher, N. Krebsfanger, O. J. Yoo and U. T Bornscheuer (1998) Enantioselectivity of a recombinant esterase from *Pseudomonas fluorescens*. *J. Mol. Catal.*, **5**, 199–202.

20. N. Monfort, A. Archelas and R. Furstoss (2002) Enzymatic transformations. Part 53. Epoxide-hydrolase-catalysed resolution of key synthons for azole antifungal agents. *Tetrahedron: Asymmetry*, **13**, 2399–2401.

21. S. Rustler, A. Müller, V. Windeisen, A. Chmura. B. C. M. Fernandes, C. Kiziak and A. Stolz (2007) Conversion of mandelonitrile and phenylglycinenitrile by recombinant *E. coli* cells synthesising a nitrilase from *Pseudomonas fluorescens* EBC191. *Enzym. Microb. Technol.*, **40**, 598–606.

22. Y. Yang, D. Zhu, T. J. Piegat and L. Hua (2007) Enzymatic ketone reduction: mapping the substrate profile of a short-chain alcohol dehydrogenase (YMR226c) from *Saccharomyce cerivisiae*. *Tetrahedron: Asymmetry*, **18**, 1799–1803.

23. F. Zambianchi, P. Pasta, G. Carrea, S. Colonna, N. Gaggero and J. M. Woodley (2002) Use of isolated cyclohexanone monooxygenase from recombinant *Escherichia coli* as a biocatalyst for Baeyer-Villiger and sulfide oxidations. *Biotechnol. Bioeng.*, **78**, 489–496.

24. R. Carr, M. Alexeeva, A. Enright, T. S. C. Eve, M. J. Dawson and N. J. Turner (2003) Directed evolution of an amine oxidase possessing both broad substrate specificity and high enantioselectivity *Angew. Chem., Int. Ed.* **42**, 4807–4810.

25. J.-S. Shin and B.-G. Kim (1999) Asymmetric synthesis of chiral amines with ω-transaminase. *Biotechnol. Bioeng.*, **65**, 206–211.

26. J. H. Ko, B. G. Kim and J.-H Ahn (2006) Glycosylation of flavonoids with a glycosyl-transferase from *Bacillus cereus*. *FEMS Microbiol Lett.*, **258**, 263–268.

27. R. Weis, P. Poechlauer, R. Bona, W. Skranc, R. Luiten, M. Wubbolts, H. Schwab and A. Glieder (2004) Biocatalytic conversion of unnatural substrates by recombinant almond R-HNL isoenzyme 5. *J. Mol. Catal.*, **29**, 211–218.

28. P. Domínguez de María, M. Pohl, D. Gocke, H. Gröger, H, Trauthwein, T. Stillger, L. Walter and M. Müller (2007) Asymmetric synthesis of aliphatic 2-hydroxy ketones by enzymatic carboligation of aldehydes. *Eur. J. Org. Chem.*, 2940–2944.

Chapter 7
An Introduction to Basic Gene Cloning for the Production of Designer Biocatalysts

7.1 Introduction

From the previous chapters, it will have become clear to those with little experience of biocatalysis that the field has undergone little short of a revolution in the last ten years, with use of recombinant organisms and enzymes derived from them becoming routine in both academic and industrial laboratories. This has largely been due to rapid developments in the area of molecular biology, but also the accessibility of this technology through convenient kits sold by most major biochemical suppliers. We will define recombinant organisms as those which express genes that are not native to the organism and have been introduced through genetic engineering. In all cases in this book, and indeed in the majority of cases in the literature, this refers to the introduction of genes encoding novel biocatalysts into the common laboratory bacterium *Escherichia coli*, although the use of other bacteria such as *Pseudomonas fluorescens*, and yeasts, notably *Saccharomyces cerivisiae* and *Pichia pastoris*, has also been common, if less frequent. *E. coli* has been a model bacterial organism in the field of biochemistry for many years. Laboratory strains of *E. coli* are non-pathogenic, easy to cultivate, fast-growing and their genetics are well understood. *E. coli* was also the first bacterium for which a whole genome sequence was determined [1]. We will therefore restrict our discussions in this chapter to cloning and expression in *E. coli* only and hence provide background only on the nature of genes from bacteria and some simple tools for exploitation of those genes alone.

Another reason for the increasing popularity of recombinant biocatalysts is quite simply the amount of genome sequence information that is appearing in the international and publicly accessible databases, as described in Chapter 2. This has meant that instead of being restricted to working with a single enzyme for a desired transformation, a whole range of possible homologues exists within nature that might be accessed very easily, giving rise to a pool of unexplored activities and selectivities that is representative of the diversity acquired through natural evolution. Another advantage of recombinant biocatalysts is that previously, when isolating an enzyme from a wild-type source, there would be a number of practical and safety issues associated with acquiring that enzyme. For example, one would have to consider the possible pathogenicity of the organism or the requirement for exotic growth conditions such as an anaerobic environment or growth at elevated temperatures. Also, the natural abundance of the enzyme in the wild-type strain may have been too low to attempt purification and even then, it may not have been easy to isolate the enzyme using conventional protein chromatography. The use of genetic engineering to optimise genes for expression by recombinant strains of *E. coli* allows for the acquisition of abundant amounts of easy-to-purify enzyme in a non-pathogenic and easy-to-grow host. Once cloned, the gene is also then amenable to engineering techniques such as rational mutagenesis and directed evolution. There are some downsides – it is possible of course that *E. coli* will either not express the foreign gene or that even if expressed, insoluble (and therefore unusable) enzyme results, but overall the developments have been hugely beneficial.

Once the laboratory is suitably equipped (see Chapter 3), all that is required to begin to create one's own recombinant biocatalyst is a sample of the DNA of the organism in question and sometimes not even that. Many industrial and academic groups are now having relevant genes encoding biocatalysts made for them by gene synthesis companies as seen in Chapter 2. However, on the basis that most recombinant biocatalysts are still made in the laboratory from samples of organismal genomic DNA, we provide a short introduction to the background to simple recombinant DNA experiments and their application to the creation of recombinant biocatalysts.

In each case, the general experimental strategy for making recombinant biocatalysts is that listed below and illustrated in Figure 7.1.

1. *Gene amplification by polymerase chain reaction (PCR).* The gene to be cloned and expressed must first be amplified from the source or template DNA using a technique known as PCR.

2. *Gene cloning (insertion into a suitable plasmid cloning vector).* The gene then needs to be cleaned up and 'ligated' into a suitable plasmid, or small circular piece of DNA, for cloning. This is then introduced into a 'cloning strain' of *E. coli* by a process called 'transformation'. The *E. coli* is then grown, generating many copies of the cloning plasmid, which is then extracted ('mini-prepped' or 'midi-prepped') for the next step.

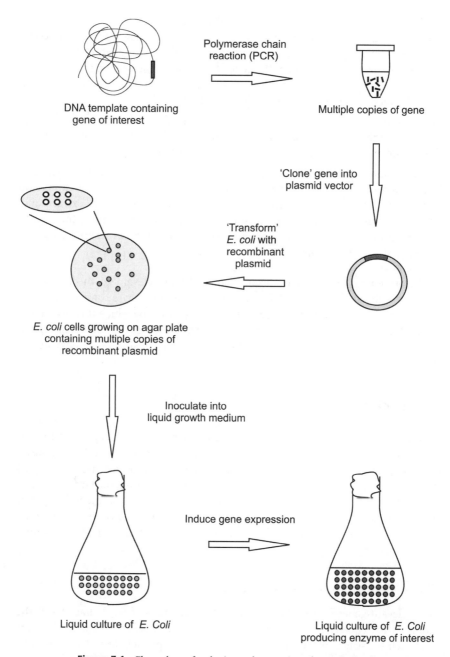

Figure 7.1 Flow scheme for cloning and expression of genes in *E. coli*

3. *'Sub-cloning' into an expression vector.* The gene is cut out of the cloning vector plasmid using restriction enzymes, and inserted into a new 'expression vector' or plasmid that possesses characteristics essential for stimulating the cellular machinery for the expression of the gene of interest i.e. the production of protein. In practice, nowadays, the gene cloning is often done straight into the expression plasmid vector, missing out the cloning vector step.

4. *Transformation of expression strain.* The expression plasmid is then used to transform into an 'expression strain' of *E. coli* that will be used as the growth and biotransformation organism – the 'recombinant biocatalyst'.

5. *Gene expression and analysis of protein products.* The recombinant biocatalyst is grown in small amounts and the optimum growth conditions for gene expression (or protein production) are assessed using SDS-PAGE analysis of cell extracts and, if possible, an activity assay. Some simple changes to, for example, the growth temperature of the expression strain may have very positive effects on the amount of expression or the solubility of the enzyme target.

Before embarking on some experimental details relevant to the cloning of genes and the analysis of DNA fragments, it is useful to revise some of the basic background about DNA structure and the mechanisms by which it is stored and replicated within bacterial cells, and how its genetic message is first transcribed, then translated into the amino acid sequence that constitute the enzymes of interest.

7.2 Background

There are many excellent textbooks which will provide a background to biochemistry and aspects of molecular biology and DNA structure and metabolism, including those by Stryer [2] and Voet and Voet [3]. For a more advanced introduction to the theories involved in genetic manipulation, the Genes series by Lewin [4] provides an excellent starting point. It is also worth mentioning that many laboratories that use molecular biological techniques possess a copy of the extensive manual of practical techniques in gene cloning techniques by Sambrook and Russell [5].

7.2.1 General – structure of DNA

The relevant genes for our discussion are the sequences of DNA that encode the amino acid sequences of enzymes. DNA is a polymer of nucleotide monomers that each consists of a deoxyribose sugar bound to a phosphate and a heterocycle known as a base. These bases can be either monocyclic *pyrimidines* [thymine (T) or cytosine (C)] or bicyclic *purines* [adenine (A) or guanine (G)] (see Appendix 2).

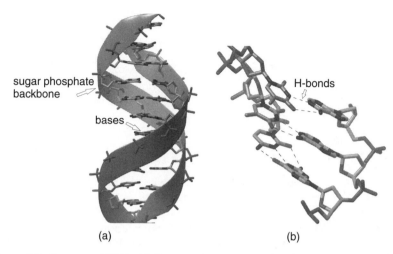

(a)   (b)

Figure 7.2 Structure of DNA. (a) Two complementary strands of DNA, one coral and one grey, make up the double helix. The backbone of each strand is made up of a chain of sugar (deoxyribose) phosphates. Attached to each sugar phosphate is an aromatic base – either a pyrimidine or a purine – the bases from complementary strands interact in the interior of the double helix through hydrogen bonding (b, dashed black lines)

The backbone of a DNA strand is comprised of the phosphate esters formed between the substituted deoxyribose sugars (Figure 7.2), in the 5′ or the 3′ position. The DNA strand is thus determined to have a 5′ or a 3′ end, analogous to the N- or C-termini of a protein. Sequences of DNA that precede the 5′ end of a gene are said to occur *upstream*; those that follow the 3′ end are said to be *downstream*. It is well known that in Nature, one strand of DNA, which might be thought of as the *coding* strand, is wound round another, the *non-coding* or complementary strand, to form a double helical structure or duplex (Figure 7.2). The interior of the DNA duplex is formed by the base residues attached to the 1′ position of the sugars.

The integrity of the double helix structure is maintained by hydrogen bonds between the bases in the interior of the double helix: A binding to T by two hydrogen bonds; and G binding to C using three hydrogen bonds (Figures 7.2 and 7.3).

Trivially, the A-T interaction can be thought of as 'weaker' than the G-C interaction and this may mean that sequences rich in A-T may be easier to separate (see the Pribnow box below). It is the sequence of only these four bases that will dictate the make-up of the genetic complement of the organism or genome. The conservation of sequence on both strands of the DNA helix means that the helix can unwind and each strand can act as a template for the synthesis of a new strand of DNA encoding a complementary sequence in a process known

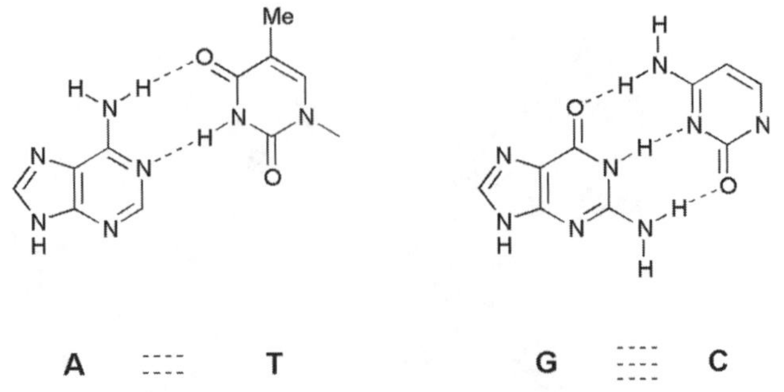

Figure 7.3 Hydrogen bonding in the DNA duplex

as replication. The enzymes that perform the polymerisation of the new DNA chain are known as DNA polymerases, and are exploited in the gene amplification process known as the PCR (see below).

7.2.2 Base pairing, codons and the genetic code

The sequence of base pairs in a gene dictates the sequence of amino acids in the resultant protein through a series of rules enshrined within the *genetic code* (see Appendix 3). This means that any series of three bases will encode one amino acid in the final protein sequence. For example the ATG *codon*, a sequence that is found at the beginning of many bacterial DNA sequences that encode proteins, encodes for the amino acid methionine. There are sixty-four different codons i.e. sixty-four ways in which to arrange a three-letter code consisting of any possible four letters, yet only twenty amino acids are commonly found within proteins. This is because most amino acids are encoded by more than one codon – for example, the amino acid alanine has four codons. The genetic code is thus said to be *degenerate*. In addition to the codons that encode the amino acids, there are three, TGA, TAA and TAG, that encode the message 'stop'.

The gene sequence along one strand of DNA may be read in one of three 'open reading frames'. This means that given the sequence ... ATGGTTCCTGTT ... , one may interpret the code starting from the first A, to give ATG, GTT, CCT (encoding Met, Val, Pro) or the T that is second in the sequence to give TGG, TTC, CTG (encoding Trp, Phe, Leu) or starting from the G that is third in the sequence. The consequences of reading the wrong reading frame when interpreting DNA sequence can be devastating for an experiment therefore, and care must be taken not to engender 'frame shifts' when, for example designing primers for PCR (see Section 7.6). Frame shifts will also be caused by the unexpected deletion or insertion of base pairs that can sometimes happen during PCR or other DNA manipulations. These will usually be revealed by DNA sequencing (see Section 7.6).

Another consideration arises from the degeneracy of the genetic code described above. As some amino acids are encoded by more than one codon (leucine and arginine are both possibly encoded by six possible three-base-pair codons), it may be that one or more of these is preferentially used by one organism. For example, of the four codons encoding proline in *E. coli*, the CCG codon accounts for 55 % of the prolines encoded; whereas the CCC codon only accounts for 10 % (a table of complete information for *E. coli* can be viewed at http://www.sci.sdsu.edu/~smaloy/MicrobialGenetics/topics/in-vitro-genetics/codon-usage.html). It may be, however, that the organism from which the gene is obtained uses very different proportions of possible codons for encoding certain amino acids. This may be for instance because the ratio of GC base pairs to AT base pairs in the genome sequence is greater than normal, as is found in strains of actinomycete bacteria such as *Streptomyces* and *Rhodococcus*. The *codon usage* ratio for an organism becomes available of course once the genome has been sequenced. Hence, when designing synthetic genes whose sequences originate in other organisms for expression in *E. coli*, it may be beneficial to optimise the *codon usage* for expression in the new strain.

7.2.3 Transcription and translation

In simple terms, the sequence of base pairs that constitute the genome will dictate the complement of enzymes and other proteins that constitute the make-up of the cell. Readers will know that, in order to synthesise proteins on the ribosome within the cell, this DNA message must first be *transcribed* into a message on a different nucleic acid molecule, known as messenger RNA (mRNA) in a process known as *transcription*. The message is transcribed from the ATG at the beginning, encoding methionine, until the message reaches one of the codons that signals 'stop'. The double helix of DNA is unwound and an enzyme called *RNA polymerase* catalyses the synthesis of the mRNA chain complementary to the sequence of the DNA chain, yet replacing the base thymine (T) with a different base, uracil (U). The transcribed message is then recognised by a different molecule of RNA, transfer RNA (tRNA), which when bound with the mRNA and the ribosome all constitute a biochemical complex for the *translation* of the mRNA message into the growing protein chain (Figure 7.4). The protein that leaves the ribosome is then folded into the stable conformation that for our purposes will constitute the active form of the enzyme of interest.

7.2.4 Bacterial genomes and the structure of a typical bacterial gene

In bacteria such as *E. coli*, the genome of the organism is contained within a long sequence of circular DNA, known as a bacterial chromosome, or within shorter

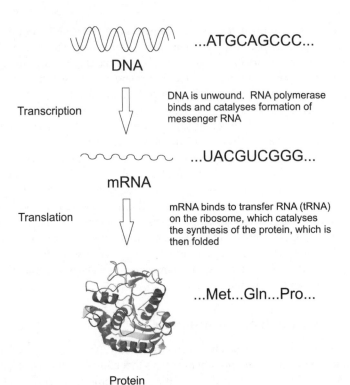

DNA

...ATGCAGCCC...

Transcription

DNA is unwound. RNA polymerase binds and catalyses formation of messenger RNA

mRNA

...UACGUCGGG...

Translation

mRNA binds to transfer RNA (tRNA) on the ribosome, which catalyses the synthesis of the protein, which is then folded

...Met...Gln...Pro...

Protein

Figure 7.4   *Transcription* of the DNA sequence into the mRNA message and *translation* into the amino acid sequence

circular pieces of DNA called plasmids. The genes that encode enzymes are laid down in a sequence that may be physiologically relevant. For example, genes involved in a particular metabolic or biosynthetic pathway may be *clustered*, that is laid down in sequence and their transcription controlled by a single region of DNA called a *promoter* (see the structure of a gene, below). Groups of genes clustered in this way are called *operons*. This provides the advantage that, once, for example, a carbon source has been detected by the cell, all of the enzymes that are required for the relevant catabolic pathway are switched on at approximately the same time, giving the cell rapid access to the energy provided by that carbon source.

It is useful at this stage to look at the major characteristics of a sequence of DNA that constitute a gene. The first gene in an operon, like all the others, will most often contain an ATG codon which encodes for the amino acid methionine at the start of the *coding sequence*. This is illustrated in Figure 7.5, in the sequence upstream of the gene *CggR* from a species of *Bacillus*, which is the first gene in an operon coding for the metabolism of glucose in that organism. The start codon for the

```
CTTTGCGGTGATTAACATCCTTTGTTTCCTGTTTGTCGTTACGATCTGTCCAGAAACGAAGAACAAATCGC
TCGAGGAAATTGAAAAGCTTTGGATAAAATGAAAACGCTTTAATGAAACAGCCCTTTCTACGGGAAGGGCT
GTTTATATTGGGATGCACCATTTGGCGCTTTCTGTATAAGATAAAGATATATAGGATAAAATATTGCTGGA
TAAAACGACGCGGCATGAAAACTCTGCGAATATTGTCGATGAATTGGCTCTTAACAGTTGAATAAACAATT
CCACCCTGTTAAAATAATTAAAGAAAGCAGAAATGATTTTTTTTGGCTATGACGGGACGTTTTTTGTCATA
GCGGGACATATAATGTCCAGCAAAAAAGGAAGGAACGTTTGAGTCATG
```

Figure 7.5 Nucleotide sequence upstream of the gene *CggR* in *Bacillus* sp

gene is shown in bold and underlined at the end of the sequence. The *transcription start site* (TSS) is shown in green in the string of six A bases 20 bases upstream of the start codon. The sequence of base pairs between the TSS and the start codon, whilst transcribed, is not translated into amino acids later. Approximately 35 base pairs upstream of the transcription start site, is a base sequence -**TTGTCA**- (in red) which is recognised by the enzyme RNA polymerase that will later catalyse the transcription reaction. Approximately 10 base pairs upstream of the transcription start site is an additional consensus sequence -**TATAAT**- (in blue) known as a *Pribnow box*. It is thought that the RNA polymerase binds to the -35 recognition sequence, and then moves along to the Pribnow box, where the DNA double helix is unwound in preparation for polymerisation of the mRNA chain, starting at the transcription start site. This is facilitated by the density of A and T bases in the Pribnow box, which participate in only 2-hydrogen bond interactions with their partner on the other strand.

At the 3′ end of the operon (not shown), there needs to be a signal to RNA polymerase to stop transcribing the DNA into mRNA, and this is often seen as a GC-rich region of DNA known as a *terminator* sequence. Such sequences often form a loop structure, which stimulates the RNA polymerase to dissociate from the DNA strand.

Whilst only the first gene in the operon sequence will have the -35 and -10 consensus sequence, each gene will also have an additional consensus region upstream from the ATG start codon that encodes a *ribosome binding site* in the resultant mRNA. A number of different sequences for the ribosome binding site, also known as 'the Shine-Dalgarno sequence' have been described. For example, in a gene sequence encoding a C-C bond lyase from *Rhodococcus* [6], the 5′ region of which is shown below, the ribosome binding site is indicated by the GAGGAA sequence underlined and in bold that ends five bases upstream of the ATG codon. The coding sequence is shown in italics and the amino acid sequence shown below the gene:

```
CGCTCCTCGGTGACTCGAACGGAGGAAGTTCGATGAAGCAATTGCCA
                               M  K  Q  L  A
```

At the end of the gene will be the 'stop' codon, either TGA, TAG or TAA, which after being transcribed to mRNA and bound to the ribosome, will indicate to the ribosome to cease the incorporation of amino acids into the nascent protein. At the end of the gene encoding the same C-C bond lyase the stop is signalled by TGA underlined and in bold below:

```
GGAATGGAGTCCGAACAGTGACCAGCACAG
  G   M   E   S   E   Q stp
```

Knowledge of the structure of a prokaryotic gene will be useful when it comes to designing *oligonucleotide primers* for gene amplification by PCR (see below) and subsequent gene cloning. The cloning of a gene requires that it will, once amplified from a source of DNA, be inserted into a small and easy-to-manipulate DNA molecule that will be stable over a long period of time. These are most commonly small circular transmissible units called plasmids.

7.2.5 Plasmids

Plasmids at their simplest are relatively small, circular pieces of DNA that are made up of sequences of genes and regulatory elements. In some bacteria, some characteristics of the organism – perhaps the ability to grow on a certain carbon source or pathogenicity traits – have the genes for the relevant enzymes encoded on isolated plasmids, rather than within the larger bacterial genome. Being small and stable, they are known to be readily transmissible between bacteria, hence providing a mechanism for one bacterium to acquire a physiological property from another, such as antibiotic resistance or the ability to degrade said carbon source. It is their size, comparative simplicity and ability to be transferred between bacteria that have made plasmids such suitable vectors for performing genetic engineering.

We have alluded in previous chapters to *commercially available plasmids*. By this, we mean that many biochemical companies sell plasmids that have been engineered for the purposes of genetic manipulation experiments or, in the context of this discussion, the introduction of 'foreign' genes into *E. coli*. These plasmids have many features that enable them to be used for such experiments. These are shown in the representative plasmid illustrated in Figure 7.6.

7.2.6 The origin of replication (ori)

The origin of replication is a region of sequence found in double-stranded DNA such as the chromosome of *E. coli*, and recruited for commercial plasmids, where unwinding of the DNA double helix begins in preparation for the binding of DNA polymerase and hence the replication of the genetic sequence that constitutes the plasmid. The origin of replication in commercially available plasmids can permit the replication of a large number of plasmids within cells that are transformed

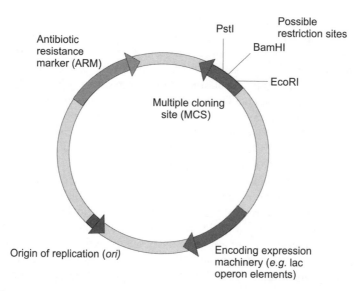

Possible restriction sites

PstI

BamHI

EcoRI

Antibiotic resistance marker (ARM)

Multiple cloning site (MCS)

Origin of replication (*ori*)

Encoding expression machinery (*e.g.* lac operon elements)

Figure 7.6 Diagram of a plasmid showing multiple cloning sites for digestion by restriction enzymes and antibiotic resistance marker (ARM)

with it (so-called 'high-copy number'). Some commercial plasmids possess two ori sequences, but only one of these is relevant to routine cloning experiments of the type described herein.

7.2.7 The multiple cloning site - restriction sites

Plasmids often have one or more *multiple cloning sites*. These are short regions of DNA sequence that include a number of nucleotide sequences (usually four or six base-pairs) called 'restriction sites' suitable for controlled digestion by restriction enzymes (see below). It is these restriction sites that will enable the plasmid to be selectively cut and new gene sequences to be inserted or ligated. The differences in base-pair sequence found in different restriction sites (see Appendix 7) means that different restriction enzymes can be used to cut the plasmid very specifically in different places, giving a large measure of control in engineering the structure of the new plasmid containing the gene of interest.

7.2.8 Antibiotic resistance marker (ARM)

Another important feature of commercial plasmids is the *antibiotic resistance marker* (ARM, Figure 7.6). During the course of your cloning and expression experiments it will be important to distinguish between those bacteria which contain the plasmid that you are using, and those that do not, the latter being useless to you and thus constituting an inconvenient 'background' for the cloning experiment. The antibiotic resistance marker in the plasmid encodes one or other factor, perhaps an enzyme that degrades an antibiotic for example, that will allow only those cells that contain the plasmid of interest to survive when the *E. coli* cells are grown in media that contain the antibiotic. A common antibiotic resistance marker encodes a β-lactamase enzyme that confers resistance to the antibiotic ampicillin. *E. coli* cells that contain a plasmid that expresses the β-lactamase are able to grow therefore when ampicillin is added to the solid or liquid growth medium; cells without that plasmid will not grow. Other common antibiotics used in this way are kanamycin, chloramphenicol, streptomycin and carbenicillin.

There is a selection of more basic plasmids that is available for simple cloning experiments. These feature little more than a circular piece of DNA, some suitable *restriction sites* (see below), an ARM and an origin of replication that will allow simple cutting and pasting of a foreign DNA sequence into the plasmid. These might be thought of as *cloning plasmids*. *Expression plasmids*, which have additional features designed to induce *E. coli* to express the gene of interest, are described below, and in Section 7.9.

7.2.9 A genetic system for induction and control of gene expression

In addition to the origin of replication, ARM and multiple cloning site, a plasmid that is used for *expressing* foreign genes will have to possess additional genes that encode a mechanism of inducing and controlling the expression of the foreign gene. One of these systems of genes is based on the so-called *lac* promoter from *E. coli*, and is induced by growth on the carbohydrate lactose, or its structural analogue isopropyl-β-D-thiogalactopyranoside (IPTG). This is detailed in Section 7.9. As part of this system, the expression plasmid will have a *promoter sequence*, to which RNA polymerase binds and an *operator*, a short DNA sequence which, in the absence of an inducer, interacts with a protein that prevents RNA polymerase binding to the promoter, and thus inhibiting transcription.

7.2.10 Manipulation of DNA – restriction enzymes and restriction sites

The revolution in DNA manipulation for the purposes of genetic engineering has come about through the recruitment of enzymes whose natural role is various aspects of DNA processing. For example, we will see that PCR methods for gene amplification are possible through the exploitation of the natural action of DNA polymerases. It is equally possible to 'cut and paste' sequences of DNA to each other giving rise to the various methods of DNA recombination that are the foundation of simple genetic engineering tools for biocatalysis. The 'pasting' enzymes are called DNA *ligases* and the process involved in joining two double-stranded pieces of DNA together is called *ligation*. The enzymes that are used to cut DNA are types of hydrolase called nucleases. These can be either *exonucleases*, which cut DNA chains at the ends or *endonucleases* which cut double-stranded DNA in the middle of a strand. It is the latter enzymes, termed *restriction endonucleases*, or, more trivially, *restriction enzymes* that constitute much of the toolbox in simple molecular biology protocols. The restriction endonucleases that have most commonly been exploited in biotechnology are either Type II or Type II S enzymes. Their names, such as *Bam*HI or *Eco*RI feature prefixes that arise from the organisms whence they have been derived e.g. *Eco*RI comes from *E. coli*.

Many biochemical companies sell a wide range of restriction enzymes that are each distinguished by their ability to cut DNA only at certain four or six base-pair sequences (see Appendix 7). These are known as *restriction sites* and feature in the multiple cloning sites of the commercial plasmids described above. A long sequence of DNA may therefore in theory be cut into known sizes using enzymes that are specific for certain restriction sites and cutting and highly selective pasting of new gene sequences into the plasmid may be accomplished.

Let us consider a simple example. Figure 7.7 shows a representative commercial plasmid (a small circular piece of DNA that is used for purposes of gene cloning).

At one position is a restriction site, **CCATTG**, which is acted on by the enzyme *Nco*I. This means that, when incubated with this enzyme, the plasmid will be cleaved once at this site to yield a linear plasmid of 5500 base pairs. The cleavage will yield overhanging single-stranded DNA at each end of the DNA duplex (linearised plasmid) that are termed 'sticky ends', so-called because pieces of DNA that have been cleaved with *Nco*I may be inserted into the plasmid by adhering to these overhangs. There is another restriction site in the plasmid, which features a sequence **CATATG** that is recognised by the enzyme *Nde*I. If the plasmid is cut with two enzymes, *Nco*I and *Nde*I, then the digestion will result in a slightly shorter piece of linear DNA, which has two *different* sticky ends. This will allow the addition of a new piece of DNA, called the *insert*, such as the gene of interest, in one direction only if the gene to be inserted has itself been engineered with the *Nco*I and *Nde*I sequences and digested with these enzymes. It should be noted that other restriction enzymes, such as *Sma*I cut DNA to leave so-called 'blunt ends', rather than sticky ends (see Appendix 7) and can also be exploited in alternative cloning techniques.

The insertion of a new piece of DNA into a plasmid will result in a new plasmid entity, a recombinant plasmid that is often termed a *construct*. It is these plasmid constructs that are the route to new recombinant biocatalysts. In order to do this, the insert itself must be engineered to possess the sticky ends that are appropriate

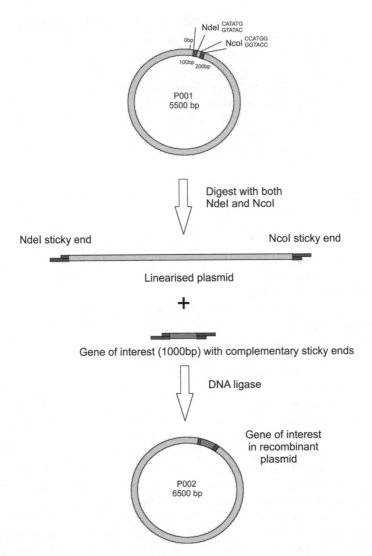

Figure 7.7   Diagram illustrating a simple cloning experiment using a representative commercial plasmid

for inserting the gene into the new plasmid. The natural gene sequence encoding the enzyme required will *not* contain these sequences in the right place. They must be engineered into the sequence at the stage of gene amplification, which is performed using a process known as PCR, which is the starting point for the creation of any recombinant biocatalyst.

7.3 Gene Amplification by PCR

The unwinding of the double helix structure of DNA allows replication of the genetic sequence as described above. The change from double- to single-stranded DNA also occurs at high temperature, and is reversible. This physicochemical behaviour of the DNA helix at different temperatures can be exploited for the amplification of DNA sequences using a technique called PCR (Figure 7.8).

When heated to 95 °C, the double helix of DNA *melts* into two individual strands. In PCR, the double-stranded DNA of interest (the *template*) is melted in the presence of short synthetic DNA sequences or *oligonucleotide primers* that have been designed to bind to the coding and complementary strand and that encode, respectively, the N-terminus and the C-terminus of the protein. These primers are in higher concentration than the DNA template. DNA polymerase and deoxynucleotide triphosphate monomeric units (dNTPs) that will act as the monomers for DNA polymer formation are also included in the PCR reaction mixture. When the temperature is lowered, the primers anneal to the complementary template sequence, typically at the ends of the gene for which they have been designed. This is called *annealing*. The temperature of the PCR reaction is then raised to a level at which the activity of the DNA polymerase is optimum. The polymerase acts to complete the gene by incorporating the nucleotide monomers that have been added to the reaction in a process known as *extension*, which always proceeds in

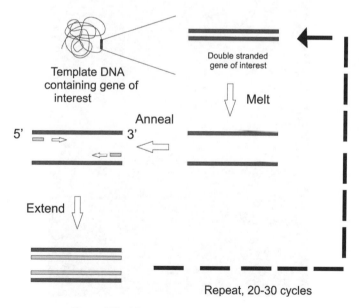

Figure 7.8   The polymerase chain reaction (PCR)

the 5' to 3' direction, on either the coding or non-coding strand. PCR uses DNA polymerases that have optimum activity at elevated temperatures, and have been isolated from thermophilic organisms such as *Thermus aquaticus*, which gives its name to one of the most commonly used PCR polymerases, *Taq*.

The melt–anneal–extend sequence described constitutes one *cycle* of the PCR. The process is conducted in a simple machine called a thermal cycler (Figure 7.9) which is able to raise and lower the temperature of incubation very rapidly. The template annealed to primer constitutes another double-stranded piece of DNA – this can be melted, primers will anneal to each of the four strands, and four double strands result in the action of the polymerase. Thus, the amount of the required gene doubles for every cycle. Typically 25 or 30 cycles of melting, annealing and extending are performed, resulting in an exponential amplification of the target DNA or gene.

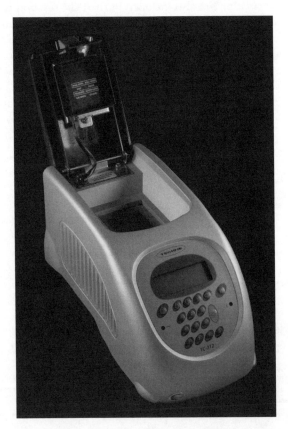

Figure 7.9   A thermal cycler (PCR machine)

We provide below a suggested protocol for starting a PCR reaction for gene amplification. In addition to giving a starting point for those of no experience, it will give an opportunity to discuss how PCR reactions may be adjusted or improved to give better results.

7.3.1 PCRs – practise and considerations

As shown above, PCRs are carried out in small benchtop machines known as thermal cyclers (Figure 7.9). Essentially, these are small incubating ovens in which it is possible to adjust the temperature for melting, annealing and extension steps of PCR and also to cool the reactions when the process has been completed. PCRs are usually conducted in microlitre volumes in small versions (200 or 500 µL) of plastic-capped tubes (Figure 7.10) in which reactions usually in the range of 10–50 µL are contained.

PCRs, at their most basic, will contain:

- A DNA template that contains the gene sequence of interest
- Oligonucleotide primers that are specific for the gene of interest, one 'forward' and one 'reverse'
- DNA polymerase
- dNTPs (deoxynucleotide triphosphate monomers, which will act as a reservoir of A, T C and G for the growing DNA chain)
- A buffer solution
- Deionised water
- Magnesium ions

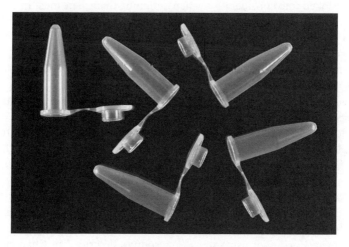

Figure 7.10   PCR tubes

7.3.2 The DNA template

We will restrict our discussion, for simplicity, to DNA templates arising from prokaryotes – eubacteria or archaea. In most cases, the template DNA will be in the form of either the 'genomic' or 'chromosomal' DNA of the bacterium. Possible sources of genomic DNA are described in Chapter 2. This will represent in most cases the entire genetic complement of the organism, and should contain the sequence encoding the biocatalyst of interest. There may be exceptions if the gene sequence desired is encoded on a small plasmid within the organism. Other templates that may be suitable for PCR may include cloning or expression plasmids that have previously been constructed by yourself or other groups. As PCR is extremely sensitive – theoretically only one molecule of the required DNA needs to be present to generate a large amount of DNA – it may also be possible to use a sample of the bacterium containing the gene, such as that which may be present in a small sample of the lyophilised material obtained from a culture collection (see Chapter 2). The required concentration of the template within the PCR may be different in each case, and it may be possible to improve the results obtained by varying this parameter. A good starting concentration would be 25 ng of genomic or plasmid DNA in a 50 µL reaction.

7.3.3 Oligonucleotide primers

One of the major considerations before beginning a PCR is the design of oligonucleotide primers that will be complementary to the 5' and 3' ends of the gene of interest. The nature of the exact sequence of these primers will be fundamental to the overall success of the experiment and a variety of factors should be taken into consideration when designing them. The most important of these is that the gene amplification must result in a length of DNA which, at its 5' and 3' ends, possesses short sequences that contain the appropriate restriction sites to generate the correct sticky ends for insertion into the expression plasmid.

It is first necessary to examine the gene of interest for restriction sites that may be cut by the restriction enzymes for which appropriate sequences have been chosen as sticky ends. If, for example, the gene of interest contains the restriction sequence for NcoI within the gene, then the gene may be cut in the middle of the coding sequence during the digest reaction for ligation (see below). It is therefore prudent to search the sequence of the gene using the 'Find' facility in word, for example, to search for sticky end restriction sites within the gene. Some of these are shown in Appendix 7, but any catalogue from a well-known molecular biology reagents supplier will contain these details. There are also a number of websites that will help identify unwanted cut sites including Webcutter (http://users.unimi.it/~camelot/tools/cut2.html). Simply pasting the gene sequence (acquired using the web resources described in Chapter 2) into the box provided and pressing 'Analyze sequence' will reveal any of the common restriction sites within the gene of interest. If the gene is at any point to be treated with a restriction enzyme for which there is a restriction site within the gene, then an alternative primer containing a site for an alternative enzyme will have to be found. The second restriction site should also be borne in mind when analysing a diagnostic digest of the plasmid that results from gene cloning, as more than one fragment corresponding to the gene of interest will result (see below).

We will now consider a simple example of gene cloning using the same theoretical plasmid shown in Figure 7.6. We will be inserting the new gene into this plasmid by making use of the *Nde*I and *Nco*I restriction sites in the multiple cloning site of the plasmid. We will therefore need to engineer a restriction site for the *Nde*I enzyme at the 5' end of the gene, to be inserted and an *Nco*I site at the 3' end.

The Nde1 restriction site is: CATATG

                                        GTATAC

Leaving sticky ends thus: CA--   TATG

                                   GTAT   --AC

The CATATG site must be engineered into the forward primer for PCR therefore. The ATG motif can be used to directly substitute for the ATG codon encoding methionine at the start of the sequence. The *Nco*I restriction site is CCATGG and must be added to the 5' end of the reverse primer as described below.

Below is the gene sequence encoding a lyase enzyme that will be the target for PCR under consideration. The gene is 774 base pairs long and encodes an enzyme of 258 amino acids. Both the start codon for methionine (ATG) and the stop codon (TGA) in the gene are in bold and underlined. The DNA sequences that precede and follow the gene, in this case on the bacterial chromosome, are known as the 'flanking sequences' and these may aid in primer design for PCR.

```
ccgctcctcg gtgactcgaa cggaggaagt tcg**atg**aagc_aattggccac ccccttccag
gagtactcac agaagtacga gaacatccgc ctcgaacgag acggcggcgt cctcctggtc
accgtccaca ccgaaggcaa gagcctggtg tggacctcaa ccgcacacga cgagctggcc
tactgcttcc acgacatcgc gtgcgaccgg gagaacaagg tcgtcatcct caccggcacc
ggcccctcgt tctgcaacga gatcgacttc acctcgttca acctcggcac cccgcacgac
tgggacgaga tcatcttcga aggccagcgt ctgctcaaca acctgctgag tatcgaggtg
ccggtcatcg cggcggtcaa cggaccggtg accaaccacc cggagatccc cgtcatgtcg
gacatcgtcc tcgccgcgga gtccgccacc ttccaggacg gaccgcactt cccttccggc
atcgtgcccg gggacggcgc ccacgtggtg tggccgcacg tgctgggctc gaaccgtgga
cgctacttcc tgctgaccgg ccaggaactc gatgctcgca ccgccctcga ctacggcgcg
gtcaacgagg tcctgtccga gcaggagctg ctgccccggg cctgggagct cgcccgcggt
atcgcccgaga aaccgctcct ggcccgccgg tacgcccgca aggtgctgac ccgtcagctg
cggcgggtca tggaagccga cctgagtctc ggcctcgcgc acgaagcgct cgccgccatc
gatctgggaa tggagtccga acag**tga**cca gcacagcacc ctcgaccgta gccgagacgg
```

The forward primer can be represented as the 5′ sequence of the gene as read off the page with the addition of the restriction site for *Nde*I, plus a few extra nucleotides at the start – this ensures that the restriction enzyme, as it binds the DNA outside the restriction sequence to stabilize itself, will work properly The length of primers required for the reaction may vary dependent on the gene in question, but a good length to start with is approximately 18–30 bp, with the catatg (underlined) site near the 5′ end of the primer.

A suggested forward primer therefore would be:

```
FOR:5'- gga gga agt cat atg aag caa ttg gcc acc - 3'
```

For the reverse primer, the primer sequence needs to be the *reverse complement* of the gene sequence that is written on the page as it will need to bind to the 5′ end of the gene sequence on the non-coding or complementary strand.

The sequence at the 3′ end of the gene reads:

```
5'- gcc gcc atc gat ctg gga atg gag tcc gaa cag tga cca gca cag - 3'
```

The reverse of this is:

```
3' - gac acg acc agt gac aag cct gag gta agg gtc tag cta ccg ccg - 5'
```

The complementary sequence, which is that found on the non-coding strand is:

```
5' - ctg tgc tgg tca ctg ttc gga ctc cag tcc cag atc cat ggc ggc - 3'
```

Add the *Nco*I restriction site (underlined) near to the 3′ end of this sequence as read to give a suggested reverse primer:

```
REV: 5' - ctg cca tgg tca ctg ttc gga ctc cag tcc - 3'
```

Once you have designed primers, it is worth checking that the *melting tempera-tures* (T_m) of each primer are within 5 °C of each other. There are a lot of different ways to do this, but the most simple is to use formulae such as those found at http://www.promega.com/biomath/calc11.htm#melt_results, which allow for the simple pasting of oligonucleotide sequences for such calculations. It is also worth noting that annealing may be improved by one or two C or G bases at the 3′ end of the primer, which is the region that will bind first to its complementary sequence.

Oligonucleotide primers can be purchased from a range of biochemical suppli-ers from whom it is now routine to order these primers online. As well as providing an analysis tool during ordering which would check the melting temperature of the primers, the sequences can also be examined for undesirable characteristics, such as the potential to form secondary structure or to dimerise, both of which should be avoided for a successful PCR. It is normal to request in the order of tens of nanomoles of primer material, which is delivered as a lyophilised sample in a plastic tube that must be diluted with water before use as directed in a PCR as described below.

With the advent of high-throughput molecular biology, many tools are available for automatic design of primers, including http://biotools.umassmed.edu/bioapps/primer3_www.cgi; http://bibiserv.techfak.uni-bielefeld.de/genefisher2/ and http://tools.invitrogen.com/content.cfm?pageid=9716, which provides the easiest to follow guide for beginners interested in designing primers for prokaryotic gene amplification, including the option to automatically insert the restriction sites of choice.

7.3.4 Setting up the PCR

There are many ways in which to carry out even a simple PCR, as there are many parameters that may be adjusted in order to optimise the performance of the reaction. These may include the concentration of template or primers, the concentration of magnesium ions, the addition of DMSO or detergents such as betaine as additives, and, crucially, the temperature at which the *annealing* step is performed, and the length of time that is applied for the melting and extension processes. We therefore provide a simple recipe for a first PCR and some conditions with which to program the thermal cycler.

7.3.5 A recipe for a PCR using genomic DNA as a template

To a 500 μL PCR tube, add:

DNA template at a concentration of 25 ng μL^{-1}	1 μL
Forward primer at a concentration of 20 μM	1 μL
Reverse primer at a concentration of 20 μM	1 μL
dNTPs (2 mM each in solution)	5 μL
MgSO$_4$ (25 mM solution, this is sometimes included in the buffer)	1 μL
DNA polymerase buffer (10x concentration)	5 μL
DNA polymerase (add last)	1 μL
Distilled water	34 μL
To a final volume of:	50 μL

Even a simple thermal cycler will have several wells in which to insert PCR tubes, so it may be worth setting up a multiple of reactions with variations of concentration in magnesium ions, or template, or perhaps with and without the addition of 1 or 2 μL of DMSO.

7.3.6 A simple program for a PCR using genomic DNA as a template

There are four stages to the PCR reaction: initial melt; 30 or so cycles of melt, anneal and extend as described above; a final extension; and cooling the reaction for storage. Below is a suggested program for the thermal cycler:

1. Initial melt 95 °C for 180 s
2. Thirty cycles of the following stages: (a) Melt; 95 °C, 60 s
                                              (b) Anneal; 55 °C, 60 s
                                              (c) Extend; 72 °C, 90 s

3. Final extension 72 °C for 180 s
4. Cool to 4 °C (if running overnight).

In the event of a failed reaction, it is recommended that the physical parameters be altered, in conjunction with varying the chemical composition of the reactions as described above, in order to attempt to get a product. One of the most successful of these is often to change the annealing temperature, perhaps by lowering (in some cases as low as 35 °C) or raising (perhaps as high as 65–70 °C), but a longer melt and/or extension time can also prove beneficial. An annealing temperature of about 10–15 °C lower than the melting temperature of the primers is a good place to start. 'Gradient' PCR machines are also available that allow the optimisation of the annealing temperature within one experiment. If the PCR uses template DNA that has come from an organism for which the DNA is rich in G and C bases, the annealing temperature often needs to be raised. Organisms frequently encountered in biocatalysis that display this feature include actinomycete species of bacteria such as *Streptomyces, Rhodococcus* and *Nocardia*. The success or otherwise of a PCR will be determined by analysis of the reaction mixture using a type of electrophoresis described below.

As PCR has developed as a technique, its applications have become wide-ranging and techniques ever more sophisticated. Whilst the above guide is merely an introduction, there are several good textbooks that include more detailed aspects of PCR and their uses [7], and a search of the internet for 'PCR' will raise many useful pages with techniques and useful tips.

7.4 DNA Fragment Analysis by Agarose Gel Electrophoresis

Fragments of DNA such as the genes amplified by PCR can be analysed by electrophoresis, but in a different format to that used for protein analysis described in Chapter 6. The DNA is usually analysed using a horizontally formatted gel that has been immersed in a buffer in a suitable tank that is fitted with electrodes such that a voltage may be set up across the gel. The gel in the case of DNA is made from agarose, essentially a refined form of laboratory agar, and the pieces of DNA migrate according to size as measured by the number of base pairs in the DNA fragment (Figure 7.11).

A 'ladder' of base pairs of known size is also included in the gel to act as a reference. The DNA can be visualised by UV light in a transilluminator after the gel has been stained with an aqueous solution of ethidium bromide or Sybrsafe®. (NB Ethidium bromide is a known carcinogen and teratogen.) Ethidium bromide-type reagents are strongly UV active and are known to intercalate between the stacks of DNA bases and the DNA fragments on the gel show up therefore as bright bands in the appropriate size regions when placed on the transilluminator.

7.4.1 Running an agarose gel

Apparatus and reagents

In order to run an agarose gel for DNA analysis you will need:

- An agarose gel electrophoresis tank and power pack
- A plastic gel mould
- A plastic gel comb with multiple narrow lanes
- Masking tape
- Automatic pipettes and tips
- Agarose (0.48 g for a 60 mL gel)
- 200 mL TAE buffer (see Appendix 5)

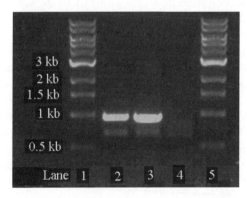

Figure 7.11 An agarose gel, developed with ethidium bromide, showing DNA fragments (lanes 2 and 3) against a ladder of DNA fragments of a standard defined size

- Agarose gel loading buffer (see Appendix 5)
- Sybrsafe ® (0.5 μL for a 60 mL gel)

NB. *Caution*: Wear disposable gloves from the outset and for the entire procedure. The Sybrsafe® reagent is genotoxic and will contaminate the gel and equipment with which it comes into contact. It may be of course that others in the laboratory have previously used the gel moulds, combs, glassware and pipettes for handling Sybrsafe®, and these are best considered as already contaminated. It is best to isolate a small area of the laboratory for agarose gels and Sybrsafe® for this reason, and to reserve one or two automatic pipettes and reusable glassware specifically for this area. Also, at the end of the experiment, any non-reusable waste such as gloves and old gels should be sequestered in a clearly labelled separate receptacle for Sybrsafe®-contaminated waste.

1. Prepare the gel mould for the agarose. This usually involves taping the ends of the plastic mould to stop the agarose from leaking out whilst it is still molten. Make sure that the taped seals are water-tight (Figure 7.12).

2. Mix 0.48 g of agarose with 60 mL of TAE buffer. This will make a 0.8 % agarose gel – a standard gel for the analysis of genes of the size that will encode the majority of biocatalysts i.e. 500–3000 bp. The agarose will not dissolve immediately. Heat the agarose suspension either in boiling water or in a microwave oven until the agarose is seen to dissolve.

3. Carefully transfer 2 μL of the neat, bright orange Sybrsafe® reagent to the gel mixture and swirl to mix. The solution will become slightly yellow.

4. Pour the molten mixture into the mould and immediately place the gel comb in the appropriate position at the end of the gel (Figure 7.12). The gel will set around the comb to create wells, in a similar fashion as for SDS-PAGE (Chapter 5).

5. Mix 5 μL of the completed PCR mix with 1 μL of the agarose loading buffer (this could be up to 50 μL of PCR mix if, say, purifying the whole sample for a cloning experiment). The buffer contains a blue dye that will allow you to watch the progress of the electrophoresis. Also, mix 5 μL of the 1 kB ladder mix with 1 μL of the loading buffer. This will serve as a standard on the agarose gel.

6. Remove the tape from the agarose gel mould – the gel should now be set. A good rule of thumb is that the set gel will appear opaque. Insert the gel into the running tank, making sure that the end with the comb is at the end for the *negative* (black) electrode. DNA is negatively charged overall and will migrate towards the *positive* (red) electrode end of the gel. Pour sufficient TAE buffer into the tank to just cover the surface of the gel. Remove the gel comb from the gel and ensure that the lane holes are open and have been filled with the buffer.

7. Using an automatic pipette, transfer 6 μL of the ladder sample plus loading buffer into the first well on the left at the top of the gel. This must be done gently,

as the sample can squirt out of the top of the lane and contaminate other lanes. After you have loaded the ladder, load the other samples into the gel, using a fresh pipette tip for each sample (Figure 7.12).

8. When you have completed the loading, cover the tank with the lid and, again making sure that the samples have been loaded at the *negative* (black) electrode end, switch on the power. Set the voltage to 100 V and allow the gel to run until the blue dye has travelled nearly towards the end of the gel. This may take an hour or possibly a little longer (Figure 7.12).

9. Once the run has finished, switch off the power and remove the gel. Place the gel on the transilluminator. *Caution:* If you are using an open transilluminator without a screen, you must wear a UV-protective face-shield and check that your labcoat/gloves cover any exposed skin so as to avoid burning. You should see the DNA fragments appear as bright bands on a dark background (Figure 7.11). The size of the gel fragment can be compared with the size of the fragments on

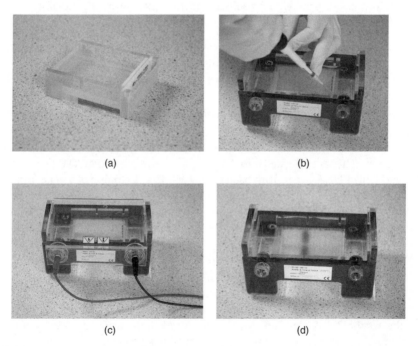

(a)   (b)

(c)   (d)

Figure 7.12 Preparation of an agarose gel for analysing PCR reactions and restriction digests. The gel mould is sealed with masking tape at each end and the molten agarose is poured inside. A plastic comb is inserted at this stage to form wells (a). The gel is covered in TAE buffer and the samples of DNA mixed with loading buffer are injected into the wells (b). The gel tank is connected to a power pack and run until the dye has moved at least halfway along the gel (c, d)

the ladder on the left. In the example shown in Figure 7.11, it can be seen that the amplified PCR product is about 1000 bp (1 kbp).

Once the gene has been amplified by PCR, it should be cleaned up, in order to remove the contaminants that have resulted from the PCR, including the template DNA, and then it may be cloned or ligated into a suitable plasmid vector.

7.5 Gene Cloning

The term gene cloning is used trivially to describe a number of different processes, but it is most useful to think of it as developing the gene of interest into a state such that it can be easily manipulated, by making many copies of it. When papers in the biocatalysis area describe a gene encoding an enzyme as having been cloned, it is usually understood that that gene has been amplified, cleaned up, ligated into a plasmid vector suitable for transformation of a suitable host strain (usually *E. coli*) and analysed to determine the sequence of the gene.

The process of gene cloning subsequent to PCR will usually consist of the following steps:

1. *PCR product clean-up.* This is either done by merely cleaning the PCR product directly from the PCR tube, or by cutting the PCR product out from an agarose gel with a scalpel, and preparing this for cloning. Commercial kits are available for both of these clean-up procedures.

2. *Digestion of plasmid vector and PCR product.* Both plasmid and gene must be digested using suitable restriction enzymes, in order to generate the appropriate complementary sticky ends such that the gene can be ligated into the plasmid.

3. *Ligation.* The gene is ligated into the plasmid of choice using an ATP-dependent enzyme called DNA ligase from bacteriophage T4. The cloning of a gene into an expression plasmid was usually done in two stages: the first to clone the gene into a cloning plasmid that had a high 'copy number' – this would allow easy access to the gene of interest; followed by sub-cloning into an expression plasmid. However, it is routine now to clone the gene of interest directly from the genomic DNA via PCR into the expression plasmid.

4. *Transformation and 'mini-prepping' of plasmids.* A cloning strain of *E. coli is* transformed with the ligation mixture and the strain is grown on agar plates containing an appropriate antibiotic. A small liquid culture is grown using one of the resultant colonies and this is processed (mini-, midi- or maxi-prepped according to scale) using a commercially available kit to give a solution of many copies of the recombinant plasmid.

5. Once the sequence of the gene has been analysed and confirmed to be correct by comparison with a sequence available in the molecular biology databases (if available), the gene may be considered to have been successfully cloned.

7.5.1 PCR product clean-up

Most routine manipulations of DNA in the laboratory are conducted using kits that are available from a range of biochemical/molecular biology suppliers. The PCR product can either be cleaned up using the procedure supplied in a PCR product clean-up kit – this basically involved taking the residual PCR reaction mixture and transferring the contents to a purpose-prepared mini-column that is then washed with a series of solutions, finally eluting the DNA required in pure water. An alternative is to run all of the PCR product on a preparative-scale DNA gel and, using the transilluminator to visualise the product material, to cut this band out with a scalpel, weigh the amount of gel produced, and again to use a purpose-designed kit for the isolation of the DNA material from the gel sample. This 'gel extraction' is sometimes to be recommended, particularly when the template for the PCR reaction is itself a recombinant plasmid. It can be that the template plasmid is purified along with the PCR product, thence being carried over into the ligation reaction and then the transformation mixture, and can lead to re-isolation of the original plasmid.

7.5.2 Digestion of the plasmid and PCR product

The primers designed for PCR will have resulted in a product that has, within either end, a restriction site that is suitable for cleaving by restriction enzymes in order to yield 'sticky ends' that will be suitable for ligation. In order to tidy up the ends to generate suitable sticky ends, these ends must be digested with the restriction enzymes of choice. In addition, the plasmid that has been chosen to carry the gene of interest (the 'vector') must be digested in order to generate the complementary sticky ends that will allow the gene to be inserted, resulting in the final complete plasmid.

The restriction digest reactions only require the DNA to be digested, the restriction enzyme of choice, and a suitable buffer, the last of which is usually supplied with the restriction enzyme from the company. It may also be necessary to add a small amount of another reagent, possibly dithiothreitol or bovine serum albumin (BSA), if recommended by the company literature. One important facet of restriction digests is the buffer. Not all restriction enzymes work equally well in the same buffer, and if performing a 'double digest' such as in the example below, you will need to consider whether the buffer of choice for each enzyme, as advised in the company literature, is suitable for both enzymes. Fortunately, the biochemical catalogues usually provide a table that allows one to easily find a suitable buffer for a digest that will include, say *Nco*I and *Nde*I, and recommend a buffer for these experiments. If compatible buffers cannot be found, it may be necessary to run sequential digests with the two enzymes under different conditions.

7.5.3 A double digest of a PCR product using *Nde*I and *Nco*I

1. Transfer 5 µL of the clean PCR product to a 500 µL PCR tube. To this, add 12 µL of deionised water, 2 µL of the recommended buffer for double digest (if using NEB buffers, the suitable buffer is no. 4), and 1 µL each of *Nde*I and *Nco*I. Incubate at 37 °C for 2 h. (NB There are new fast enzyme formulae which allow for much faster digestion.)

2. Clean up the digestion mixture using a PCR-product clean-up kit as described above. The product can be checked using an agarose gel if necessary to confirm the presence of a fragment of the anticipated length.

7.5.4 A double digest of a plasmid using *Nde*I and *Nco*I

1. Transfer 1 µL of the commercial plasmid Novagen pET26b(+) (this is supplied at 1 µg µL^{-1} concentration in glycerol) to a 500 µL PCR tube. To this, add 15 µL of deionised water, 2 µL of the recommended buffer for double digest (again, NEB no. 4 would be appropriate), and 1 µL each of *Nde*I and *Nco*I. Incubate at 37 °C for 2 h.

2. Clean up the digested plasmid to remove restriction enzyme and additives using a PCR-product clean-up kit as described above. The digested plasmid, which is now described as 'linearised' can be checked using an agarose gel to confirm that digestion has taken place. The digested plasmid should run at the predicted size of the plasmid – in this case approximately 5500 bp – if the plasmid appears 'smaller', for instance in the 3000 bp region, this is usually indicative of the plasmid still being in a state of supercoiling, and hence undigested.

7.5.5 Ligation

The digested plasmid and insert gene must now be ligated to form the recombinant plasmid. This is done using an ATP-dependent DNA ligase enzyme, which is again obtained from a commercial source together with a requisite buffer in which the performance of the enzyme is optimum. DNA ligase, like restriction enzymes, should be kept cold at all times when not in use, and it is recommended that it is stored in a cool-box like the other molecular biology reagents.

1. To a 500 µL PCR tube, add 1 ng of the clean digested plasmid and 3 ng of the clean digested PCR product. Add a volume of ligase buffer equal to one-tenth of the final reaction volume (ligase buffer contains magnesium chloride, dithiothreitol and the ATP that is essential for the reaction). In the event of an unsuccessful ligation, the ratio of the insert:vector can be varied, although a ratio of 3:1 is a good starting point.

2. Finally, add 1 µL of the DNA ligase enzyme. Incubate at 16 °C overnight. The incubation can of course be performed in a PCR machine that has merely been

programmed to run in isothermal mode for the length of time required. There are also kits available for 'quick ligation' which allow for much shorter incubation times.

7.5.6 Transformation of a cloning strain of *E. coli*

The ligation reaction should result in at least some intact recombinant plasmid. It is now necessary to 'persuade' the *E. coli* cells to take up this plasmid and to synthesise enough copies of it that may be isolated using mini- or midi-prep techniques (see below). We have considered the techniques of 'transforming' *E. coli* in an earlier chapter. A cloning strain of *E. coli* should be used for the initial transformation using the ligation mix, such as Top10 from Invitrogen or Novablue Singles from Novagen.

Apparatus and reagents

- 100 mL LB agar
- Sterile plastic Petri dishes
- 50 μL aliquot of competent cells of cloning strain of *E. coli*
- 1 mL stock solution of antibiotic (e.g. kanamycin)
- Ice bath
- Automated pipettes and sterile tips
- Heating block set to 45 °C
- Glass plate spreader
- 70 % ethanol solution

1. Prepare 100 mL of LB agar (see Appendix 4), sterilise by autoclaving and allow to cool. When the bottle has cooled sufficiently for it to be held in the hand, add the antibiotic to the agar, mix by swirling and pour approximately 20 mL of the agar into one of each of five sterile plastic Petri dishes.

2. Thaw a 50 μL aliquot of competent cells on ice, and to these add 1 μL of the solution of plasmid that has been adjusted to a concentration of 2 ng μL^{-1}. Mix with the end of a plastic pipette tip, then leave on ice for a further 30 min.

3. The cell suspension is 'heat shocked' in a water bath at 42 °C for 45 s, then transferred to ice for a further 2 min.

4. Add 1 mL of sterile LB medium (SOC medium, which is often supplied with commercial competent cells, is often used for this step) and grow the cells in an orbital shaker for 1 h at 37 °C. After 1 h the suspension should appear slightly cloudy. Pipette 100 μL of the suspension onto the now solid LB-ampicillin agar plates and use the spread plate technique (Chapter 3) to spread the liquid culture around the surface of the agar. Leave the plates to incubate at 37 °C overnight.

5. On the following day, you should observe a number (could be <10, but could be >1000, depending on the 'transformation efficiency' – see Chapter 4). The efficiency obtained from transformation using a ligation mix can be poor, but it is important to remember that it is only necessary to obtain one or two colonies that have taken up the recombinant plasmid.

The colonies obtained from the transformation of the ligated plasmid should contain the recombinant plasmid, but there are other possibilities that require that the insertion of the gene has occurred. For example, it may be possible that the intact commercial plasmid has contaminated the transformation reaction. Two major techniques are used to confirm that insertion of the gene has taken place – *colony PCR* uses the cell colonies directly as templates for an amplification reaction, using primers usually designed against either the 5' and 3' end of the inserted gene, *or*, against one of these ends and a short sequence of DNA which is just outside the relevant restriction site. This last option ensures that any amplification results from a stretch of DNA that includes both the commercial plasmid *and* the insert. Also, the plasmid DNA can be digested with the two restriction enzymes that recognise the insertion sites – this will require the isolation of the recombinant plasmid as described below.

7.5.7 Mini-, midi-, and maxi-prepping of plasmid material

The colonies obtained from transformation of *E. coli* using the ligation mix are suitable for growing small cultures for the specific purpose of making plasmid stocks for further transformations. The technique for doing this, dependent on scale, is known as 'mini-prepping' (or 'midi-/maxi-prepping'). Once again, kits are available for this technique from a range of biochemical suppliers, and will include all necessary solutions, tubes and columns containing filters which will bind and release the DNA during the course of the procedure. To generate material for the mini-prep, it is necessary to grow a 2 or 5 mL culture from one or more of the colonies obtained.

The practical details of the mini-prep procedure are described in the instructions that come with the kits. A microcentrifuge will be required for the procedure, that starts with the harvesting, by centrifugation, of the small cell cultures that have been grown overnight for the purpose. The cell material is then resuspended in a small amount of a cell lysis solution, centrifuged again, and the supernatant loaded onto a small filter in a column that fits within a microcentrifuge tube. A series of wash steps bind and then release the plasmid DNA, which is finally eluted into a volume of 50 or 100 μL of distilled water, which, given that a transformation of *E. coli* will usually require not more than 1 μL of a plasmid solution, will provide a more than adequate supply of the relevant recombinant plasmid for a project. Plasmid DNA from mini-preps can be stored at −20 °C or −80 °C and can be freezed and thawed several times with little or no loss of plasmid stability.

7.5.8 Analysis of the PCR product – confirmation of gene cloning

The gene of interest can be considered to have been cloned only once the plasmid has been analysed by both a restriction digest and by sequencing, which will reveal the sequence of the gene ligated within the recombinant plasmid.

7.5.9 Analysis by restriction digest

The restriction sites that have allowed the insertion of the gene into the recombinant plasmid will now allow analysis of the plasmid to confirm that the insert has been cloned successfully.

1. To a 500 μL PCR tube, add:

Recombinant plasmid	5 μL
NdeI	1 μL
NcoI	1 μL
Restriction digest buffer	2 μL
Distilled water	11 μL
To a final volume of:	20 μL

2. Incubate the digest for 2 h at 37 °C. Take 5 μL of the digest mix, add 1 μL of the agarose gel loading buffer and run a 0.8 % agarose gel of the digest mix. View the gel on the transilluminator.

A successful result is shown in Figure 7.13. The digest should have resulted in two products: a band at approximately 5500 bp which represents the linearised

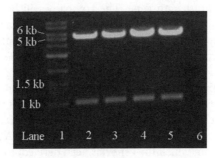

Figure 7.13   An agarose gel illustrating a successful analytical restriction digest. Lane 1 contains the ladder of DNA fragments of known length; lanes 2–5 contain restriction digest mixtures, showing the plasmid (approximately 5500 bp) and insert (1000 bp – 1 kb)

host plasmid and a band at 1000 bp, in the example shown, that represents the gene product. It is useful to include in one of the gel lanes a sample of the digested host plasmid that has not been used for cloning. It may be possible that the host plasmid has merely recircularised, without the inclusion of the required gene (in which case, the ligation has not worked) or even that no plasmid is observed at all, which may represent a flaw in the transformation protocol.

7.6 Analysis by DNA Sequencing

A successful digest does not in itself provide sufficient evidence of a successful gene cloning experiment. It may be that mutations have arisen during the PCR or that one has failed to spot a restriction site within the gene to be cloned. The gene must therefore have its sequence verified by DNA sequence analysis.

DNA sequencing is routinely performed by a host of companies, which merely require details of the plasmid used and a sample of the plasmid to be sequenced, usually in the form of some of the solution that has resulted from the mini-prep. The company would usually want to know the concentration of the plasmid provided, which can be easily determined using a spectrophotometer reading at 260 nm. Plasmid preps derived by mini-prep from 2 or 5 mL culture routinely contain of the order of $10-500\,ng\,\mu L^{-1}$, and the company would usually only require $5-10\,\mu L$ of this solution.

The methods by which the DNA sequence is determined are not relevant to the present discussion (for details see, for example, reference [4]), but the process is dependent on sequence-specific oligonucleotide primers in a similar fashion to PCR. If the company is familiar with the plasmid that you have used for the cloning experiment, they should be able to advise on the primers necessary for the sequencing reaction. The reactions are usually unable to read more than a few hundred bases, so in the case of genes longer than 500–800 bases, it may be necessary to request more than one sequencing reaction in order to obtain the sequence of the full length gene. The company will be able to design new primers for sequencing based on the sequence that they have already acquired. The gene information is usually sent back in the form of either text files or files that are derived directly from the sequencing machine, the results of which can be observed manually in a program such as CHROMAS or GeneEdit. In the event of an unexpected variation in the sequence from what has been expected, it is often possible to interpret the raw data and to correct some faults that have arisen due to the automated interpretation of the raw data by the program.

The raw DNA sequence may be examined against the DNA sequence contained within the genome from which the gene has been derived, to check that the gene has the correct sequence. In the event of 'point mutations' or single base changes, it is possible that these have arisen not through mutation but rather by an error in the genome sequence itself. If the mutations are 'silent' – that is, the change in base does not affect the sequence of amino acids – this should not cause too much concern, but should an amino acid in the sequence be changed, the possibility exists that this amino acid may be involved in substrate binding or catalysis, and the PCR and cloning experiment should be repeated, or a different clone from the ligation/transformation selected. The amino acid sequence encoded by the DNA sequence may be obtained easily using the 'Translate' tool on the EXPASY website as described in Chapter 2. This only requires cutting and pasting the DNA sequence obtained from sequencing into the appropriate window, and pressing 'Translate'. Six possible interpretations of the gene sequence will be presented – these represent three open reading frames for each of the 5′–3′ and 3′–5′ directions. One of these should correspond directly to the amino acid sequence of the enzyme that is expected. Occasionally, some of the translated gene has the correct sequence, but a single base insertion or deletion causes a 'frame shift' resulting in the wrong sequence from that point.

7.7 Troubleshooting the Gene Amplification and Cloning Process

The practical procedures for the simple cloning of bacterial genes is simple, provided one has access to the necessary reagents, kits and equipment. The experiments can prove frustrating in the case of failure however, as there are many points, from the PCR reaction onwards, which may have contributed specifically to the failure of the experiment. In the case of the experiment not working it is useful to consider a number of these eventualities, and what might be done to address them.

1. *Failed PCR reaction.* It would be fair to say that most genes from bacterial genomes should probably be amenable to PCR dependent amplification. Some possible reasons for the reaction not working were alluded to in Section 7.3. In addition to these, if PCRs do not work, first double-check the primer sequences and then of course check that the correct primers have been used in the reaction mix (the primers all come in identical vials, and may have similar names or numerical descriptors). If the primer sequences are acceptable, both physical and chemical parameters in the PCR can be altered as described previously. It may also be worth using fresh DNA polymerase, as the activity of the enzymes used for molecular biology can deteriorate rapidly if not treated carefully and kept cold.

2. *Gene fails to clone (as judged by agarose gel analysis of mini-prepped plasmids).* Unfortunately the gene cloning process can go wrong in a number of places. First, it may be worth checking that the clean PCR product is actually present in the restriction digest/ligation reaction, by running a gel of the product that has been

either excised from a gel or merely isolated from a PCR reaction using a clean-up kit, as sometimes the PCR product gets lost in the clean-up.

Second, as with DNA polymerase described above, the activity of the restriction enzymes should be verified, perhaps by running a simple digest of the plasmid vector – fresh enzyme should of course be used if you can confirm that one or both of the enzymes is inactive. Also, it may be that the restriction digests of plasmid and insert have not been run for a sufficiently long time, or at the required temperature.

Third, the activity of the DNA ligase may be a problem, as with the other enzymes. If the robustness of the rest of the protocol has been validated, the use of fresh DNA ligase may be considered. The ATP in the ligation buffer will be inactivated by freezing and thawing so fresh buffer may also help. It may also be worth adjusting the ratio of insert:vector in the ligation reaction as described above, and also possibly to try different temperatures of incubation.

Fourth, the competency of the cells used for transformation can be checked (particularly of course in the event of no, or very few, transformants). Again, the competence of cells can fail if they have either been prepared or stored incorrectly and a simple transformation of, for example, an established plasmid for which the transformation efficiency is known to be good, can confirm whether those competent cells are satisfactory.

Ultimately, the processes of PCR and gene cloning are really just organic chemical reactions for which the optimal reaction conditions need to be established on a case-by-case basis. Most PCR and gene cloning experiments work therefore, and it is more usual to find problems in the optimisation of gene expression (below).

7.8 Ligation-Independent Cloning

The sheer amount of new gene sequences now available through genome sequencing and metagenomics projects has acted unsurprisingly as the stimulus for the development of high-throughput procedures for gene cloning and expression. These are particularly effective for large amounts of bacterial genes of the kind often sought by those involved in biocatalysis. Given the dependence of standard gene cloning protocols on case-by-case primer design and the activity of a number of expensive enzymes, new approaches to cloning that are independent of both restriction site design and the activity of ligases have been combined using a technique known as ligation-independent cloning (LIC) [8, 9]. In LIC, the gene is amplified by a PCR reaction which adds generic nucleotide ends ('LIC ends') onto the gene of interest which will aid in the ligation and cloning protocol as described below. As the ends are generic it is necessary only to design primers for the amplification of any prokaryotic gene without having to consider the integration of restriction sites within the primer sequences themselves. This is also helpful in automating the design of primers using simple computer programs. As the name suggests, the technique is also independent of the use of DNA ligase enzymes, exploiting instead the ability of longer sequences of single-stranded DNA to anneal spontaneously. Methods of LIC include the LIC-PCR (below) and also company-specific technologies such as Gateway® (Invitrogen) and InFusion^{TM} (ClonTech).

The theory behind LIC-PCR is outlined in Figure 7.14. The method will create single-stranded extensions on both the gene amplified by PCR and the LIC-plasmid vector such that annealing will be spontaneous.

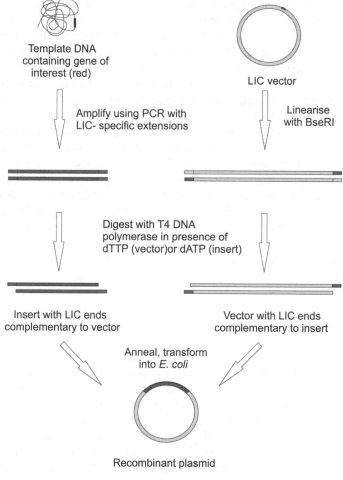

Figure 7.14   Ligation-independent cloning (LIC-PCR)

Hence, PCR primers for each gene are designed with generic LIC extensions that are merely added to the regular primer sequence. Once the gene has been amplified, it is cleaned up, and incubated with an enzyme, T4 DNA polymerase, and dATP. The polymerase, which is in this instance acting as a $3'$- $5'$ exonuclease, digests both strands of the amplified gene until it encounters an A, at which point the polymerase stops digesting the sequence and reveals a single strand on each end of the gene. Next the LIC-plasmid is digested by a restriction enzyme at a specific site to yield a 'blunt ended' linear plasmid (i.e. no single-stranded overhangs). This linearised plasmid is then also incubated with T4 DNA polymerase, but in the presence of dTTP. The $5'$ end of each strand is digested back to the first T base in the sequence, revealing single-stranded LIC sequence which is complementary to that revealed by the action of the T4 polymerase on the gene sequence to be inserted. Simple incubation of both linearised, T4-treated plasmid and insert results in spontaneous annealing of the complementary single strands, giving rise to a recombinant plasmid, which is then transformed into a high competency strain of *E. coli* to give colonies that should contain that plasmid. These colonies can be grown and their plasmids removed by mini-prepping procedures as described for regular cloning, to assess the success, or otherwise, of the experiment.

A number of molecular biology suppliers provide LIC-compatible plasmids, including Novagen and Stratagene. These companies also provide manuals detailing the use of those particular plasmids and necessary considerations in the design of primers for cloning genes into them. The LIC plasmids, as with other standard 'cut-and paste' expression vectors often also feature useful characteristics such as the encoding of protease-cleavable sites in the gene products which also facilitate the removal of affinity chromatography tags such as the poly-histidine tag described in Chapter 6. The LIC protocols, whilst accelerating and simplifying the cloning process are not without fault however, and are still subject to the quality of enzymes and competent cells used in the process addressed in troubleshooting considerations above.

7.9 Gene Expression

Once the gene of interest has been cloned and the sequence confirmed, one can begin experiments designed to find out the optimum conditions for expression of that gene. It is expression of the gene by the host strain of *E. coli* that will result in large amounts of the required enzyme being produced. Gene expression in bacteria is controlled by a whole host of factors that we will only address in the detail relevant to the context of biocatalysis, however, it is sufficient to say that genes that encode enzymes in bacteria will be expressed or 'switched on' in response to a change in physical or chemical environmental factors, such as the availability or lack of a carbon source, a change in temperature, or the relative availability of water. We have already seen in wild-type systems that the expression of genes encoding enzymes can be either *constitutive*, apparently requiring no specific chemical inducer, or *inducible*, such as the range of enzymes that are produced by *Pseudomonas putida* as a result of a change in gene expression in response to growth on the terpene camphor. It is the manipulation of these responses that has allowed a wide range of simple expression systems to be constructed for use in microbiology and biotechnology.

There are many such expression systems and we will address only one, that which exploits the so-called *lac* operon of *E. coli* and the inducer of gene expression, IPTG, an analogue of the natural inducer, lactose.

7.9.1 The lac-operon and IPTG as an inducer in *E. coli* expression systems

E. coli is able to use the carbohydrate disaccharide lactose as a carbon source. In order to enable the bacterium to utilise lactose, it has encoded within its genome an enzyme, β-galactosidase, which is encoded by the gene *lacZ*, which is capable of hydrolysing lactose to its constituent monosaccharide sugars, galactose and glucose. In order to metabolise lactose, the *lacZ* gene must be 'switched on' so that it may be expressed, resulting in the production of the β-galactosidase. The biochemical and genetic system employed by *E. coli* for sensing the need to express the *lacZ* gene can be simplified as illustrated in Figure 7.15. The transcription of genes involved in lactose metabolism is under the control of a region of DNA known as the *lac* promoter. When lactose is not available, a protein, the *lac* *repressor*, encoded by the gene *lacI*, binds to a sequence of DNA upstream of the promoter of the *lacZ* gene, called the *operator*. When the repressor protein is bound to the operator, it prevents the binding of RNA polymerase to the promoter, and thus inhibits the transcription of genes required for lactose metabolism, including the β-galactosidase. When lactose is at a sufficiently high concentration, it binds to the *lac* repressor, and induces a change in conformation of that protein that results in its release from the *lac* operator.

This frees the lac promoter and allows the binding of RNA polymerase and the transcription of the *lacZ* gene, leading to the production of the β-galactosidase. This relatively simple system of chemical induction is exploited in many commercial plasmids for the heterologous expression of genes in *E. coli*.

Many recombinant plasmids, in conjunction with expression strains, exploit some of the machinery of the *lac* operon for gene expression. In the popular pET series of expression plasmids developed by F. W. Studier and co-workers and commercialised by Novagen (summarised at http://www.merckbiosciences. co.uk/docs/ndis/C183-001.pdf), for example, the *lac* promoter does not feature, and is replaced by a promoter for an RNA polymerase from a virus-like organism, a bacteriophage called T7. The promoter is situated next to the *lac* operator

(a) In *E. coli*

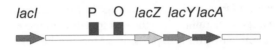

(b) In engineered expression plasmids exploiting
the *lac* operon for recombinant gene expression

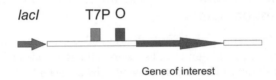

Gene of interest

Figure 7.15 (A) Structure of the *lac* operon in the chromosome of wild-type *E. coli*. (b) In the pET range of commercial plasmids, the promoter P is replaced by the promoter T7P, which recognises the T7 polymerase, encoded within the genomes of certain expression strains of *E. coli*

upstream of the multiple cloning site into which the foreign gene will be inserted, and any foreign gene thus comes under the control of the *lac* operator, and will be transcribed by the T7 polymerase when circumstances allow. The T7 promoter will only function in conjunction with the specific RNA polymerase from that phage, and will not function with the *E. coli* RNA polymerase, thus introducing an excellent measure of control in the transcription of the foreign gene. The gene encoding the phage T7 polymerase is not encoded on this series of plasmids but has rather been engineered into the chromosome of *E. coli* and is a feature of the commercial strains of that organism that are used for cloning and expression (DE3 strains, described in Chapter 4). The *lacI* gene, encoding the lac repressor protein, is also encoded in the engineered *E. coli* chromosome. The gene encoding T7 polymerase on the *E. coli* chromosome is under the control of a promoter–operator system called *lacUV5*, which responds to induction by lactose, as in wild-type *E. coli*.

If lactose is added to the cells of the expression strain containing the pET plasmid, the following events ensue. Lactose causes the release of the lac repressor protein from the *lacUV5* operator, the *lacUV5* promoter is exposed and native *E. coli* RNA polymerase then binds to the promoter and catalyses the transcription of the T7 polymerase. The T7 polymerase is then able to bind to the T7 promoter on the pET plasmid, which has also been exposed following lactose-induced release of the *lac* repressor. The foreign gene in the pET plasmid is then transcribed by T7 polymerase.

In practise, in order to prolong gene expression in the interests of inducing *overexpression*, or production of very large amounts of the required enzyme, lactose is not routinely used as the inducer, but rather the non-metabolisable analogue IPTG. The latter is somewhat expensive and also toxic to cell growth, but is still routinely used as an inducer in gene expression in *E. coli*, as it can be very successful in achieving the overproduction of recombinant enzymes in short growth times.

7.9.2 Simple strategies for improving soluble gene expression

In Chapter 4 we introduced the concept of designer biocatalysts and described methods for using plasmids that had been acquired from collaborators to perform transformations of commercially available expression strains of *E. coli* in order to create recombinant catalysts. Recombinant gene expression in *E. coli* is routinely achieved using very simple protocols that allow for adjustments in different variables in an effort to optimise the overproduction of the enzyme of interest. All that is required is that the expression strain of *E. coli* is transformed with the recombinant plasmid. We have already discussed the improvement in performance of recombinant biocatalysts in a previous chapter, which assumed that the recombinant plasmid had been received from a collaborator.

In order for the biocatalyst to be functioning at an optimal level, it needs to have folded correctly within the cell and a good indication that this has happened is if the expressed gene product is soluble. There are many reasons why a gene product will be insoluble – this may result in the formation of membrane-enclosed vesicles called *inclusion bodies*, which will not allow the exploitation of the biocatalyst in the recombinant system. The solubility can be improved in many cases merely by changing one of a number of chemical or physical variables. The extent of soluble expression can be determined trivially using SDS-PAGE. In the following example, we will consider the effect of the temperature of incubation following the induction of expression.

7.9.2.1 Optimising gene expression through small-scale test cultures

Apparatus and reagents

- Petri dish of LB agar bearing colonies of recombinant *E. coli* transformed with the expression plasmids containing the gene of interest
- Inoculation loop
- 10–50 mL sterile vials
- Aliquot of relevant antibiotic solution at 1000x required concentration
- Aliquot of 1 M IPTG

274   *Practical Biotransformations: A Beginner's Guide*

- Orbital shakers at different temperatures (perhaps 16 , 30 and 37 °C) – for testing expression at different temperatures
- Apparatus and reagents for performing SDS-PAGE (see Chapter 6)

1. First, a small (2 mL) starter culture of the recombinant strain of *E. coli* is grown overnight from a plate colony, with shaking at 37 °C in the presence of a suitable amount of the appropriate antibiotic.

2. Then 50 μL of this starter culture is used to inoculate each of four small cultures consisting of 5 mL of LB broth containing the appropriate amount of antibiotic. The series of cultures could be intended to study one particular variable, say the concentration of IPTG added, or the temperature at which cells are grown post-induction. The cells are grown with shaking at 37 °C until they have reached an optical density (A_{600}) of 0.5.

3. To each of the four small cultures, add IPTG from a 1 M stock solution in water, such that a final concentration of 1 mM in the cultures is achieved. Place one of the small cultures at 16 °C and one each at 20, 30 and 37 °C. The lower temperature incubations can be grown overnight, but the 30 and 37 °C samples can be harvested after 3–5 h.

4. Harvest the cell samples by centrifugation after an appropriate incubation time and resuspend the cells in 0.5 mL of either water or a simple buffer at neutral pH. Sonicate the cell suspensions for 30 s at high power and centrifuge again to obtain two fractions for each temperature culture sample: one represents the 'soluble fraction'; the other, a pellet of insoluble material that can be reusupended in 0.5 mL of buffer for purposes of analysis.

5. Against an appropriate series of low molecular weight markers, analyse the samples using SDS-PAGE (described in Chapter 6), running the soluble and insoluble fractions of each temperature experiment in adjacent lanes. After staining and destaining of the gel, it should be possible to see the band corresponding to the enzyme of interest at the appropriate molecular weight region. It will be seen that the gene product will be distributed between the soluble and insoluble fractions (Figure 7.16).

It may be observed that incubation at a lower temperature results in greater levels of soluble expression, and that further, larger scale growths should exploit the successful expression conditions revealed by this small scale experiment.

There are other physical and chemical parameters that can be investigated in this way:

- Concentration of the inducer (0.1–2 mM).
- The expression strain (there are many strains of *E. coli* available from biochemical suppliers that often result in improvement in soluble gene expression over standard strains; see Chapter 4).

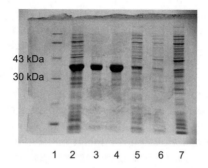

Figure 7.16   An SDS-PAGE gel illustrating the results of an experiment designed to investigate the conditions for optimum gene expression. Lane 1 contains low molecular weight markers. Lanes 2, 3 and 4 show the levels of expression in the insoluble fraction of cells grown at 16, 30 and 37 °C, respectively. Lanes 5, 6 and 7 show levels of expression in the soluble fraction of cells grown at 16, 30 and 37 °C, respectively. The soluble expression of the protein with a molecular weight of approximately 40 kDa is greatly improved at 16 °C

- The time of induction – at any time in the growth curve between early exponential phase (OD = 0.3) to late exponential phase/stationary phase (OD > 1.0) in shake flask cultures grown on LB broth.

If the investigation of these and other variables does not result in successful soluble gene expression, it may be that the gene needs to be engineered into a different expression plasmid, or that an entirely different expression system needs to be used. This can particularly be the case when attempting to express genes from eukaryotic organisms in *E. coli*, a prokaryotic host. It is common in both academia and industry to exploit expression systems in yeasts such as *Saccharomyces cerivisiae* (baker's yeast) or *Pichia pastoris* or fungi such as *Aspergillus oryzae* for the expression of more challenging gene products. It is also true that the protein required may, for a variety of reasons, elude soluble expression altogether and that an alternative target of similar activity or properties should be sought.

7.9.2.2 Auto-induction media (AIM)

The means of induction in standard commercial vectors are, as discussed, those which are based either on physical or chemical means through the addition of inducers such as IPTG the analogue of lactose in *lac*-repressor-T7 promoter-controlled systems. IPTG is expensive and toxic to cells however, and methods which avoid its use are potentially valuable from both safety and economic perspectives. Studier has recently described an excellent range of media for *auto-induction* that is reported to work well for recombinant strains of *E. coli* transformed with pET-type and other common vectors [10]. In auto-induction, the cells are grown on a defined medium with both glucose and lactose as sole carbon sources. Glucose actually inhibits the expression of genes under the control of the *lac* promoter, and the organism will use this preferentially as a carbon source over lactose. However, once the cells have utilised glucose and built up their biomass as a direct result, they then turn to lactose as the carbon source, the transcription of genes under the control of the *lac* promoter is switched on, and the cells start to produce the enzyme of interest in the desired manner. The advantage of AIM in practice is the very high cell densities that may be achieved, and hence a much higher yield of enzyme. Improvements in the solubility of expressed gene products are also possible. The disadvantages are the careful preparation of the solutions needed for the AIM protocols, compared with the simple preparation of LB broth, and longer growth times on the whole, although the greater yields can make this approach very worthwhile.

7.9.2.3 Solubility screens

Soluble expression can also be increased in some cases by the choice of buffer used for the resuspension and disruption of the cells. These conditions may take into account salt concentration, buffer pH and the presence or absence of small amounts of solvents or detergents. There are commercial screens available and also some useful publications including Lindwall and co-workers who describe the application of a range of thirty buffers applied to improve the solubility of expression of human tubulin protein in *E. coli* [11]. Whilst the preparation of the buffers is somewhat time-consuming, it may be beneficial in trying to improve the activity of some biocatalyst strains, or the recovery of isolated enzymes, if not through directly affecting the expression of the desired gene.

7.10 Conclusion

In this chapter, we have introduced the background to aspects of molecular biology fundamental for creating recombinant biocatalysts. We have also introduced a simple series of techniques that might enable the worker inexperienced in gene cloning to start some simple experiments using basic facilities and equipment. We have emphasised that developments in molecular biology have had a profound and lasting effect on applied enzymology and biocatalysis, and the area of recombinant biocatalysis and protein engineering should not be ignored by anyone beginning to work within this field. The scope and length of this book preclude any detailed attempts to discuss the huge range of available techniques, and indeed the suite of techniques required to embark upon studies of *in vitro* evolution of enzyme activity (see Chapter 8) that are now central to industrial enzymology. However, it is hoped that it provides a useful introduction to simple molecular biology principles that will help the uninitiated to engage with the relevant literature and with co-workers actively involved in biocatalyst development.

References

1. F. R. Blattner, G. Plunkett 3rd, C. A. Bloch, N. T. Perna, V. Burland, M. Riley M, J. Collado-Vides, J. D. Glasner, C. K. Rode, G. F. Mayhew, J. Gregor, N. W. Davis, H. A. Kirkpatrick, M. A. Goeden, D. J. Rose, B. Mau and Y. Shao (1997) The complete genome sequence of *Escherichia coli* K-12. *Science*, **277**, 1453–1474.
2. L. Stryer, J. M. Berg and J. L Tymoczko (2002) Biochemistry, 5th Edn, W.H. Freeman, San Francisco.
3. D. J. Voet and J. G. Voet (2004) Biochemistry, 3rd Edn, Wiley-VCH, Weinheim.
4. B. Lewin (2007) Genes IX (or earlier editions), Jones and Bartlett Inc., Boston.
5. J. Sambrook and D. W. Russell (2001) Molecular Cloning: A Laboratory Manual, Cold Spring Harbour Laboratory Press, Cold Spring Harbour.
6. G. Grogan, G. A. Roberts, D. Bougioukou, N. J. Turner, and S. L. Flitsch (2001) The desymmetrisation of bicyclic β-diketones by enzymatic *retro*-Claisen reaction. *J. Biol. Chem..*, **276**, 12565–12572.
7. M. J. McPherson and S. G. Moller (2006) PCR (The Basics), Taylor and Francis, London.
8. C. Aslandis and P. J. de Jong (1990) Ligation-independent cloning of PCR products (LIC-PCR). *Nucl. Acids Res.*, **18**, 6069–6074.
9. C. Li and R. M. Evans (1997) Ligation independent cloning irrespective of restriction site compatibility. *Nucleic Acids Res.*, **25**, 4165–4166.
10. F. W. Studier (2005) Protein production by auto-induction in high-density shaking cultures. *Protein Expression Purif.*, **41**, 207–234.
11. G. Lindwall, M.-F. Chau, S. R. Gardner and L. A. Kohlstaedt (2000) A sparse matrix approach to the solubiilzation of overexpressed proteins. *Protein Eng.*, **13**, 67–71.

Chapter 8
Engineering Enzymes

8.1 Introduction

Enzymes contain a vast and diverse wealth of catalytic reaction chemistries, activities and regio- and enantioselectivities within their natural repertoire. However, even though the amount of available gene sequences available for biocatalysis has increased exponentially over the past few years, it is still true to say that the application of enzymes is nevertheless inherently limited by the natural scope of substrate specificity and stability available within wild-type enzymes. As a consequence, much of the current literature in biocatalysis, and, indeed many of the best examples of the use of biocatalysis in an industrial setting exploits enzymes that have therefore been in some way *engineered*. Such engineering helps to expand the scope of biocatalytic activity, perhaps by creating enzymes which display enantioselectivities or regioselectivities that are distinct from those displayed naturally, by conferring greater thermostability or tolerance to specific solvents or in order to improve tolerance of high substrate concentrations that would be necessary to render industrial processes economically viable.

The basic elements of engineering an enzyme reside within the ability to change its amino acid sequence. Methods which look to changing the amino acid sequence must necessarily target the *gene* encoding the enzyme of interest; hence methods that alter the gene sequence are known collectively as *mutagenesis*. There have historically been two distinct general strategies that have been used for generating mutant gene sequences; those of *rational* and *random* mutagenesis:

1. *Rational mutagenesis.* Changes to amino acids may be performed *rationally*, that is, if the structure and mechanism of an enzyme are known, strategic decisions may be made as to changing one or more amino acid residues by changing one three-base-pair codon in the gene, so as to change, for example, the electrostatic or steric interactions between enzyme and substrate in the active site.

Practical Biotransformations: A Beginner's Guide   © 2009   Gideon Grogan

2. *Random mutagenesis or 'directed evolution'*. Changes to the amino acid sequence may be made randomly, using high-throughput mutagenesis techniques that alter any number of codons within a gene. Very large amounts of randomly generated variants, known as *libraries*, will result, which may be screened for improved performance, resulting in improved enzymes for which the molecular basis of improved properties will not be confirmed until after more detailed investigation. The advantage of random mutagenesis strategies is of course that it is not necessary that the structure of the enzyme under investigation is known.

We will not provide a detailed experimental guide to performing the techniques of protein engineering, as to do so would require a much larger book. However, we offer a brief guide on the considerations that might inform such experiments, and a description of some of the techniques that are used to accomplish them.

8.2 Site-Directed or Targeted Mutagenesis as a Tool for Investigating Enzyme Mechanism or Altering Catalytic Attributes

The rational mutagenesis of amino acids based on a knowledge of structure or mechanism is targeted at specific residues and is hence popularly known as *site-directed mutagenesis* (SDM). Given the superficial simplicity of the experiment, a surprisingly large amount of information about enzyme mechanism might be gleaned from engineering a change of one amino acid in an enzyme sequence. It may be possible, for example, to study the effect of the hydroxyl group of a tyrosine residue by mutating it to phenylalanine, or investigate the effect of removing steric bulk from within an active site by changing the aromatic side-chain of a phenylalanine to the smaller methyl group of an alanine. It is also possible to change the electrostatic environment of an active site, perhaps by changing a positively charged lysine to a negatively charged glutamic acid. Mutants that have been generated using SDM are usually designated using a notation which places the wild-type amino acid first, followed by the amino acid to which it has been changed, thus Y329F or Tyr329Phe, indicating that the tyrosine at position 329 has been mutated to phenylalanine.

SDM experiments are technically trivial, but are largely limited at this stage to the repertoire of twenty amino acids that are encoded by the genetic code. Progress is being made on the introduction of non-canonical amino acids into enzymes, but these techniques are far from routine as yet [1]. The other constraints on informed, rational mutagenesis of an enzyme are that the three-dimensional structure of the enzyme, or a closely related enzyme of similar activity are known, and that the gene encoding the enzyme has been cloned and is thus available in a suitable plasmid that can be manipulated.

The first application of SDM to the study of enzyme mechanism was reported in 1982 [2]. The enzyme tyrosyl tRNA synthetase, which transfers the amino acid tyrosine to its cognate transfer RNA molecule in the process of translation, was mutated based on knowledge of its structure to reveal, in the first instance, some of the mechanistic determinants of substrate binding in the enzyme. The technique was difficult and time-consuming to perform in 1982, but it is nowadays a straightforward technique that uses the polymerase chain reaction (PCR – see Chapter 7) as a basis. Whilst the technique is trivial, it should be remembered that the effects of changing even one amino acid in a protein of three hundred may be more far-reaching than one might imagine. As well as possibly resulting in a enzyme that is merely inactive, it could be, for example, that a change in a single amino acid may disrupt an important hydrogen bond or salt bridge in an enzyme that leads to structural instability – the protein may thus be expressed in inclusion bodies (see Chapter 7) as a result of a failure to fold properly, and would not be expressed in a soluble form.

8.2.1 A SDM experiment. Considerations and practise

A contemporary method for SDM using PCR is summarised in Figure 8.1. The process starts with the expression plasmid into which the gene of interest has been cloned. First, a PCR is conducted using short oligonucleotide primers that span the site of mutation. The whole plasmid is amplified essentially, but the template remains. It is very important that the template, which does not harbour the intended mutation, is removed or else it will contaminate the plasmid mini-preps that result from the end of the SDM experiment, and parent plasmid that bears only the wild-type gene may be recovered. The removal of the template plasmid is performed by incubation of the PCR product with the restriction endonuclease DpnI. DpnI is a nuclease that specifically recognises methylated DNA. DNA is methylated naturally in bacteria such as *E. coli*, so only the template plasmid, which has come from amini-prep of an *E. coli* culture, will be digested away and only the mutant plasmid will remain. The mixture after DpnI digestion is then used to transform a cloning strain of *E. coli*, during which process the nicked plasmid product of PCR is repaired to give an intact circular plasmid. Mini-preps of plasmids isolated from the subsequent transformants should harbour the plasmid with the intended mutation, which can be confirmed by DNA sequencing.

Before embarking on the SDM experiment, there are a number of considerations that need to be addressed:

1. *Examining the protein structure for amino acid targets for mutation.* With an ever-increasing amount of web resources available for bioinformatics and protein

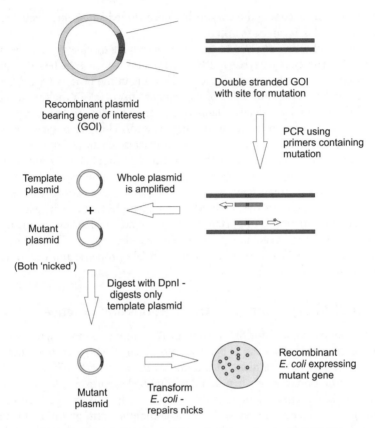

Figure 8.1 A contemporary method for site-directed mutagenesis using PCR (Quikchange Stratagene)

structure, it is not necessary to work in a structural biology laboratory to be able to examine the detail of enzyme structures contained within the protein databank, and whilst sophisticated modelling programs can be used to suggest which mutations might be made within an active site in order to obtain a desired effect on activity, a study of the actual enzyme structure, in complex with its substrate or an inhibitor is most informative. Programs such as Coot [3] are easy to use protein modelling programs that are used for building and refining protein structures, but may also be used simply for examining structures from the Protein Data Bank. There are numerous other simpler freeware applications for examining protein structure including Firstglance in Jmol and Protein Explorer, which require various plug-ins. It is also possible to view structures using programs on for example, the RCSB pdb website itself (http://www.rcsb.org/pdb/home/home.do)

including Quickpdb. Each protein structure is accorded a four character accession code, such as 1o8u, and the pdb files contained within the databank are text files of the coordinate locations of each atom within the structure. It is these pdb files that are read by the programs above.

2. *Which mutation should be made?* A decision must be made as to the identity of the amino acid to replace the one that exists in the wild-type sequence. This will depend on what aspect of the role of that amino acid is under investigation. If this is a role in side-chain hydrogen-bonding example, the mutation may be to remove that hydrogen-bond participant. For example, serine (S), whist it cannot be mutated to a side-chain ethyl group, could be mutated to the methyl group of alanine (A). It is relatively more convenient to investigate the contribution to substrate binding or catalysis by the hydroxyl group of tyrosine (Y), for example, as this may be mutated to phenylalanine (F). Where near-isosteric amino acid replacement can be made, these are usually best. For example, glutamic acid (E) and aspartic acid (D) can be mutagenised to their amide homologues, glutamine (Q) and asparagine (N). Steric interactions may be crudely examined using a range of hydrophobic amino acid residues, valine (V) or isoleucine substituting for alanine (A) or phenylalanine (F) for example. Default mutation to glycine is usually avoided as this small side-chain (R = H) can introduce unwanted mobility in the protein backbone, so a default mutation to alanine (R = Me) is often preferred. Indeed the effect of various amino acids on enzyme action is often investigated using a technique that has been named 'alanine scanning', where each potential amino acid of interest is changed to alanine. With the advent of high-throughput methods, it is now also possible to convert any amino acid to all other nineteen amino acids using a technique known as saturation mutagenesis, the principle of which is outlined later. This will give rise to a small library of variants, some of which will not express, or be expressed in soluble form, but will yield valuable information nonetheless from those mutants that express successfully.

The new amino acid may be encoded by more than one three-letter-base codon, so only one of these codons will be selected for incorporation into the new DNA sequence. For instance, one may wish to change a lysine residue in a sequence to an alanine. If lysine (K) is encoded by AAA in the wild-type sequence, it could be changed to GCT, GCC, GCA or GGG in order for the SDM experiment to successfully incorporate alanine (A) in the required position. It is true that *E. coli*, in which the genetic engineering is to be performed, uses 'preferred' codons, but in practice, any of these codons would probably suit.

3. *Mutagenic oligonucleotide primer design.* It will then be necessary to design oligonucleotide primers that cover the site of mutation and that include the new codon. An example of this is shown in Figure 8.2. In the carbon-carbon bond lyase enzymes used as an example in the database search in Section 2.8 (Figure 2.6)

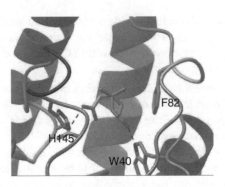

Figure 8.2   The structure of the active site of OCH containing the bound product of reaction and revealing a possible catalytic role for histidine residue in position 145, which is hydrogen-bonded to the pendant carboxylic acid moiety of the product (pdb code 2szo)

and the cloning experiment described in Section 7.3, the structure of one of the enzymes, bound to its product reveals a possible catalytic role for histidine residue in position 145. The role could be perhaps as a catalytic base in the activation of a water molecule for attack at one of the carbonyl groups of the substrate, using the lone-pair on the nitrogen of the imidazolium ring to abstract a proton from a water molecule. In the enzyme product complex, this histidine is observed to be hydrogen-bonded to the carboxylic acid moiety of the product (Figure 8.2). In order to change this histidine residue to another, which would test the ability of the enzyme to catalyse the reaction in the absence of this base, a mutation to alanine is suggested.

The amino acid and corresponding base pair sequence around the histidine residue of interest are shown below:

```
N --- Trp Thr Ser Thr Ala His Asp Glu Leu Ala Tyr ---C
5'--- TGG ACC TCA ACC GCA CAC GAC GAG CTG GCC TAC ---3'
3'--- ACC TGG AGT TGG CGT GTG CTG CTC GAC CGG ATG ---5'
```

We therefore design mutant oligonucleotide primers that change the codon for His (CAC) to one that encodes Ala (GCA):

```
N --- Trp Thr Ser Thr Ala Ala Asp Glu Leu Ala Tyr ---C
5'--- TGG ACC TCA ACC GCA GCA GAC GAG CTG GCC TAC ---3'
3'--- ACC TGG AGT TGG CGT CGT CTG CTC GAC CGG ATG ---5'
```

The primers for the experiment will have the same sequence as those written above with the top 5'-3' one being the 'forward primer' and the bottom 3'-5' one, the 'reverse primer'. As these primers will merely amplify the pre-existing plasmid, and there is no new ligation or cloning step, there are no considerations here with respect to restriction sites, unless of course the mutation now introduces a new restriction site into the gene sequence itself. The primers, once acquired, are then used in conjunction with the template plasmid containing the wild-type gene in a PCR that will amplify the whole plasmid, but containing the new codon.

8.2.2 Protocol for a simple SDM experiment

In order to conduct an SDM experiment, it is possible to purchase all the reagents through kits, such as the commonly used Quikchange™ kit from Stratagene. The experiment may also be performed using the individually purchased reagents, as the core of the SDM experiment is really just a PCR, followed by an enzymatic digestion, and then a transformation into a competent strain of *E. coli*. The protocol below is based on that provided by Stratagene (http://www.stratagene.com/products).

Apparatus and reagents

- Expression plasmid bearing wild type gene
- Mutagenic primers (forward and reverse)
- PCR machine
- Pfu Turbo DNA polymerase
- PCR ingredients (as detailed in Section 7.3)
- Enzyme DpnI
- Cloning strain of *E. coli* (Top10 or Novablue Singles, etc.)

1. First, set up a PCR according to the following recipe:

To a 500 μL PCR tube, add:

Wild-type plasmid template at a concentration of 25 ng μL^{-1}	1 μL
Forward mutagenic primer at a concentration of 125 ng μL^{-1}	1 μL
Reverse mutagenic primer at a concentration of 125 ng μL^{-1}	1 μL
dNTPs (2 mM each in solution)	1 μL
MgSO$_4$ (25 mM solution)	1 μL
DNA polymerase buffer (10x concentration)	5 μL
Pfu Turbo DNA polymerase (at 2.5 U μL^{-1})	1 μL
Distilled water	38 μL
To a final volume of:	50 μL

Then run a standard PCR program in the thermal cycler as follows. (Also run a blank in which there are no oligonucleotide primers.)

(i)   Initial melt 95 °C for 30 s
(ii) Twelve to eighteen cycles of the following stages: (a) Melt; 95 °C, 30 s
                                                            (b) Anneal; 55 °C, 60 s
                                                            (c) Extend; 68 °C, 300 s
(iii) Final extension 72 °C for 180 s
(iv) Cool to 4 °C.

(NB In the extend stage: usually 60 s for each kb of plasmid length.)

2. The mixture of wild-type plasmid template and linear, variant plasmid i.e. the PCR product, are digested with the DpnI enzyme. To 50 μL of the cooled PCR mixture is added, 1 μL of DpnI enzyme at $10\,U\,mL^{-1}$. The digest reaction is mixed carefully and centrifuged for 1 min and then incubated at 37 °C for 1 h.

3. Transform an aliquot of *E. coli* cloning strain (XL1B supercompetent cells are provided with the Stratgene kit) with 1 μL of the DpnI digest mix using the simple transformation protocol described in Chapter 7. Incubate the resultant plates at 37 °C overnight. Isolate the plasmids from a representative number of the colonies that result by mini-prepping and submit these for sequencing, paying particular attention to the region that corresponds to the desired mutation. The mutant plasmids can then be used to transform an expression strain of *E. coli* that was transformed with the wild-type, and expression trials conducted (as described in Section 7.9) to check that the variant proteins are expressed in soluble form.

8.2.3 Saturation mutagenesis

It may be of interest to change the amino acid at one site to all other possible proteinogenic amino acids. This is fairly easily accomplished using a technique called saturation mutagenesis. In this technique, in the design of the oligonucleotide primers used in the mutagenic PCR reaction, the codon for the amino acid to be substituted is replaced with an 'NNK' codon, where N corresponds to any base (A, T, C or G) and K to G or T. This means that all possible amino acids are encompassed within 32 codon possibilities. This can be accomplished using techniques such as cassette mutagenesis, or more conveniently, using a Quikchange kit [4].

8.2.4 Examples of SDM and its uses in applied enzymology

The impact of rational mutagenesis has been far-reaching within applied enzymology and still exerts much force when it comes to analysing enzyme mechanisms or even changing the attributes of enzymes such as their stability and enantioselectivity. The possibility of creating novel enzyme activities also exists, as illustrated below.

8.2.5 Engineering greater stability to oxidation in the protease subtilisin

Subtilisin is a serine protease which, in addition to having a range of applications in biotransformations reactions, is also used in washing powder formulations. As a serine protease, it has an active site catalytic triad that is composed of a serine residue (at position 221) and also an aspartic acid (position 32) and histidine (position 64). As it is used in washing powder formulations, the enzyme needs to be stable under the oxidising conditions that will be encountered when the enzyme cones into contact with bleach oxidants such as hydrogen peroxide. Some amino acid side chains are of course more susceptible to damage by oxidation than others, notably those containing sulfur, such as cysteine and methionine. In the active site of subtilisin is a methionine residue at position 222. It was postulated [5] that mutating the methionine to other residues may decrease the susceptibility of the enzyme to inactivation through oxidation. The methionine was hence mutated to alanine, serine and cysteine. Whilst the mutants were overall, comparably or less active than the wild-type, the stability of all three mutants when exposed to 0.1 M hydrogen peroxide was markedly improved. This was even more so for the alanine and serine variants, which retained activity even having been exposed to concentrations of 1.0 M hydrogen peroxide.

8.2.6 Using SDM to release steric constraints in the active site of P450cam

We have already seen that cytochrome $P450_{cam}$ is an oxidoreductase that is capable of stereoselectively hydroxylating the monoterpene $(1R)$-$(+)$-camphor on the 5-*exo* position (Section 1.6). The enzyme has no activity toward an alternative substrate, diphenylmethane, however (Figure 8.3).

A modelling analysis of the structure of P450cam suggested that both steric bulk and important hydrogen bonding interactions between one aromatic residue, tyrosine 96, and camphor within the active site might be relieved by mutation and hence aid in the accommodation of larger substrates. Tyr96 was therefore mutated to alanine [6] and the Y96A mutant was found now to accept diphenylmethane as a substrate, with a turnover of $6\,s^{-1}$.

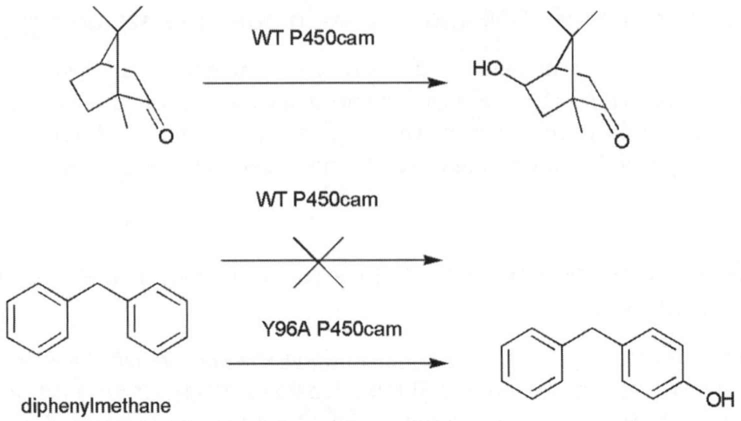

Figure 8.3   Using SDM to release steric constraints in the active site of P450cam

8.2.7 Using SDM to invert the enantioselectivity of vanillyl alcohol oxidase

Vanillyl alcohol oxidase is a flavin-dependent enzyme that catalyses the enantios-elective addition of water to a quinine methide intermediate that is formed by the deprotonation of *para*-substituted phenols such as 4-(methoxymethyl)phenol. The enantioselectivity of the enzyme is attributed to the activation of a water molecule at only one face of the intermediate by an aspartic acid residue at position 170, the water then being used to hydrate the double bond of the intermediate (Figure 8.4) [7].

A related enzyme, *para*-cresol methyl hydroxylase catalyses an equivalent hydration, with opposite enantioselectivity, and features an acid residue, Glu427, on the opposite face of the substrate. It was proposed that simply removing Asp170 from VAO by mutating to alanine, plus introducing an acidic residue at a site spatially equivalent to Glu427 in PCMH, in position 457, would allow activation of water at the opposite face of the substrate and thus a reversal in enantioselectivity of the enzyme. This was indeed the case, with a D170S/T457E mutant displaying an enantioselectivity of 80 % for the (*S*)-enantiomer of the product, compared with 94 % (*R*)- for the wild-type.

8.2.8 Using SDM as a handle for chemical modification

SDM can also be used as a tool to introduce specific chemical functionality into the active site of an enzyme, such that it might be covalently modified. Subtilisin from *Bacillus lentus* has no cysteine residues in its sequence. Cysteine, which bears a thiol side-chain, is an interesting 'handle' for chemical modification as it is

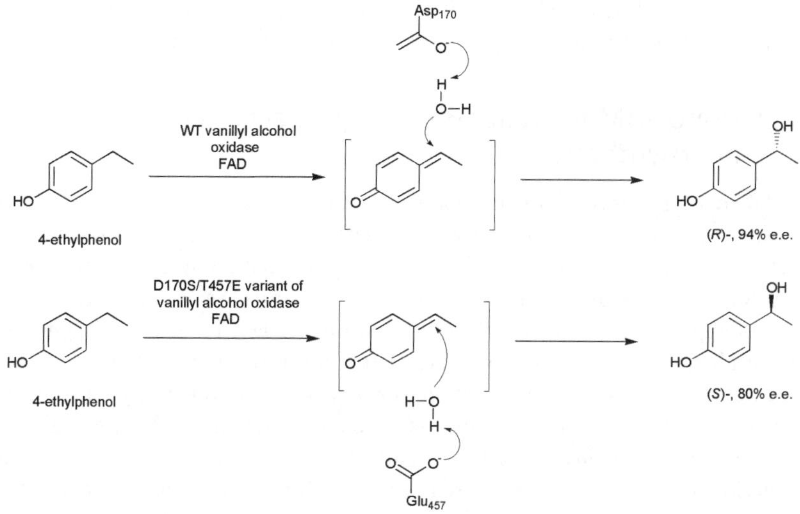

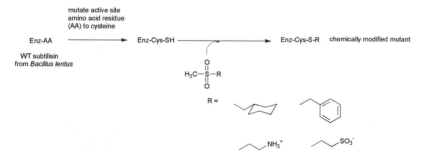

Figure 8.4 Using SDM to invert the enantioselectivity of vanillyl alcohol oxidase

reactive to attack by nucleophiles and hence, to chemical derivatisation. In order to attempt to change the chemical environment within the active of subtilsin, an active site asparagine was mutated to cysteine, then the enzyme was reacted with a series of sulfnoyl reagents bearing R groups of different functionality, to form a range of 'chemically modified mutants' (CMMs) in which the active site now featured a hydrophobic group (phenyl) or positively charged or negatively charged (SO_3^{2-}) group (Figure 8.5) [8].

One of these CMMs, wherein a cyclohexane ring had been introduced into the active site of subtilisin, altered the active site environment, such that the modified

Figure 8.5 Using SDM as a handle for chemical modification in subtilisin from *Bacillus lentus*

enzyme now catalysed the hydrolysis of a model peptide substrate with a k_{cat}/K_m that was 3.2-fold greater than that measured for the wild-type enzyme.

8.2.9 Using SDM to create new enzyme activities – the glycosynthase

A detailed appreciation of the mechanism of an enzyme can help in the design of new enzyme activities using single point mutations. In Chapter 5, we examined the mechanism of retaining β-glycosidases, the enzymes that catalyse the stereospecific hydrolysis of glycosidic bonds between saccharide units (Section 5.5). It was seen that, after protonation of the glycosidic bound oxygen by a catalytic glutamic acid residue, a nucleophilic glutamate at the 'bottom' of the active site attacks the anomeric position of the acceptor sugar to give a covalent glycosyl enzyme intermediate. The formation of this intermediate is a barrier to the use of glycosyl hydrolases in the *synthetic* direction, as any glycoside bond formed between two saccharide residues will be hydrolysed as a result of the continuing presence of the nucleophilic residue. Withers and co-workers [9] postulated that replacing the nucleophile in glycosyl hydrolases with a residue such as alanine would result in an enzyme that would not be able to hydrolyse a disaccharide that was formed in the active site, provided that the formation of a glycosidic bond was encouraged by the use of an activated sugar donor that had a good leaving group in the anomeric position, such as fluoride. Using a β-glycosidase from an *Agrobacterium* species, the nucleophilic Glu358 was mutated to alanine, and it was found that this variant enzyme, a *glycosynthase*, when used with activated fluorinated sugar donors, was indeed able to catalyse the formation of saccharide bonds with little or no hydrolysis of the product (Figure 8.6).

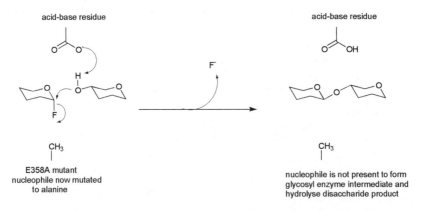

Figure 8.6   Engineering a glycosynthase enzyme from a retaining β-glycosidase

Even the small number of choice examples above illustrates the potential power of rational SDM for the improvement or alteration of enzyme properties based on established conclusions on structure and/or mechanism. However, it is clear that not all enzymes would be amenable to alteration in this way and that the current understanding of enzymatic catalysis is not yet sophisticated enough to allow the rational alteration of enzymatic activity in a generic manner. The development of improved properties through *random* mutagenesis has allowed access to improvements in a change in enzyme structure or mechanism that could quite simply not be predicted.

8.3 Engineering Using Random Mutagenesis Directed Evolution of Enzymes

The field of biocatalysis has been revolutionised in the last ten years by the application of random mutagenesis methods, popularly known as *directed evolution*, to the improvement or alteration of enzyme performance. Such methods are dependent on obtaining an enzyme of improved properties through indiscriminate base-pair changes throughout the length of the gene that encodes the enzyme of interest. There are numerous methods of introducing random mutations in the gene sequence, some of which are briefly described below. However, first, it is important to mention that, in contrast to rational or structure-guided mutagenesis, random mutagenesis experiments will result in huge *libraries* of variants, from which the enzyme of altered activity must be identified. Random mutagenesis methods, therefore, whilst opening up a huge range of possibilities for the simple alteration of enzyme activity, also raise fundamental and essential considerations of either *screening* or *selection* of that new activity from the library.

8.3.1 Screening

Screening of a library refers to the individual assay of each enzyme variant for an altered activity. The library of mutant plasmids that arises from a random mutagenesis experiment will be transformed into an expression strain of *E. coli*, giving rise to a number of transformant colonies on a plate, each of which may represent a clone containing a different variant. It may be necessary to grow each variant colony and test the enzyme from each for altered selectivity using a relatively low-throughput method such as HPLC or GC (particularly if the targeted property for improvement is, for example, enantioselectivity) and so it is obvious that only a small fraction of possible variants can be screened using these methods. However, even experiments that are only able to screen such small representative fractions of libraries have been successful in unearthing mutant enzymes that display greatly improved properties. It is of course more useful to have a colorimetric screen of some kind, for example, where the agar plates used for plating out variant transformants contain a substrate whose transformation by an improved enzyme at a certain threshold results in a colour change in or around the colony, allowing the rapid sieving of the best mutants in a preliminary round of screening. Examples of screening techniques are seen in the examples given below.

8.3.2 Selection

It would of course, be much more preferable to have a 'selection' assay for the improved activity. In a selection, the acquisition of an enzyme capable of a certain transformation will confer on the relevant clone the ability to survive on an agar plate that contains a certain substrate, which may otherwise be poisonous to cells. Mutants that do not express a gene encoding an enzyme capable of metabolising such a poison will simply not grow. To use a simple example, a selection from a library of randomly mutated β-lactamases might employ a high concentration of penicillin on the agar plates used for transformation, meaning that only mutant β-lactamases that displayed improved catalytic efficiency will confer on the relevant clones the ability to survive the screen. This example is of course fairly contrived in the context of applied biocatalysis, wherein the properties to be improved will more likely be enantioselectivity, thermal stability or solvent tolerance, and in the first of these examples, the design of an appropriate selection would be extremely challenging, although it has been used.

8.3.3 Methods for generating genetic diversity and libraries of variants

Some of the methods used for generating the genetic diversity that will lead to the formation of mutant libraries are outlined below.

8.3.4 Random introduction of single base changes leading to single amino acid substitutions: Error-prone PCR (epPCR)

It is well known that some DNA polymerases such as the Taq polymerase frequently used in PCR protocols, has an inherent error rate (approximately 1 in 10^4) in the incorporation of nucleotides in the growing DNA chain, leading to misincorporation of the wrong base. In some cases, owing to the degeneracy of the genetic code, this will have no effect on the overall amino acid sequence of the enzyme, as the mutations will be *silent*. For example, a change from AGT, encoding serine, to AGC, will still encode serine, however, a change to AGA will then cause the sequence to encode arginine.

The error-rate of DNA polymerase may be increased in the PCR reaction, by increasing the concentration of magnesium ions in the reaction, or the concentration of GTP (one of the deoxynucleotide triphosphate monomers). As a result, misincorporations into the new DNA sequence will be encouraged. For various reasons, not all misincorporations are equally likely, but, overall, what will result from a so-called 'error-prone' PCR (epPCR) reaction is a library of gene sequences, all based on the wild-type, but all featuring one or more mutations within the sequence, each of which could be different (Figure 8.7). Statistically, the consequences are huge. If we assume briefly (taking into account the silent mutations described above) that every single base substitution gave rise to a change in one amino acid in the sequence, a single epPCR reaction on a 900 base pair gene encoding an enzyme of 300 amino acids could give rise to 20^{300} variants. In practise of course, the number is somewhat lower, but the statistics are

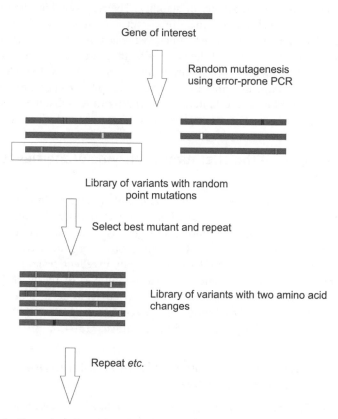

Gene of interest

Random mutagenesis using error-prone PCR

Library of variants with random point mutations

Select best mutant and repeat

Library of variants with two amino acid changes

Repeat *etc.*

Figure 8.7 Generation of mutant enzyme libraries using random mutagenesis achieved through error-prone PCR

illustrative of the possibilities afforded in terms of exploring the *sequence space* within protein sequences. Once a variant with a single amino acid has been generated, this mutant plasmid may then be used in another cycle of random mutagenesis, leading to a double mutant of even more greatly improved properties. In practise many iterations are usually performed until the required performance of the new enzyme is attained, as shown in Figure 8.7.

In addition to the sheer number of variants that can be generated, it is the unexpected structural consequences of random mutation that have caught the attention of researchers. It has been common in random mutagenesis studies of the type described that the mutations that have caused an improvement in activity are often not within the active site, but are at sites remote from that, perhaps even at the periphery of the globular enzyme. Rationalising the improvement of enzyme activity in these cases is somewhat difficult therefore, but may be illustrative of changes in weak interactions that stabilise the protein structure giving rise to increased or reduced flexibility or stability. The relatively high frequency of mutations in improved mutants that are distant from the active site is, in part, merely a statistical phenomenon, explained by the fact that there are simply more amino acids that are not part of the active site, than there are of those that are. However, the consequences of such remote mutations giving rise to such catalytic improvement is striking nonetheless and reveals in some small part how little we know about the subtle chemical interactions within proteins that have resulted in some cases in perfect catalysts through millions of years of natural evolution.

8.3.5 Examples of the alteration of enzyme properties using epPCR

Some of the earliest examples of the applications of directed evolution to enzyme improvement using epPCR were carried out in the laboratory of Frances Arnold, who first showed that such random mutagenesis techniques could be used to improve the stability of the protease subtilisin in organic solvents such as dimethylformamide [10]. There have been many excellent reviews in the last decade that summarise these and other early examples of the work [11, 12], including enzyme thermostability and enantioselectivity as detailed below. Some methods for both directed evolution and the screens or selection strategies used have also been collected together in technical manuals [13–15].

8.3.5.1 Improving the enantioselectivity of an esterase from *Pseudomonas aeruginosa*

In one of the first experiments designed to improve the enantioselective properties of an enzyme through epPCR-mediated directed evolution, Reetz and co-workers chose as their model an esterase enzyme from *P. aeruginosa* [16, 17]. This esterase was chosen as it performed the enantioselective hydrolysis of the racemic ester shown in Figure 8.8 very poorly, with an E value of only 1.0. The esterase gene was subjected to epPCR and the library of variant enzymes that resulted was assayed against separate enantiomers of the substrate in multi-well format using the hydrolysis of *para*-nitrophenyl acetate as the basis of the assay. One round of epPCR resulted in a mutant for which the enantiomeric ratio E was improved to 2.1. Further rounds of epPCR gave variants that were even more substantially improved and after four iterations, a variant that exhibited an E value of 11.3 was obtained.

Saturation mutagenesis of the amino acid residues identified as 'hot-spots' for mutation resulted in even better mutants, illustrating the power of combining directed evolution methods for identifying such amino acids, which could then be refined using rational methods. A five-point mutant, S149G/S155F/V47G/V55GS164G, that had been arrived at using a mixture of directed evolution and saturation mutagenesis exhibited an E value of 25.8. An examination of the protein structure subsequently revealed that the mutations that had been made were, in many cases, remote from the active site and would simply not have been targeted as possible residues that would significantly affect enantioselectivity. Early examples of directed evolution studies such as this realised the significance that *remote* mutations might have on the catalytic properties, perhaps through indirect effects on active site structure caused by altered flexibility in the tertiary structure of enzymes by the breaking or making of new weak interactions such as hydrogen bonds. A number of the residues in the most active mutants were mutations to glycine, a residue that allows flexibility owing to its small R group.

8.3.6 Improvements in the activity of the monoamine oxidase from *Aspergillus niger*

Improvements in enzyme activity have not by any means been restricted to enzymes of 'simple' mechanism such as serine hydrolases, and indeed a number of oxidoreductases have also been improved using these techniques, including the cyclohexanone monoxygenase from *Acinetobacter* (see Chapters 5 and 6) and also the monoamine oxidase from *A. niger*, one application of which was described

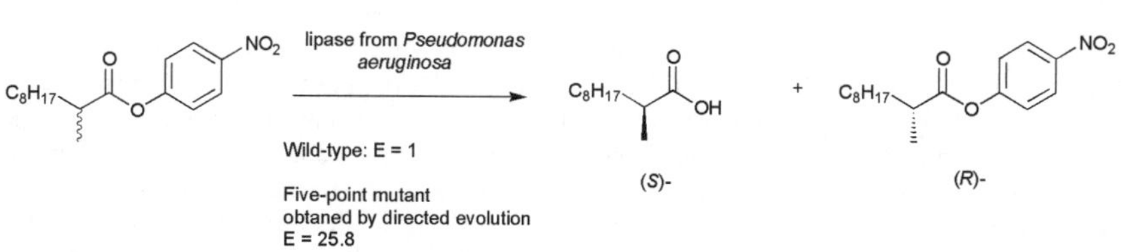

Figure 8.8 Improving the enantioselectivity of an esterase from *Pseudomonas aeruginosa*

in Chapter 6. As described therein, this enzyme, originally from the fungus *Aspergillus*, catalyses the oxidation of simple amines such as α-methylbenzylamine to their corresponding imines, which are hydrolysed *in aqua* to give the corresponding ketones. These enzymes can then be exploited in chemoenzymatic oxidation–reduction cycles when used in conjunction with a chemical reductant for the deracemisation of chiral amines. The wild-type MAO, expressed in *E. coli*, displayed poor activity toward the target substrate. Random mutagenesis was applied to the gene encoding ANMAO (using an alternative method that exploits a 'mutator' strain of *E. coli* in transformation) and a number of variants that exhibited improved activity toward the target substrate [18]. In contrast to the screening of individual variant colonies adopted in the directed evolution of the esterase above, the improved activity of the amine oxidases was conducted using a high-throughput plate-based colorimetric screen. The hydrogen peroxide that was released during the amine oxidase reaction is used by an enzyme, horseradish peroxidase (HRP), contained within the agar, to oxidise a HRP substrate that turns the colonies that express more active amine oxidase variants blue (Figure 8.9).

Agar plate copies could be made that were tested against separate enantiomers of the chiral amine substrate in order to rapidly identify those which would be most enantioselective. Iterations of the random mutagenesis, using epPCR, have resulted in further improved mutants that are capable of accepting a wide range of primary, secondary and tertiary amines that are not accepted as substrates by the wild-type enzyme [19, 20], such as the example given in Figure 6.32. The

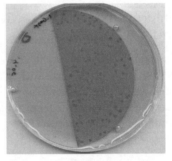

Figure 8.9   Plate screen for mutants of amine oxidase from *Aspergillus niger* obtained by directed evolution. The plate of transformed *E. coli* is 'copied' onto two sections of a membrane, which are each then tested against different enantiomers of the substrate. The contrast in colours in colonies on the two plates is indicative of the enantioselectivity of the mutant enzymes in the colonies (photograph courtesy of Professor Nick Turner). M. Alexeeva, A. Enright, M. J. Dawson, M. Mahmoudian and N. J. Turner, Deracemization of α-Methylbenzylamine Using an Enzyme Obtained by In Vitro Evolution, Angewandte Chemie International Edition, 2002, **41**, 3177–3180. Copyright Wiley-VCH Verlag GmbH & Co. KGaA. Reproduced with permission.

recently solved structure of ANMAO has revealed that, again, whilst some of these mutations are within the active site and their efficacy might be partially explained by a release of steric constraints within the active site, some exert their effect more remotely, and may be due perhaps to the opening of access channels through the exterior of the protein to the active site where the reaction chemistry is catalysed.

8.3.7 Gene site saturation mutagenesis (GSSM)

GSSM represents a special case of high-throughput point mutation, which, whilst performed in a high-throughput manner, is rational in that the strategy targets the systematic replacement of every amino acid in a protein sequence with every other amino acid. The result is a large library of single amino acid point mutants in which the result of saturation mutagenesis at every position is represented. The technique was first described by Diversa, which applied it to a nitrilase enzyme, in the interests of developing a catalyst for the selective hydrolysis of the glutaronitrile derivative shown in Figure 8.10 [21].

The nitrilase has 330 amino acids in its sequence, and thus the library generated consisted of 10 528 variants. This large number of variants could be screened rapidly by using a mass spectromety based screen, in which the hydrolysis of one or other prochiral nitrile group would result in products of a different molecular weight (Figure 8.10). A number of the point mutants obtained were shown to improve the enantioselectivity of the enzyme – many of these were mutants of the same position, which had been changed to a variety of amino acids – a result that would not have been observed with standard epPCR as described above, as more than one change to a single codon is rarely observed using this method.

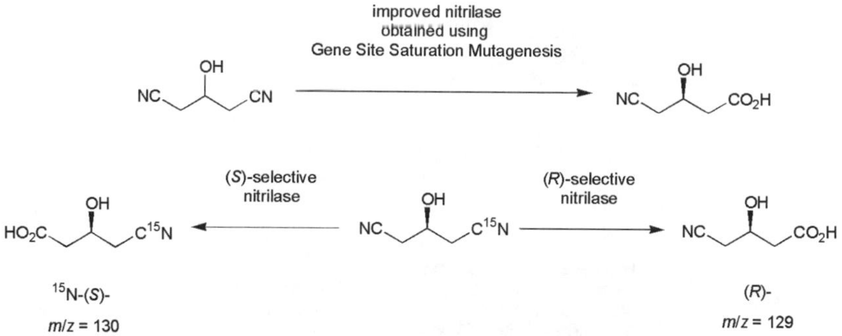

Figure 8.10 Gene site saturation mutagenesis for the improvement of a nitrilase

8.3.8 DNA shuffling or molecular breeding™

The idea of exploring protein sequence space can be further demonstrated by one of the other major methods of generating sequence diversity in random mutagenesis experiments, that of *DNA shuffling*. There are many variants of this, some of which are patent protected by companies such as Diversa, who are experts in the area. The original idea grew out of celebrated research published by Stemmer in the mid-1990s [22] and has also hence been referred to as *Stemmer shuffling*. The principle of these techniques is quite simple (Figure 8.11).

A family of genes encoding, for example, a set of lipases, will, whilst exhibiting many similarities, notable in mechanism and perhaps overall protein fold, display

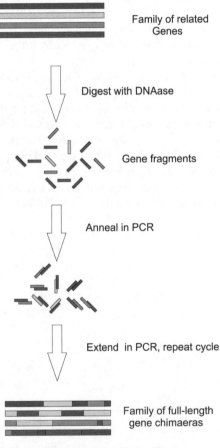

Family of related
Genes

Digest with DNAase

Gene fragments

Anneal in PCR

Extend in PCR, repeat cycle

Family of full-length
gene chimaeras

Figure 8.11   DNA shuffling

different properties with respect to stability or enantioselectivity, which it may be in the interests of the researcher to combine. DNA shuffling 'shuffles' a pack of DNA sequences of a minimum level of sequence identity (about 60 %), resulting in a library of chimeric sequences made up of gene fragments from different parent genes from the original pool. In practise, the genes are first digested together, then fragments are annealed to each other in a PCR reaction, exploiting the similarity of some of the gene sequences to each other and hence their complementarity. DNA polymerase then extends the fragments to full length genes and the process is reiterated. As with random point mutagenesis, the statistical possibilities afforded by such techniques are staggering, and indeed also raise the possibility for shuffling genes encoding enzymes of *different chemical reactivity* in an effort to increase the diversity of biocatalysts available for industrial processes.

An early example of DNA shuffling as applied to industrial enzymes was performed by Maxygen on the enzyme savinase, a type of subtilisin from the bacterium *Bacillus* [23]. Twenty-six other wild-type genes encoding subtilisin homologues that displayed a variety of different activity characteristics were employed as parent sequences for DNA shuffling, in an effort to improve the activity of savinase in four respects: activity, thermostability, solvent stability and pH optima. It was found that, in addition to creating progeny that had acquired one improved property from a parent sequence, such as improved activity at pH 5.5, some enzymes had acquired perhaps two inherited properties such as a shift in pH optima plus greater thermal stability. The advantage of this technique therefore is that complementary properties that might not ordinarily be found combined in a single natural enzyme may be combined in artificial variants.

In some cases, gene shuffling has been used as a complement to epPCR dependent mutagenesis, where the family of parent genes for shuffling is itself the library of variant genes that has resulted from an epPCR experiment. The epoxide hydrolase from *Agrobacterium radiobacter* (EchA) catalyses the enantioselective hydrolysis of racemic epoxides such as *para*-nitrophenyl glycidyl ether (PNGE) (shown in Figure 8.12), with an E value of 3.4 [24]. epPCR was first used to generate a library of variants, fifteen of which were discovered to display an improved E value of up to 14.0. An agar-plate based screen was first used to identify the best variants, for which the enantioselectivity was then measured using well-plate based

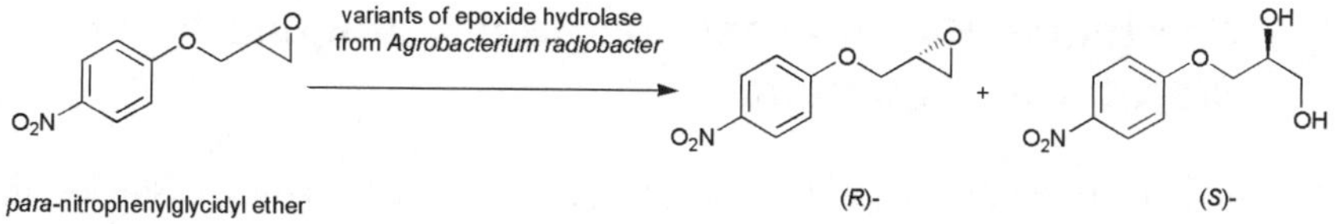

para-nitrophenylglycidyl ether   (R)-   (S)-

Figure 8.12   EchA catalyses the enantioselective hydrolysis of *para*-nitrophenyl glycidyl ether

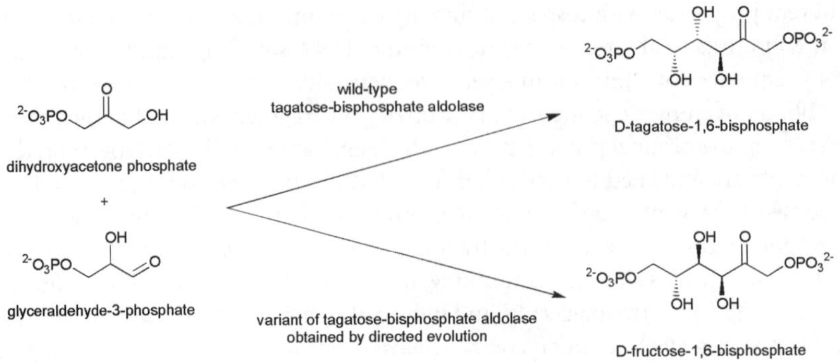

Figure 8.13 Tagatose-1,6-bisphosphate aldolase evolved to catalyse the cleavage of fructose-1,6-bisphosphate

screens that monitored transformation of PNGE. On sequencing, these variants were found to possess between one and six amino acid substitutions. These fifteen variants were then used as a family for gene shuffling, which resulted in variants with improved E values of up to 44. The mutation sites in improved mutants in this case were mostly either in, or had direct effects on, active site residues.

As well as hydrolases, DNA shuffling has also found applications in the directed evolution of carbon-carbon bond forming enzymes. The aldolase enzyme tagatose-1,6-bisphosphate (TBP) aldolase was evolved by Berry and co-workers [25] to catalyse the cleavage of fructose-1,6-bisphosphate (FBP) (Figure 8.13).

Screening of variants was carried out using a plate-based assay that coupled the formation of the reaction products to the formation of a purple dye, via the activity of an NAD-dependent dehydrogenase. Shuffling was used in the first instance on the wild-type gene encoding TBP aldolase to yield variants, which were altered on only one amino acid position. The best variants were then pooled in a further round of shuffling to give variants altered in up to three amino acid positions and for which the specificity of the enzyme had been converted from a 99:1 specificity for TBP cleavage to a 4:1 preference for FBP cleavage. The variant enzymes were able to catalyse the reverse, carboligation reaction to give different product stereochemistry, dependent on the variant used.

8.4 Combining Rational and Random Mutagenesis for Biocatalyst Improvement

The tools and techniques described for rational mutagenesis leading to single-site mutants have been available for nearly thirty years and are now routine in biocatalysis laboratories. Even the techniques involved in random mutagenesis have now been applied to biocatalysis for over a decade and new examples are routinely encountered in the organic chemical literature. Informed by the experience of the biocatalyst engineering seen in recent years, some consensus is perhaps emerging as to the preferred methods for addressing the engineering of enzyme activity for an industrial application.

Whilst the frequency of the observation of mutations remote from enzyme active sites has been noted, it is now recognised that the majority of single-site mutations that give rise to *the most significant* changes in the desired enzyme property occur within a short distance from the active centre of the enzyme, as would be expected [26, 27]. Many of the contemporary studies in directed evolution are therefore focusing on engineering mutations that are based initially on structural observations. High-throughput cloning and screening technologies are then applied to the production of focused libraries of variants which are only mutated at selected positions in the enzyme active site. This might include *saturation mutagenesis* at, say, three positions in the active site, where mutant libraries are generated with each of three positions potentially mutated to the other nineteen amino acids. A large range of variants is still generated, but such experiments give libraries of perhaps more meaningful variants in the context of the chemical problem, and have been shown to produce enzymes of greatly improved properties.

8.4.1 Combinatorial active site mutagenesis test (CASTing)

CASTing is a name applied to a technique in directed evolution which first targets a small number of regions in the active site to generate sub-libraries, which then themselves act as the basis for random mutagenesis at the other site. The technique was applied to both lipases [28] and also the epoxide hydrolase from *A. niger* [29], which was known to catalyse the hydrolysis of the phenylglycidyl ether, similar to that shown in Figure 8.12 (although without the *para*-nitro substituent) with an enantiomeric ratio value of only 4.6. Six regions around the active site, represented by two or three amino acids in close proximity in sequence were chosen as the basis for the sub-libraries A to F. Sites A, B and C were subjected to random mutagenesis first, but only mutations at site B yielded mutants of three-point mutations displaying improved enantioselectivity of *E* value 14. The best mutants from this sub-library were therefore used as the basis for random mutagenesis in region C. This resulted in a further improved variant displaying an *E* value of 21, with a further two point mutations in region C. Further mutations in regions D, E and F ultimately resulted in a mutant displaying an *E* value of 115, with nine-point mutations different from the wild-type enzyme.

8.4.2 Directed evolution using protein sequence activity relationships (ProSAR)

Given the technical infrastructure and expense involved in high-throughput technologies for random mutagenesis and the evauation of variants, it is no surprise that many of the key new approaches in the area that have been developed in industry and companies such as Diversa and Codexis have been at the forefront of applying random mutagenesis technology to the improvement of enzymes with industrial applications firmly in mind. Codexis have recently reported the evolution of a halohydrin dehalogenase (HHDH), which is applied in the synthesis of the side-chain of the important hyopcholestemic drug atorvastatin using computationally assisted directed evolution methods that have their roots in quantitative structure–activity relationships that have been used to optimise small molecular structures [30]. HHDH catalyses the interconversion of vicinal halohydrins such as ethyl-4-chloro-3-hydroxybutyrate and its relevant epoxide derivative (Figure 8.14). When incubated with cyanide ions, the enzyme can accept cyanide as a nucleophile for attack at the epoxide to give the vicinal hydroxynitrile. A range of established mutational techniques were used to generate diverse libraries, the results of which were analysed by the ProSAR technique using a linear regression analysis that was able to identify the nature of the contribution of any single mutation to activity.

Thus, in a large variant library it was possible to identify deleterious mutations, which were then discarded, beneficial mutations, which were stored, and possibly beneficial mutations, which were re-analysed. Mutations which exerted a beneficial effect were accumulated within the best mutant through their inclusion in variants used as the basis for further rounds of mutation. This approach militates against the inclusion of non-beneficial mutants and also acts to combine the mutations that have proven most beneficial. Ultimately, variants were identified that contained at least thirty-five point mutations in the enzyme compared with the wild-type, with eight of the twenty residues that occurred within a 7.5 Å shell surrounding the active-site bound product mutated in improved variants, again supporting the notion that mutations that occur closer to the active site exert more profound effects on activity, on the whole. These enzymes were able to catalyse the transformation of ECHB with a 4000-fold improved performance over the

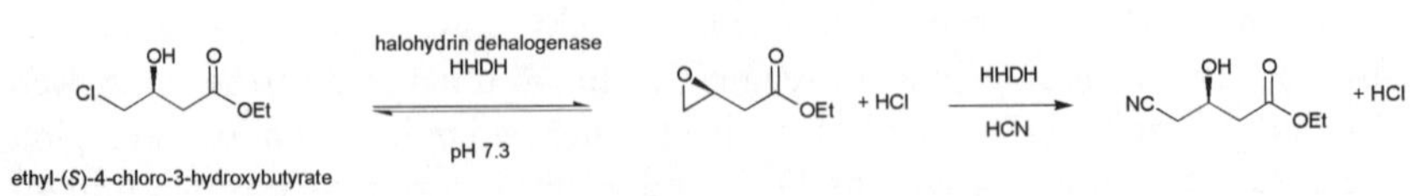

Figure 8.14 Halohydrin dehalogenase (HHDH) catalyses the interconversion of vicinal halohydrins such as ethyl-4-chloro-3-hydroxybutyrate, and allows the incorporation of cyanide to form cyanohydrins. This enzyme was used as the basis for ProSAR-driven evolution by Codexis

wild-type enzyme. The computational aspects of this approach are suggestive of a greater contribution of *in-silico* design methods in the future improvement of existing enzymes, or, indeed, the design of entirely new enzyme activities using pre-existing enzyme scaffolds.

8.5 Exploiting Catalytic Promiscuity for Creating New Enzyme Activities

The diversity of biocatalytic reactions available for application in organic synthesis is growing enormously, as a result of the rise in genome sequencing and the development of the rational and random mutation technologies and computational approached described above. Another aspect of diversity that is attracting increasing level of attention within the biocatalysis community is that of *catalytic promiscuity*. This can loosely be defined as the ability of an enzyme to catalyse a chemical reaction additional to that to which it is usually attributed. A simple example of this might be the observation that protease enzymes such as chymotrypsin are capable of catalysing the cleavage of carboxyl esters. There is some confusion about the nomenclature used in describing this phenomenon, but some useful descriptors have been established by Hult and Berglund in a recent article [31]. Whilst some enzymes may, to some degree, be described as promiscuous in the sense that they recognise different substrates, or use the equivalent mechanisms/active site residues to catalyse the hydrolysis of different bonds (amide or ester), there are more stark examples of catalytic promiscuity where enzymes catalyse reactions that are formally unrelated to the ones for which they are known.

A good example of this is illustrated by the carbon-carbon bond lyase enzyme hydroxynitrile lyase, which is known to catalyse the enantioselective addition of cyanide to aldehydes and ketones to give optically active cyanohydrins, as seen in Section 5.6. It has recently been observed that this enzyme is capable of catalysing the Henry or nitroaldol reaction [32]. The enzyme is able to deprotonate the methyl group of a molecule such as nitromethane, such that the resultant carbanion equivalent may be added to an aldehyde such as benzaldehyde, in a similar fashion to the addition of cyanide to the same substrate (Figure 8.15).

Many other examples of catalytic promiscuity exist within the literature [33]. It is thought that some of these are explicable through tracing the evolutionary origin of the enzyme activity in question. One theory suggests that a low level of promiscuous activity in an enzyme can assist the evolution of that enzyme into one that has optimum levels of activity for the new reaction. For example a phosphotriesterase (PTE) enzyme that catalyses the degradation of the herbicide paraoxon is thought only to have acquired this ability relatively recently, as previously, the pesticide substrate would not have occurred in the environment.

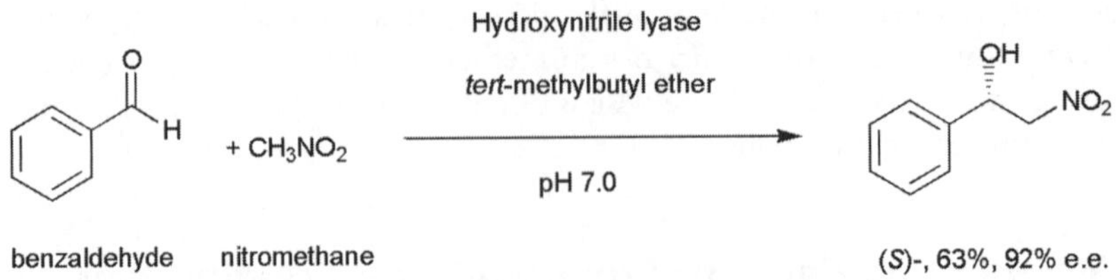

Figure 8.15 Catalytic promiscuity in hydroxynitrile lyase, which catalyses the addition of nitromethane to aldehyde substrates such as benzaldehyde

The PTE also displays a residual promiscuous ability to catalyse the hydrolysis of lactone substrates [34], and indeed, it was shown that microbial lactonases isolated from the environment related to the PTE did indeed have a promiscuous PTE activity. It may be therefore that an enzyme displays promiscuous activity for a reaction that was catalysed by an evolutionary progenitor prior to the protein fold being recruited for a new chemical role. Overall, the studies reveal that there may be more scope for the engineering of novel chemical reactivity within enzymes than has previously been thought, if a commonality of mechanism, such as the stabilisation of reactive intermediates, can be identified between a reaction catalysed by an enzyme, and a different chemical transformation.

8.6 Designing Enzymes in *Silico*

Finally, as the number of structures of enzymes increases and more is learned about the molecular and structural determinants of substrate recognition and transformation by enzymes, there is an increasing move towards the computational design for enzyme activities that as yet have no equivalent in natural enzyme chemistry [35]. Having identified the necessary three- dimensional requirements for the Kemp elimination (Figure 8.16), Baker and co-workers designed a number of suitable protein scaffolds based on pre-existing enzymes for the catalysis of

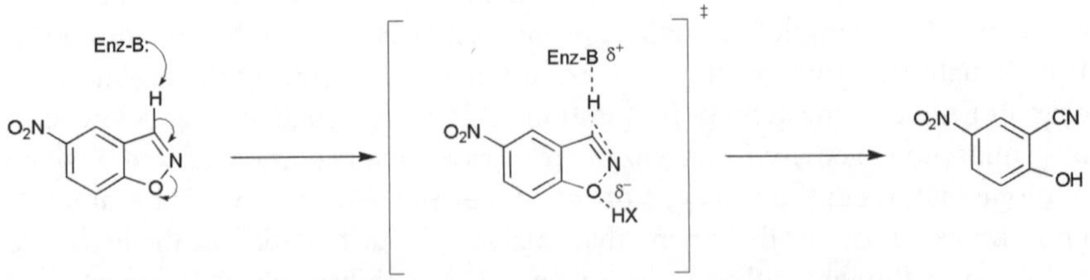

Figure 8.16 Kemp elimination catalysed by a computationally designed biocatalyst

this reaction, each based on the inclusion of an efficient base for abstraction of a proton from the carbon atom of the oxazole substrate.

After having designed and identified *de novo* enzyme catalysts for performing these reactions with up to 100 000-fold greater rate enhancements, these were then improved using directed evolution approaches to give enzymes which displayed 200-fold increases in k_{cat}/K_m for the desired transformation. The powerful combination of *in silico* design and directed evolution suggests that protein evolution strategies in future may be credibly targeted towards entirely new enzymes for which no natural enzymes currently exist.

8.7 Conclusion

As the application of enzymes continues to grow within both academia and industry, it is evident that, in order to satisfy the stringent requirements of industrial processes with respect to catalyst turnover and productivity, protein engineering techniques will be essential to the growing incorporation of this technology into synthetic chemical processes. The engineering described in this chapter can, in some instances, require a profound understanding of protein structure and molecular biology techniques for the manipulation of proteins, as well as the computational methods which can be used to optimise the findings of random mutagenesis experiments such as those described within. In the search for novel reactions and improved activities, it is evident that our appreciation of the subtle interactions that govern properties such as stability and selectivity is still growing, but recent developments suggest that a greater knowledge of protein structure and dynamics, in close collaboration with projected increases in computational power, offer the very real possibility of *de novo* enzyme design for the development of better enzyme catalysts in the future.

References

1. L. Wang and P. G. Schultz (2002) Expanding the genetic code. *Chem. Commun.*, 1–11.
2. G. Winter, A. R. Fersht, A. J. Wilkinson, M. Zoller and M. Smith (1982) Redesigning enzyme structure by site-directed mutagenesis: tyrosyl tRNA synthetase and ATP binding. *Nature*, **299**, 756–758.
3. P. Emsley and K. Cowtan (2004) Coot: model building for molecular graphics. *Acta Crystallogr., Sect D*, **60**, 2126–2132.
4. D. L. Steffens and J. G. K. Williams (2007) Efficient site-directed saturation mutagenesis using degenerate oligonucleotides. *J. Biomol. Techniques*, **18**, 147–149.
5. D. A. Estell, T. P. Graycar and J. A. Wells (1985) Engineering an enzyme by site-directed mutagenesis to be resistant to chemical oxidation. *J. Biol. Chem.*, **260**, 6518–6521.

6. S. M. Fowler, P. A. England, A. C. G. Westlake, D. R. Rouch, D. P. Nickerson, C. Blunt, D. Braybrook, S. West, L.-L. Wing and S. L. Flitsch (1994) Cytochrome P450cam monooxygenase can be redesigned to catalyse the regioselective aromatic hydroxylation of diphenylmethane. *J. Chem. Soc. Chem. Commun.*, 2761–2762.

7. R. H. H. van den Heuvel, M. W. Fraaije, M. Ferrer, A. Mattevi and W. J. H. van Berkel (2000) Inversion of stereospecificity of vanillyl alcohol oxidase. *Proc. Natl Acad. Sci. USA*, **97**, 9455–9460.

8. P. Berglund, G. DeSantis, M. R. Stabile, X. Shang, M. Gold, R. R. Bott, T. P. Graycar, T. H. Lau, C. Mitchinson and J. B. Jones (1998) Chemical modification of cysteine mutants of subtilisin *Bacillus lentus* can create better catalysts than the wild-type enzyme. *J. Am. Chem. Soc.*, **119**, 5265–5266.

9. L. F. Mackenzie, Q. Wang, R. A. J. Warren and S. G. Withers (1998) Glycosynthases: mutant glycosidases for oligosaccharide synthesis. *J. Am. Chem. Soc.*, **120**, 5583–5584.

10. K. Chen and F. H. Arnold (1993) Tuning the activity of an enzyme for unusual environments: sequential random mutagenesis of subtilisin E for catalysis in dimethyl-formamide. *Proc. Natl Acad. Sci. USA.*, **90**, 5618–5622.

11. F. H. Arnold (2001) Combinatorial and computational challenges for biocatalyst design. *Nature*, **409**, 253–257.

12. N. J Turner (2003) Directed evolution of enzymes for applied biocatalysis. *Trends Biotechnol.*, **21**, 471–478.

13. F. H. Arnold and G. Georgiou (2003) Directed Evolution Library Creation: Methods and Protocols, Humana Press, New York.

14. F. H. Arnold and G. Georgiou (2003) Directed Enzyme Evolution: Screening and Selection Methods, Humana Press, New York.

15. J.-L. Reymond (Ed.) (2006) Enzyme Assays: High Throughput Screening, Genetic Selection and Fingerprinting, Wiley-VCH, Weinheim.

16. M. T. Reetz, A. Zonta, K. Schimossek, K. Liebeton and K.-E. Jaeger (1997) Creation of enantioselective biocatalysts for organic chemistry by in vitro evolution. *Angew. Chem., Int. Ed. Engl.*, **36**, 2830–2832.

17. K. Liebeton, A. Zonta, K. Schimossek, M. Nardini, D. Lang, B. W. Djikstra, M. T. Reetz and K.-E. Jaeger (2000) Directed evolution of an enantioselective lipase. *Chem. Biol.*, **7**, 709–718.

18. M. Alexeeva, A. Enright, M. J. Dawson, M. Mahmoudian and N. J. Turner (2002) Deracemisation of alpha-methylbenzylamine using and enzyme obtained by directed evolution. *Angew. Chem., Int. Ed.*, **41**, 3137–3139.

19. R. Carr, M. Alexeeva, A. Enright, T. S. C. Eve, M. J. Dawson and N. J. Turner (2003) Directed evolution of an amine oxidase possessing both broad substrate specificity and high enantioselectivity *Angew. Chem., Int. Ed.*, **42**, 4807–4810.

20. C. J. Dunsmore, R. Carr, T. Fleming and N. J. Turner (2006) A chemo-enzymatic route to enantiomerically pure cyclic tertiary amines. *J. Am. Chem. Soc.*, **128**, 2224–2225.

21. G. DeSantis, K. Wong, B. Farwell, K. Chatman, Z. Zou, G. Tomlinson. H. Huang, X. Tan, L. Bibbs, P. Chen, K. Kretz and M. J. Burk (2003) Creation of a productive, highly enantioselective nitrilase through gene site saturation mutagenesis. *J. Am. Chem. Soc.*, **125**, 11476–11477.

22. W. P. C. Stemmer (1994) Rapid evolution of a protein *in vitro* by DNA shuffling. *Nature*, **370**, 389–391.

23. J. E. Ness, M. Welch, L. Giver, M. Bueno, J. R. Cherry, T. V. Borchert, W. P. C. Stemmer and J. Minshull (1999) DNA shuffling of subgenomic sequences of subtilisin. *Nature Biotechnol.*, **17**, 893–896.

24. B. van Loo, J. H. Lutje Spelberg, J. Kingma, T. Sonke, M. G. Wubbolts and D. B. Janssen (2004) Directed evolution of epoxide hydrolase from *A. radiobacter* toward higher enantioselectivity by error-prone PCR and DNA shuffling. *Chem. Biol.*, **11**, 981–990.

25. G. J. Williams, S. Domann, A. Nelson and A. Berry (2003) Modifying the stereochemistry of an enzyme-catalyzed reaction by directed evolution. *Proc. Natl Acad. Sci. USA*, **100**, 3143–3148.

26. S. Park, K. L. Morley, G. P. Horsman, M. Holmquist, K. Hult and R. J. Kaslauskas (2005) Focusing mutations into the *P. fluorescens* esterase binding site increases enantioselectivity more effectively than distant mutations. *Chem. Biol.*, **12**, 45–54.

27. K. L. Morley and R. J. Kaslauskas (2005) Improving enzyme properties: when are closer mutations better? *Trends Biotechnol.*, **23**, 231–237.

28. M. T. Reetz, M. Bocola, J. D. Carballeira, D. Zha and A. Vogel (2005) Expanding the range of substrate acceptance of enzymes: combinatorial active-site saturation test. *Angew. Chem., Int. Ed.*, **44**, 4192–4196.

29. M. T. Reetz, L.-W. Wang and M. Bocola (2006) Directed evolution of enantioselective enzymes: iterative cycles of CASTing for probing protein-sequence space. *Angew. Chem. Int. Ed.*, **45**, 1236–1241.

30. R. J. Fox, S. C. Davis, E. C. Mundorff, L. M. Newman, V. Gavrilovic, S. K. Ma, L. M. Chung, C. Ching, S. Tam, S. Muley, J. Grate, J. Gruber, J. C. Whitman, R. A. Sheldon and G. W. Huisman (2007) Improving catalytic function by ProSAR-driven enzyme evolution. *Nature Biotechnol.*, **25**, 338–344.

31. K. Hult and P. Berglund (2007) Enzyme promiscuity: mechanism and applications. *Trends Biotechnol.*, **25**, 231–238.

32. M. Gruber-Khadjawi, T. Purkarthofer, W. Skranc and H. Griengl (2007) Hydroxynitrile lyase-catalysed enzymatic nitrolaldol (Henry) reaction. *Adv. Synth. Catal.*, **349**, 1445–1450.

33. U. T. Bornscheuer and R. J. Kaslauskas (2004) Catalytic promiscuity in biocatalysis: using old enzymes to form new bonds and follow new pathways. *Angew. Chem., Int. Ed.*, **43**, 6032–6040.

34. L. Afriat, C. Roodveldt, G. Manco and D. S. Tawfik (2006) The latent promiscuity of newly identified microbial lactonases is linked to a recently diverged phosphotriesterase. *Biochemistry*, **45**, 13677–13686.

35. D. Rothlisberger, O. Khersonsky, A. M. Wollacott, L. Jiang, J. DeChancie, J. Betker, J. L. Gallaher, E. A. Althoff, A. Zanghellini, O. Dym. S. Albeck, K. N. Houk, D. S. Tawfik and D. Baker (2008) Kemp elimination catalysts by computational enzyme design. *Nature*, **453**, 190–195.

Appendices

A.1 Structures of the proteinogenic amino acids

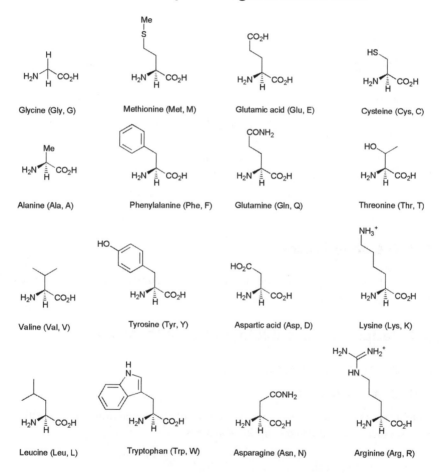

Glycine (Gly, G) Methionine (Met, M) Glutamic acid (Glu, E) Cysteine (Cys, C)

Alanine (Ala, A) Phenylalanine (Phe, F) Glutamine (Gln, Q) Threonine (Thr, T)

Valine (Val, V) Tyrosine (Tyr, Y) Aspartic acid (Asp, D) Lysine (Lys, K)

Leucine (Leu, L) Tryptophan (Trp, W) Asparagine (Asn, N) Arginine (Arg, R)

Practical Biotransformations: A Beginner's Guide © 2009 Gideon Grogan

Isoleucine (Ile, I)	Histidine (His, H)	Serine (Ser, S)	Proline (Pro, P)

A.2 Structures of bases found in nucleic acids

Adenine, A	Thymine, T	Uracil, U

Guanine, G	Cytosine, C

A.3 The genetic code

Table A3.1 shows the genetic code in terms of DNA codons. Encoded amino acids are shown in three-letter code in bold.

A.4 Recipes for microbiological growth media

Liquid media for bacterial growth

Luria-Bertani (LB) broth

Suitable for the growth of recombinant strains of *E. coli* and other gram-positive bacteria such as *Pseudomonas* and *Rhodoccoccus*.

- Sodium chloride $10 \, \text{g L}^{-1}$
- Yeast extract $5 \, \text{g L}^{-1}$
- Tryptone $10 \, \text{g L}^{-1}$

Add distilled water to the required volume and autoclave.

Table A4.1

	T	C	A	G
T	TTT (**Phe**)	TCT (**Ser**)	TAT (**Tyr**)	TGT (**Cys**)
	TTC (**Phe**)	TCC (**Ser**)	TAC (**Tyr**)	TGC (**Cys**)
	TTA (**Leu**)	TCA (**Ser**)	TAA (**Stop**)	TGA (**Stop**)
	TTG (**Leu**)	TCG (**Ser**)	TAG (**Stop**)	TGG (**Trp**)
C	CTT (**Leu**)	CCT (**Pro**)	CAT (**His**)	CGT (**Arg**)
	CTC (**Leu**)	CCC (**Pro**)	CAC (**His**)	CGC (**Arg**)
	CTA (**Leu**)	CCA (**Pro**)	CAA (**Gln**)	CGA (**Arg**)
	CTG (**Leu**)	CCG (**Pro**)	CAG (**Gln**)	CGG (**Arg**)
A	ATT (**Ile**)	ACT (**Thr**)	AAT (**Asn**)	AGT (**Ser**)
	ATC (**Ile**)	ACC (**Thr**)	AAC (**Asn**)	AGC (**Ser**)
	ATA (**Ile**)	ACA (**Thr**)	AAA (**Lys**)	AGA (**Arg**)
	ATG (**Met**)	ACG (**Thr**)	AAG (**Lys**)	AGG (**Arg**)
G	GTT (**Val**)	GCT (**Ala**)	GAT (**Asp**)	GGT (**Gly**)
	GTC (**Val**)	GCC (**Ala**)	GAC (**Asp**)	GGC (**Gly**)
	GTA (**Val**)	GCA (**Ala**)	GAA (**Glu**)	GGA (**Gly**)
	GTG (**Val**)	GCG (**Ala**)	GAG (**Glu**)	GGG (**Gly**)

Terrific broth (TB)

Suitable as for LB – may result in improved cell densities in cultures of *E. coli* particularly. TB broth is made up of solutions of yeast extract and buffer salts, which need to be autoclaved separately then mixed after autoclaving and cooling.

Yeast extract solution

- Bacto-tryptone $12 \, \text{g L}^{-1}$
- Yeast extract $24 \, \text{g L}^{-1}$
- Glycerol $4 \, \text{mL l}^{-1}$

Psi broth

For making competent cell of *E. coli* (see Chapter 4).

- Yeast extract $5 \, \text{g L}^{-1}$
- Magnesium sulfate $5 \, \text{g L}^{-1}$
- Tryptone $20 \, \text{g L}^{-1}$

Adjust to pH 7.6 with sodium hydroxide and autoclave.

Add the required volume of distilled water and autoclave.
In a separate flask, make up:

- Potassium dihydrogen phosphate, KH_2PO_4 $2.3\,g\,L^{-1}$
- Dipotassium hydrogen phosphate, K_2HPO_4 $12.5\,g\,L^{-1}$

Add one-tenth the volume of distilled water used for the yeast extract solution and autoclave. When the solutions are cool, they can be mixed together using aseptic technique.

M9 medium

M9 medium is good as a basis for growth experiments where an identified sole-carbon source is going to be used for bacterial growth. M9 consists of a number of constituents based on an 'M9 salts' salts solution, which should be mixed after autoclaving and cooling.

M9 salts (10 × concentration)

- Disodium hydrogen phosphate $60\,g\,L^{-1}$
- Potassium dihydrogen phosphate $30\,g\,L^{-1}$
- Sodium chloride $5\,g\,L^{-1}$
- Ammonium chloride $10\,g\,L^{-1}$

Add the required volume of distilled water and autoclave. After cooling, the M9 medium is then prepared as follows:
For 1 L of M9 medium, to 100 mL of the autoclaved, cooled M9 salts add, using aseptic technique:

- 1 mL 1 M magnesium sulfate
- 10 mL of a 20% solution of glucose, added through a sterile filter (see Chapter 3)
- 10 mL of a 100 mM solution of calcium chloride that has been autclaved
- 1 mL of a 1% solution of thiamine through a sterile filter

This is then made up to 1 L with distilled water and is ready for the addition of a carbon source (see Chapter 4) and inoculation with organism.

Basal salts medium 1

For similar applications to M9. This medium is composed of two solutions, prepared and autoclaved separately and mixed after cooling, with the subsequent addition of ferrous sulfate.

Component 1 (per litre)

- Potassium dihydrogen phosphate   3.1 g
- Dipotassium hydrogen phosphate   8.2 g
- Ammonium sulfate   2.4 g
- Yeast extract   0.1 g
- Tryptone   0.1 g

Component 2 (per litre)

- Magnesium sulfate heptahydrate   0.5 g
- Manganese sulfate monohydrate   0.05 g
- Calcium chloride dihydrate   0.01 g
- Sodium molybdate monohydrate   0.01 g

To the mixed cooled solution, add $0.05\,\text{g L}^{-1}$ of ferrous sulfate, dissolved in water and added through a sterile filter. The complete medium is again then suitable for the addition of a sole carbon source.

Basal salts medium 2

A simpler and cheaper alternative to basal salts medium 1.

Solution 1 (per litre)

- Sodium dihydrogen phosphate   4.0 g
- Dipotassium hydrogen phosphate   2.0 g
- Ammonium chloride   3.0 g
- Yeast extract   0.1 g

Solution 2 (per litre)

- Magnesium sulfate heptahydrate   0.5 g
- Calcium chloride dihydrate   0.01 g

To the mixed cooled solution, add $0.05\,g\,L^{-1}$ of ferrous sulfate, dissolved in water and added through a sterile filter. The complete medium is again then suitable for the addition of a sole carbon source.

Auto-induction medium

Suitable for the growth of recombinant strains of *E. coli* to high cell density according to methods described by Studier (Chapter 7, reference [10]). The medium ZYM-5052 has proven successful for many applications in our laboratory. ZYM 5052 is made up of a solution of salts, yeast extract and carbon sources including glucose and lactose and also includes a trace metal solution. To prepare 1 L of ZYM-5052 medium the following reagents are prepared separately to stock concentrations, autoclaved separately and then added to an appropriate volume of distilled water that has been autoclaved in the culture flask. The final concentrations of each ingredient should be:

- NZ amine $0.1\,g\,L^{-1}$ (can be purchased from Sigma-Aldrich)
- Yeast extract $0.5\,g\,L^{-1}$
- Sodium dihydrogen phosphate $3\,g\,L^{-1}$
- 25 mM potassium dihydrogen phosphate $3.4\,g\,L^{-1}$
- Ammonium chloride $2.7\,g\,L^{-1}$
- Disodium sulfate $0.7\,g\,L^{-1}$
- Magnesium sulfate $0.24\,g\,L^{-1}$
- Trace metals at $0.2 \times$ final concentration
- Glycerol $5\,g\,L^{-1}$
- Glucose $0.5\,g\,L^{-1}$
- Lactose $2\,g\,L^{-1}$

The trace metal solution consists of a 1000x stock solution in 60 mM HCl containing:

50 mM iron (III) chloride
20 mM calcium (II) chloride
10 mM manganese (II) chloride
10 mM zinc sulfate
2 mM cobalt (II) chloride
2 mM copper (II) chloride
2 mM nickel (II) chloride

2 mM disodium molybdate (Na$_2$ MoO$_4$)
2 mM disodium selenite (Na$_2$SeO$_3$)
2 mM boric acid (H$_3$BO$_3$)

Medium for growth of *Streptomyces*

Soybean-meal medium

The soybean-meal medium used for the growth of a *Streptomyces* strain for hydroxylation reactions in Chapter 4 is grown on the following medium, which is thought to induce the hydroxylation activity. The recipe is taken from Chapter 4, reference [5] and used for the growth of *Streptomyces* in Chapter 4, reference [6].

- Soybean meal   5 g L^{-1}
- Glucose   20 g L^{-1}
- Yeast extract   5 g L^{-1}
- Sodium chloride   5 g L^{-1}
- Dipotassium hydrogen phosphate   5 g L^{-1}

Add the required amount of distilled water, adjust to pH 7.0 using hydrochloric acid and autoclave.

Two additional, simpler media for the growth of *Streptomyces*, derived from the NBIMCC website (http://www.nbimcc.org/cabricat/media.php) are:

Malt medium
- Glucose   4.0 g L^{-1}
- Malt extract   10.0 g L^{-1}
- Yeast extract   4.0 g L^{-1}

Add the required volume of distilled water and autoclave.

Glucose medium
- Glucose   10.0 g L^{-1}
- Yeast extract   1.0 g L^{-1}
- Meat extract   4.0 g L^{-1}
- Peptone   4.0 g L^{-1}

Add the required volume of distilled water and autoclave.

Liquid media for growth of filamentous fungi

There are many tips for growth media for filamentous fungi included on the UK National Culture Collection website (http://www.ukncc.co.uk/html/Databases/ Growth%20and%20Media.htm#media) These include various commercial preparations including malt extract, potato-dextrose and potato sucrose. One additional useful medium is the corn-steep solids medium, the composition of which is:

Corn-steep medium (per litre)

- Corn steep solids   7.5 g
- Glucose   10 g

Adjust pH to 4.75 using 2 M sodium hydroxide and autoclave.

Solid media for the maintenance of micro-organisms

Solid media, as described in Chapter 4, are usually based on the polysaccharide material agar, which can be purchased from the usual biochemical suppliers. As well as providing agar, which can be added to, for example LB broth or complete M9 medium at a level of 1–2% to make solid media such as LB agar or M9 agar, respectively, other speciality agars are supplied which are suitable directly for bacterial or fungal growth once made up according to the suppliers' instructions, autoclaved and dispensed into Petri dishes. These include **nutrient agar**, which is suitable for the maintenance of most non-fastidious bacteria used in biotransormations such as *Pseudomonas* and *Rhodococcus* and **malt extract agar** and **potato dextrose agar**, which are each suitable for the maintenance of filamentous fungi.

A recipe for oatmeal-tomato purée agar, useful for the growth and maintenance of *Streptomyces* spp. is given below

Oatmeal-tomato purée agar

- Fine-milled oatmeal $20 \, \text{g L}^{-1}$
- Tomato purée $20 \, \text{g L}^{-1}$
- Agar $15 \, \text{g L}^{-1}$

Dissolve in water and adjust pH to 7.0 before autoclaving. It may be necessary to include $5 \, \text{g L}^{-1}$ of yeast extract in the medium to improve the sporulation of some strains.

A.5 Buffers

Biological buffers

Table A5.1 lists some of the more commonly used buffers in isolated enzyme biotransformations and enzyme purifications and their recommended effective buffer ranges.

Buffers for making competent cells

Sterile TfB1 and TfB2 buffers are used in the preparation of competent cells of *E. coli* in recombinant biocatalyst experiments.

TfB1

- 30 mM potassium acetate
- 100 mM rubidium chloride
- 10 mM calcium chloride
- 50 mM manganese (II) chloride
- 15% (v/v) glycerol

Adjust to pH 5.8 with acetic acid and autoclave.

Table A5.1

Buffer	Effective pH range
Bis-tris propane	6.3–9.5
Tris-HCl	7.0–9.0
TEA (Triethanolamine)	7.3–8.3
Phosphate	5.8–8.0
MES [2-(*N*-morpholino)ethanesulfonic acid]	5.5–6.7
PIPES [Piperazine-1,4-bis(2-ethanesulfonic acid)]	6.1–7.5
MOPS [3-(*N*-morpholino)propanesulfonic acid]	6.5–7.9
HEPES [*N*-(2-hydroxyethyl)-piperazine-*N'*-2-ethanesulfonic acid]	6.8–8.2
CHES [*N*-cyclohexyl-2-aminoethanesulfonic acid]	8.6–10.0
CAPSO [3-(Cyclohexylamino)-2-hydroxy-1-propanesulfonic acid]	8.9–10.3

TfB2

- 10 mM MOPS
- 75 mM calcium chloride
- 10 mM rubidium chloride
- 15% (v/v) glycerol

Adjust to pH 6.5 with sodium hydroxide and autoclave.

Electrophoresis buffers

Resolving gel buffer

The resolving gel buffer is a solution of 1.5 M Tris plus (0.4% w/v) SDS, adjusted to pH 8.0.

Resolving gel mix for a 12% acrylamide gel

For 10 mL:

• Distilled water	3.2 mL
• Resolving gel buffer	2.5 mL
• Acrylamide stock [30% (w/v) acrylamide; 0.8% bis-acrylamide)] purchased from supplier	4.2 mL
• 10 % (w/v) ammonium persulfate solution in water	30 μL
• TEMED	16 μL

Stacking gel buffer

The stacking gel buffer contains 0.5 M Tris buffer plus 0.4% SDS (w/v) adjusted to pH 6.8.

Stacking gel mix

For 5 mL:

• Distilled water	3.2 mL
• Resolving gel buffer	1.3 mL
• Acrylamide stock [30% (w/v) acrylamide; 0.8% bis-acrylamide)] purchased from supplier	0.5 mL
• 10 % (w/v) ammonium persulfate solution in water	16 μL
• TEMED	8 μL

Loading gel buffer for SDS-PAGE

The protein sample is boiled as a 1:1 mixture with this loading buffer prior to loading on the SDS-PAGE gel.

For 10 mL:

- Distilled water   4.8 mL
- 0.5 M Tris-HCl buffer pH 8.0   1.2 mL
- Glycerol   1 mL
- 10 % (w/v) SDS in water   2 mL
- 0.1% (w/v) bromophenol blue   0.5 mL
- β-Mercaptoethanol   0.5 mL

SDS-PAGE Running buffer

The running buffer for SDS-PAGE is 100 mM Tris-HCl buffer containing 192 mM glycine, adjusted to pH 8.8, followed by the addition of 1% (v/v) 10% (w/v) SDS solution in water.

Coomassie Brilliant Blue Stain

For 1 L:

- Isopropanol   250 mL
- Glacial acetic acid   100 mL
- Water   650 mL
- Coomassie Brilliant Blue R   2 g

Stir and filter through standard filter paper before use.

SDS-PAGE gel destain solution

For 1 L:

- Isopropanol   50 mL
- Glacial acetic acid   70 mL
- Water   880 mL

TAE Running Buffer for Agarose gels

Agarose gels are run under this buffer in the gel tank. In order to make 1 L of 50x concentration stock solution:

- Tris   242 g
- 0.5 M EDTA   100 mL
- Glacial acetic acid   57.1 mL

Add water to 1 L.

Loading buffer for Agarose gels

The DNA sample is mixed with this buffer prior to loading the agarose gel. The buffer is mixed in a 1:4 ratio with the DNA sample.

The loading buffer contains: 60 mM Tris-HCl (pH 6.8), 2% (w/v) SDS, 0.02% (w/v) bromophenol blue, 10% glycerol and 5% (v/v) β-mercaptoethanol.

A.6 Ammonium sulfate fractionation Table

Table A6.1 is derived from that provided at http://www.science.smith.edu/departments/Biochem/Biochem_353/Amsulfate.htm. **I%** refers to the intial concentration of ammonium sulfate in the sample and each figure refers to the quantity, in grams, of solid ammonium sulfate that should be added to 100 mL of solution in order to achieve the required level of saturation. Thus, if starting at a concentration of 0% and wanting to gain a saturation of 80%, 51.6 g per 100 mL of solution should be added, and so on.

Table A6.1

I%	20	25	30	35	40	45	50	55	60	65	70	75	80	85	90	95	100
0	10.6	13.4	16.4	19.4	22.6	25.8	19.1	32.6	36.1	39.8	43.6	47.6	51.6	55.9	60.3	65.0	69.7
5	7.9	10.8	13.7	16.6	19.7	22.9	26.2	29.6	33.1	36.8	40.5	44.4	48.4	52.6	57.0	61.5	66.2
10	5.3	8.1	10.9	13.9	16.9	20.0	23.3	26.6	30.1	33.7	37.4	41.2	45.2	49.3	53.6	58.1	62.7
15	2.6	5.4	8.2	11.2	14.1	17.2	20.4	23.7	27.1	30.6	34.3	38.1	42.0	46.0	50.3	54.7	59.2
20	0	2.7	5.5	8.3	11.3	14.3	17.5	20.7	24.1	27.6	31.2	34.9	38.7	42.7	46.9	51.2	55.7
25	-	0	2.7	5.6	8.4	11.5	14.6	17.9	21.1	24.5	28.0	31.7	35.5	39.5	43.6	47.8	52.2

(continued)

Table A6.1 Continued

30	-	-	0	2.8	5.6	8.6	11.7	14.8	18.1	21.4	24.9	28.5	32.3	36.2	40.2	44.5	48.8
35	-	-	-	0	2.9	5.7	8.7	11.8	15.1	18.4	21.8	25.8	29.6	32.9	36.9	41.0	45.3
40	-	-	-	-	0	2.9	5.8	8.9	12.0	15.3	18.7	22.2	26.3	29.6	33.5	37.6	41.8
45	-	-	-	-	-	0	3.0	5.9	9.0	12.3	15.6	19.0	22.6	26.3	30.2	34.2	38.3
50	-	-	-	-	-	-	0	3.0	6.0	9.2	12.5	15.9	19.4	23.5	26.8	30.8	34.8
55	-	-	-	-	-	-	-	0	3.1	6.1	9.3	12.7	16.1	20.1	23.5	27.3	31.2
60	-	-	-	-	-	-	-	-	0	3.1	6.2	9.5	12.9	16.8	20.1	23.9	27.9
65	-	-	-	-	-	-	-	-	-	0	3.2	6.3	9.7	13.2	16.8	20.5	24.4
70	-	-	-	-	-	-	-	-	-	-	0	3.2	6.5	9.9	13.4	17.1	20.9
75	-	-	-	-	-	-	-	-	-	-	-	0	3.3	6.6	10.1	13.7	17.4
80	-	-	-	-	-	-	-	-	-	-	-	-	0	3.4	6.7	10.3	13.9
85	-	-	-	-	-	-	-	-	-	-	-	-	-	0	3.4	6.8	10.5
90	-	-	-	-	-	-	-	-	-	-	-	-	-	-	0	3.4	7.0
95	-	-	-	-	-	-	-	-	-	-	-	-	-	-	-	0	3.5
100	-	-	-	-	-	-	-	-	-	-	-	-	-	-	-	-	0

A.7 Restriction enzymes and restriction sites

Table A7.1 shows the abbreviations of some of the most commonly used restriction enzymes found in commercial plasmids, followed by their restriction sites. There are many more of course and comprehensive lists can be found in all catalogues of the major molecular biology reagent suppliers.

Table A7.1

Enzyme	Restriction site
BamHI	$5'$ ... G GATCC ... $3'$ $3'$ ... CCTAG G ... $3'$
BglII	$5'$ ... A GATCT ... $3'$ $3'$ ... TCTAG A ... $3'$
EagI	$5'$ ... C GGGCC ... $3'$ $3'$ ... GCCCG G ... $3'$
EcoRI	$5'$ ... G AATTC ... $3'$ $3'$ ... CTTAA G ... $3'$

(*continued*)

Table A7.1   Continued

EcoRV	5′ … G ATATC … 3′ 3′ … CTATA G … 3′
HindIII	5′ … A AGCTT … 3′ 3′ … TTCGA A … 3′
KpnI	5′ … G GTACC … 3′ 3′ … CCATG G … 3′
NcoI	5′ … C CATGG … 3′ 3′ … GGTAC C … 3′
NdeI	5′ … C ATATG … 3′ 3′ … GTATA C … 3′
NotI	5′ … GC GGCCGC … 3′ 3′ … CGCCGG CG … 3′
NruI	5′ … T CGCGA … 3′ 3′ … AGCGC T … 3′
PstI	5′ … C TGCAG … 3′ 3′ … GACGT C … 3′
SacI	5′ … G AGCTC … 3′ 3′ … CTCGA G … 3′
SalI	5′ … G TCGAC … 3′ 3′ … CAGCT G … 3′
Sau3AI	5′ … GATC … 3′ 3′ … CTAG … 3′
SmaI	5′ … CCC GGG … 3′ 3′ … GGG CCC … 3′
XbaI	5′ … T CTAGA … 3′ 3′ … AGATC T … 3′
XhoI	5′ … C TCGAG … 3′ 3′ … GAGCT C … 3′
XmaI	5′ … CCC GGG … 3′ 3′ … GGG CCC … 3′

Biotransformations–Index
